AF892351

OUR THREATENED CLIMATE

To my respected colleague

Professor Dr. Hermann Flohn

for his long-time involvement in CO_2/climate research and the decisive influence he has had on our understanding of this field.

OUR THREATENED CLIMATE

Ways of Averting the CO_2 Problem Through Rational Energy Use

by

WILFRID BACH

Center for Applied Climatology and Environmental Studies, University of Münster, Münster, West Germany

Translated by Jill Jäger

D. REIDEL PUBLISHING COMPANY

A MEMBER OF THE KLUWER ACADEMIC PUBLISHERS GROUP

DORDRECHT / BOSTON / LANCASTER

Library of Congress Cataloging in Publication Data

CIP

Bach, Wilfrid
Our threatened climate.

Translation of: Gefahr für unser Klima.
Bibliography: p.
Includes indexes.
1. Atmospheric carbon dioxide. 2. Climatic changes.
3. Climatology–Social aspects. 4. Man–Influence on nature.
5. Energy consumption. I. Title.
QC879.8.B3213 1983 363.7'392 83-19074
ISBN-13: 978-94-009-7244-5 e-ISBN-13: 978-94-009-7242-1
DOI: 10.1007/978-94-009-7242-1

Published by D. Reidel Publishing Company,
P.O. Box 17, 3300 AA Dordrecht, Holland.

Sold and distributed in the U.S.A. and Canada
by Kluwer Academic Publishers,
190 Old Derby Street, Hingham, MA 02043, U.S.A.

In all other countries, sold and distributed
by Kluwer Academic Publishers Group,
P.O. Box 322, 3300 AH Dordrecht, Holland.

All Rights Reserved
© 1984 by D. Reidel Publishing Company, Dordrecht, Holland
Original title: Gefahr für unser Klima
© 1982 by Verlag C.F. Müller, Karlsruhe
Softcover reprint of the hardcover 1st edition 1984
No part of the material protected by this copyright notice may be reproduced or
utilized in any form or by any means, electronic or mechanical
including photocopying, recording or by any information storage and
retrieval system, without written permission from the copyright owner

FOREWORD

In this book the author has succeeded in presenting the many facets of the global problems and hazards for our climate and their interdisciplinary aspects, as well as making these understandable for the non-specialist. In doing this, the author has not restricted himself to an analysis of the difficult problems but has indicated the necessity and the possibilities for rational solutions. The book, therefore, can be a valuable decision-aid for all those who directly or indirectly are in positions of responsibility at various levels of administration or in industry and business.

Well-timed precautionary measures against a global deterioration of climate are not only necessary for reasons of environmental protection. They are also an economical and political necessity. The measures include the reduction of the combustion of fossil fuels, a more rational energy utilization, as well as the establishment of a global equilibrium between forest loss and reforestation.

The Federal German Government takes these potential anthropogenic climate changes very seriously. In order to obtain better scientific information, the Federal Government has initiated an interdisciplinary national Climate Programme. At the same time, the Federal Republic of Germany supports the relevant activities within the frame of international cooperation.

I would be pleased if this book finds as large and wide a readership as befits its timeliness and significance.

Gerhart Rudolf Baum, Bonn, June 1982
Federal Minister of the Interior.

PREFACE AND ACKNOWLEDGMENTS

For many years I have tried to build a bridge between scientists and decision-makers. The receivers of scientific knowledge repeatedly point out that if research is to make a notable contribution to the solution of the many remaining human problems it must become more transparent.

Therefore, in this book I have attempted to find a fair compromise between scientific language and general comprehensibility, in order to present the CO_2/climate problem to the largest readership possible. This does not mean that cuts can be made in the exactness of the scientific presentation. Instead, it means that there are many possible ways of increasing the general comprehensibility and transparency, especially in the case of interdisciplinary themes.

This book, an updated version of the German original, presents the recent state of scientific knowledge, in as much as this is possible considering the continual flow of publications. A detailed and updated reference list of national and international publications for each chapter is intended for those who wish to go deeper into particular subjects. This is supplemented by a selection of recent handbooks and books compiled according to CO_2 and climate subjects, energy subjects and those of general interest.

A glossary with the specialist terms that have been used and appendices with additional explanations of presentations in the main text are intended to further increase the general comprehensibility. The frequent cross-references in the text and the extensive subject index should make it easier to find related themes. In many cases, an overview of the following material is given at the beginning of a chapter or section. The tables and interaction diagrams should help the reader to understand some of the rather complicated interactions.

Rather than have the individual conclusions scattered over the ends of the corresponding paragraphs and chapters, I have placed them in a complete summary at the beginning of the book. The list of abbreviations and conversion factors, likewise at the beginning, should help to avoid the frequent confusions and misunderstandings resulting therefrom.

This book aims to encourage an intensification of research efforts to remove the existing gaps in our knowledge. At the same time, the arguments should convince decision-makers that the proposed precautionary measures are not only sensible for averting a CO_2/climate problem, but are economically and politically necessary even without climatic reasons. Above all, I hope that the extensive information presented here will lead to better founded reporting in the future. Finally, I would like to express the wish that all those who are sceptical about many of the arguments presented in this book will read it especially carefully.

With the large range of subjects to be covered, such a book can only be written with the direct and indirect cooperation of many colleagues. In particular, I have benefited from the discussions and conference contributions, and the papers that were prepared for a number of international conferences which I organized on behalf of the Federal Environmental Agency, within a research programme entitled "Effects of air pollution on the global climate" and supported by the Federal Ministry of the Interior from 1978-1981.

So many have contributed to the production of this book that it is not possible to thank them all individually. For the careful review of the entire manuscript, for the thorough criticism and for the many valuable references to material and literature, I would like to thank Hermann Flohn, Hermann Hambloch, Ulrich Hampicke and Jill Jäger. Many of the ideas, concepts, arguments and suggested solutions that are brought together in this book came from my colleagues. The cooperative work or discussions have helped me considerably in the formulation and evaluation of the possible ways of avoiding a CO_2/climate problem. Here, I would particularly like to thank Robert Chervin, Paul Crutzen, Roger Dahlman, Dieter Ehhalt, W. Lawrence Gates, Andrew Gilchrist, Wolf Häfele, James Hansen, Martin Hoffert, Charles D. Keeling, William W. Kellogg, Gundolf Kohlmaier, Frederick Koomanoff, Florentin Krause, Amory B. Lovins, Michael MacCracken, Syukuro Manabe, Herbert Meinl, John Mitchell, Harry Moses, Jürgen Pankrath, V. Ramanathan, Roger Revelle, Walter O. Roberts, Michael Schlesinger, Stephen H. Schneider, Warren Washington, Richard Wetherald, Tom Wigley and Julius Werner. However, I bear the full responsibility for the contents of this book.

My co-workers Hans-Josef Jung, Heinrich Knottenberg, Wolfgang Lückert, Annette Neuhaus and Harald Westbeld were also involved in the critical review and preparation of the manuscript. Marianne Michelka prepared the illustrations accurately and legibly. All of them are gratefully acknowledged.

Mention must be made of Günter Keller and Christof Müller-Wirth of C.F. Müller Publishers, Karlsruhe, who suggested the German version of this book. I would also like to thank David Larner of the Reidel Publishing Company, Dordrecht, who lent continual support to the production of the English version of this book.

Finally, my special thanks go to Jill Jäger who very skilfully did most of the translation and to Nigel Huckstep for his contribution to it. For the technical supervision of the translation I am especially grateful to Tom Wigley, the Director of the Climatic Research Unit at the University of East Anglia, where also the camera-ready production of the book was very ably done by Susan Boland.

Moreover, this book is based on a number of research projects. These were supported directly by the Ministry of the Interior through the Federal Environmental Agency, the Ministry of Science and Technology, the Aspen Institute in Berlin, the German Research Society and the University of Münster, and indirectly by the Ministry for Economic Cooperation, the German Weather Service, the International Federation of Institutes for Advanced Studies, Sweden, the Commission of the European Communities, Belgium, and the United Nations University, Japan. I would like to thank them here for this support.

I would also like to thank Inter Nationes, Kennedyallee 91-103, 5300 Bonn 2, for making available a subsidy towards the cost of translation.

Finally, I wish to offer special thanks to my son Alexander for designing the cover of this English edition.

Anyone who has written a book knows what an enormous strain this puts on the family. My wife Anneliese, with much good will and skill protected me from all disturbances, as much as she possibly could. I know that she secretly hoped that, at least for a while, I would not begin with another manuscript and would have more time for family life. This book is a small expression of gratitude for all that they have sacrificed.

Münster, May 1983

Wilfrid Bach

TABLE OF CONTENTS

Foreword V
Preface and Acknowledgments VI
List of Figures XIV
List of Tables XIX
List of Appendices XXII
List of Abbreviations and Conversion Factors XXIV

Executive Summary 1

I. Introduction 27

II. Climate and Climatic Change 30
1. History of the Earth's Climate.......................... 30
2. Climate System.. 33
2.1 Radiation Budget of the Earth-Atmosphere System. 35
2.2 Anthropogenic Influence on the Natural Radiation Budget.. 35
2.3 Causes of Climatic Change and Possibilities for Early Detection................................. 37
3. Climate Models and Prediction........................... 39
3.1 Simple Climate Models.............................. 39
3.2 Energy Balance Models.............................. 40
3.3 Zonally Averaged Models............................ 41
3.4 Stochastic Models.................................. 41
3.5 General Circulation Models......................... 42
3.6 Climatic Predictability............................ 43
4. Climate Research Programme.............................. 44

III. Sociopolitical Aspects of the CO_2/Climate Problem 46
1. Population Growth....................................... 46
1.1 Past Growth.. 46
1.2 Future Growth...................................... 46
2. Settlement Patterns..................................... 49
2.1 Urbanisation in Industrialised Countries........ 50
2.2 Urbanisation in Developing Countries............ 50
3. Economic Development.................................... 51
4. Availability of Fossil Fuels............................ 54
4.1 Coal... 54
4.2 Oil.. 57
4.3 Natural Gas.. 57
5. Future Energy Use 58
5.1 Energy Amount and Mix 58
5.2 Inequities in Energy Use........................... 61
5.3 Market Penetration Times........................... 63
5.4 Capital Requirements............................... 64
6. Degradation of Forest and Soil.......................... 66
6.1 Deforestation...................................... 66
6.2 Loss of Soil....................................... 68

IV. Influence of Society on Climate 70
1. Climatic Effects of CO_2............................ 70
1.1 The CO_2/Climate Problem........................ 70
1.2 The Natural Carbon Cycle........................ 72
1.3 Evidence of Carbon Cycle Perturbation........... 74
1.3.1 Observed CO_2 Increase in the Atmosphere... 74
1.3.2 CO_2 Monitoring........................... 76
1.4 Causes of Perturbation.......................... 78
1.4.1 Role of the Conventional Fossil Fuels..... 78
1.4.2 Role of the Unconventional Fossil Fuels... 80
1.4.3 Role of the Biosphere.................... 82
1.4.4 Role of the Ocean........................ 87
1.4.5 Carbon Budget............................ 91
1.5 Assessment of Past CO_2 Concentration........... 93
1.5.1 Indirect Methods: Analysis of Tree Rings.. 93
1.5.2 Direct Methods: Analysis of Ice Core Samples................................. 95
1.6 Assessment of Future CO_2 Concentration......... 97
1.6.1 Carbon Cycle Models...................... 97
1.6.2 Future Atmospheric CO_2 Content........... 99
1.7 Climate Response Due to a CO_2 Increase......... 103
1.7.1 Equilibrium Response Experiments......... 104
1.7.2 Transient Response Studies............... 116
1.7.3 Critical Comparison of Model Results...... 122
2. Climatic Effects of other Anthropogenic Trace Gases. 126
2.1 Production and Concentration.................. 126
2.2 Climate-Chemistry Interactions................. 131
2.3 Assessment of Climatic Effects................. 133
3. Climatic Effects of Aerosols....................... 134
3.1 Sources, Emissions, Concentrations............. 135
3.2 Climatic Effects.............................. 135
4. Climatic Effects of Land Use Changes............... 139
4.1 Overgrazing - Desertification - Deforestation... 140
4.2 Urbanisation and Industrialisation............. 143
4.3 Energy Conversion and Use..................... 143
4.4 Water Engineering Projects..................... 144
5. Climatic Effects of Waste Heat..................... 144
5.1 Natural and Anthropogenic Energy Fluxes......... 144
5.2 Climatic Effects of Heat Emissions............. 146
5.3 Possibilities for Waste Heat Reduction......... 147
6. Assessment of Overall Effects - Critical Threshold Values - Prospects............................... 148
6.1 Individual Climate Factors and the Greenhouse Effect.................................... 148
6.2 Combined Greenhouse Effect and Critical Threshold Values for the CO_2 and Temperature Increase..... 150
6.3 Outlook...................................... 151

V. Impacts of Climate Change on Society 154
1. Climate Impact Programmes.......................... 154
1.1 Nature and Methods of Impact Studies........... 154

1.2 Climate Scenarios 156
1.3 Socioeconomic Scenarios 158
2. Ecosystems 158
2.1 Influence of Mankind on the Biomes 158
2.2 Response of the Biomes to Climate Changes 159
3. Energy Use 162
3.1 Climate and Energy Demand 163
3.2 Climate and Energy Supply 165
3.3 Climate and Energy Costs 166
4. Food Security 167
4.1 Climate and the World Food Situation 168
4.2 Climate and Harvest Yields 169
4.3 Climate and Pest Management 173
4.4 CO_2 Fertilization and Plant Productivity 174
4.5 Learning from Past Experience 175
5. Water Resources 176
5.1 Water Reserves and Water Demand 176
5.2 Water Availability 177
5.3 Global Water Problems 180
6. Fisheries 181
7. Population and Settlement 183
7.1 Examples from the Past: Decline of Civilisations - Migrations - Abandonments - Climate as a Causal Factor? 184
7.2 Possible Scenarios for the Future: Partial Disintegration of the West Antarctic Ice Sheet - Rise of Sea Level - Impact on Coastal Regions; An Ice-free Arctic Ocean - Shift of Agroclimatic Zones 185
8. Health, Disease and Welfare 193
8.1 Climate and Comfort 193
8.2 Climate and Disease 194
8.3 Climate Change and Welfare 195
9. Recreation 196

VI. Strategies for Averting a CO_2/Climate Problem 198
1. The Scope of Alternatives and Concepts for Action 198
2. Technical Fixes 199
2.1 CO_2 Removal 199
2.2 CO_2 Storage 202
2.3 CO_2 Conversion to Useful Products 204
2.4 Albedo Modification 205
3. Biological Methods 206
3.1 CO_2 Storage in Land Plants 206
3.2 CO_2 Storage in Water Plants 207
3.3 CO_2 Storage in Soil Humus 207
4. Energy Saving Through Better Energy Use 208
4.1 Future Energy Use 209
4.2 Some Selected Studies of Rational Energy Use 213
4.2.1 German Studies 213

4.2.2 European Studies.......................... 218
4.2.3 US Studies................................ 219
4.3 Efficient Energy Use in Industrialized and Developing Countries............................ 220
5. Introduction of Renewable Energy Resources........... 221
5.1 Substitution of Non-Renewable by Renewable Energy Sources.. 221
5.2 Global Potential of Renewable Energy Sources..... 223
6. Effects of Various Energy Strategies................ 226
6.1 Characteristics of Selected Scenarios............ 226
6.1.1 Synthetic Fuel Production Scenario......... 226
6.1.2 IIASA Scenarios........................... 227
6.1.3 "Zero Growth" Scenario..................... 229
6.1.4 Efficiency Scenario........................ 230
6.2 Effects of Various Energy Strategies on CO_2 and Temperature Changes............................. 232
6.2.1 Synthetic Fuel Strategy.................... 232
6.2.2 Pure Fossil Fuel Strategy.................. 233
6.2.3 Spectrum of Various Energy Strategies...... 234
7. Flanking Measures.................................. 239
8. Decision Aids...................................... 240

VII. Opportunities for the Future 244
1. The Old Credence in Progress - The New Credibility... 244
2. Man: Shaping his Own Future?........................ 245
3. Prospects: Extrapolating Old Trends.................. 247
3.1 Growth.. 247
3.2 North-South Conflict............................ 249
4. Prospects: Shifting Emphases........................ 251
4.1 Reorientation.................................... 251
4.2 Using Available Technological Know-How........... 252
4.2.1 Cogeneration.............................. 252
4.2.2 Combined-Heat-and-Power Plants............. 252
4.2.3 Neighbourhood Cogeneration Systems......... 253
4.2.4 Fluidised-Bed Combustion.................. 253
4.2.5 Central Station District Heating........... 254
4.2.6 Heat Pumps................................ 255
4.3 Examples of Product Innovation.................. 257
4.3.1 Insulation................................ 257
4.3.2 Energy-Saving Light Bulbs.................. 258
4.3.3 Fuel Economy of Light Vehicles............. 258
4.3.4 Feedstock Uses of Carbon.................. 259
4.3.5 Microelectronics........................... 259
4.4 Removal of Institutional Barriers................ 260
5. Ways out of the CO_2/Climate Threat.................. 262

Bibliography... 266

Further Reading 307

Appendices 313

Glossary 332

Author Index 341

Subject Index 350

LIST OF FIGURES

II. Climate and Climatic Change

II.1 History of mean air temperature as inferred from a variety of palaeoclimatic indicators. After: Mitchell Jr. (1977). 31

II.2 Schematic illustration of the climate system. After: WMO (1975), Schönwiese (1979), Flohn (1980). 34

II.3 The average radiation and heat balance of the earth-atmosphere system, relative to 100 units of incoming solar radiation. After: Rotty (1975) and Gates (1979). 36

II.4 Interdependence of the major components of a climate research programme. Source: US NAS (1975). 45

III. Sociopolitical Aspects of the CO_2/Climate Problem

III.1 Momentum of world population growth. Source: Mauldin (1980). 48

III.2 Rural and urban population distribution. After: U.N. (1975) and Sassin (1980a). 49

III.3 World primary energy substitution. Source: Marchetti and Nakicenovic (1979). 64

IV. Influence of Society on Climate

IV.1 The global carbon cycle (about 1978). Sources: Kester and Putkowicz (1977); Woodwell (1978); Bolin et al. (1979); Hampicke and Bach (1979); Lieth et al. (1980); US DOE (1980); Hampicke (pers. comm., 1981). 73

IV.2 Trend in atmospheric CO_2 concentration at Mauna Loa, Hawaii, 1958-1982. Sources: Keeling et al. (1976a); Pearman (1980); MacCracken and Moses (1982); Keeling (pers. comm., 1983). 75

IV.3 Stations monitoring atmospheric CO_2 trends, July 1979. Source: US DOE (1980). 77

IV.4 Global CO_2 production from fossil fuels, cement and gas flaring, 1860-1980. Sources: Keeling (1973 for 1860-1949) and Rotty (1980, 1981 for 1950-1980). 80

IV.5 The 1979 CO_2 emissions from fossil fuels and cement manufacture of $5340 x 10^6$ tons C by latitudinal bands. Source: Rotty (1981). 81

IV. 6 a) Carbon release from forest clearing and agricultural activity, 1860-1970;
b) Global carbon flux, showing terrestrial and fossil fuel components, 1860-1970.
Source: Moore et al. (1980). 83
IV. 7 Total carbon fluxes from 1900, with projections beyond 1980.
Source: Loucks (1980). 88
IV. 8 Diagram showing the patterns of the flow within the ocean. Upwelling occurs in the equatorial zone; downwelling occurs in the temperate zone and polar zone. The ocean area in each zone is shown as a percentage of the total ocean area.
Sources: Dietrich et al. (1975); Broecker et al. (1980). 90
IV. 9 Graph showing the relationship between the atmospheric CO_2 level reconstructed from Camp Century, North Greenland, core measurements (upper curve) and the corresponding global change in surface air temperature estimated from current climatic models (lower curve).
Sources: Berner et al. (1980); Thompson and Schneider (1981). 96
IV.10 Projections of atmospheric CO_2 concentration for different growth concepts of fossil fuel use. 99
IV.11 Industrial CO_2 production for various logistic growth rates. The ultimate production is fixed at 8.2 times the amount of CO_2 in the pre-industrial atmosphere.
Source: Keeling (1980). 101
IV.12 Predicted increase in the atmospheric CO_2. Curves are shown for the four simulated fossil fuel consumption patterns shown in Fig. IV.11.
Source: Keeling (1980). 102
IV.13 Geographical distribution of monthly mean surface air temperature (^oK) in February. Top: the model simulation. Bottom: the observed distribution.
Source: Manabe and Stouffer (1980a, b). 107
IV.14 Latitude - height distribution over the northern hemisphere of the zonal mean temperature (^oC) in response to a doubling of CO_2.
After: Manabe and Wetherald (1980). 109
IV.15 Latitude - time distributions of sea ice thickness(m). Top: $1xCO_2$ experiment. Bottom: $4xCO_2$ experiment. Shading indicates the regions where sea ice thickness exceeds 0.1 m.
Source: Manabe and Stouffer (1980a). 110

IV.16 Horizontal distribution of the rates of change of precipitation minus evaporation (P-E) in response to a doubling of the CO_2 content. Units are cm/day.
Source: Manabe and Wetherald (1980). 111
IV.17 Recorded and estimated temperature changes ($^{\circ}C$) for the N-hemisphere plus the 180-year temperature cycle as inferred from Greenland ice cores.
After: Broecker (1975); Bernard (1980). 117
IV.18 CO_2 warming versus noise level of natural climate variability. The effect of other trace gases is not included.
Source: Hansen et al. (1981). 119
IV.19 Global temperature trend obtained from a global climate model for several assumed radiative forcings.
Source: Hansen et al. (1981). 122
IV.20 Schematic illustration of the ocean-atmospheric feedback processes by which a CO_2 increase warms the surface. All of the numbers correspond to hemispherically averaged conditions and apply to a doubling of CO_2.
Source: Ramanathan (1981). 123
IV.21 Global temperature change of model climates due to changes in CO_2 concentration.
Source: Boville and Döös (1981). 124
IV.22 Climate-chemical interactions due to trace gases.
Source: Ramanathan (1980). 132
IV.23 Stratospheric and surface air temperature change in the tropics after the Mt. Agung eruption on Bali in 1963. Dots are based on calculations with a one-dimensional radiative model.
Source: Toon and Pollack (1980). 137
IV.24 Precipitation calculations for various values of surface albedo in the Sahara at 18°N.
Source: Charney (1975). 141
IV.25 Power densities related to man's activities ca. 1970 and projected.
Source: Bach (1980b). 147
IV.26 Potential radiative perturbations of climate.
Source: Hansen et al. (1981). 149
IV.27 Critical threshold values for a CO_2 and temperature increase related to model-dependent sensitivity parameters.
Source: Flohn (1981b). 151

V. Impacts of Climate Change on Society

V. 1 System showing the CO_2/climate-interactions. With additions from: Markley and Carlson (1980). 155

V. 2 World biome types in relation to mean annual precipitation and temperature. The dot-and-dash line encloses a wide range of environments in which either grassland or one of the types dominated by woody plants may form the prevailing vegetation.
Source: Whittaker (1975). 160

V. 3 Estimate of the average annual precipitation distribution for a warm climatic period in comparison with recent times. Hatched areas have an increase, dotted areas have a decrease, and blank areas have no change in precipitation.
Source: Sergin (1980). 162

V. 4 Heating demand and mean annual temperature. Dotted area lies between 2°C and 18°C.
Source: McKay and Allsopp (1980). 164

V. 5 Areas favourable for exploitation of solar, biomass and ocean-thermal energy.
Source: McKay and Allsopp (1980). 165

V. 6 Simulated five-state weighted average corn yields using 1973 technology and harvested average: Ohio, Indiana, Illinois, Iowa, Missouri.
Source: McQuigg et al. (1973). 170

V. 7 Global water problems: D (drought), S (water shortage), F (floods), P (industrial pollution), HQ (health- and water quality problems).
Source: Lindh (1981). 181

V. 8 Ice shield and ice shelves in West Antarctica. Double arrows point to critical glacier flows.
Adapted from: Mercer (1978) and Flohn (1980). 187

V. 9 German coastal areas affected by a potential sea level rise. Heavy line denotes a 5 m and dotted line a 10 m rise. 191

V.10 Projected changes of annual surface air temperature (left, ^{o}C) and annual precipitation (right, % of present) for an ice free Arctic Ocean.
Source: Flohn (1981b). 192

V.11 Distribution of world population in the annual $\overline{T}_{min}$ and $\overline{T}_{max}$ diagram. Blank area: total land surface. Lightly shaded area: extension of permanent habitats. Heavily dotted area: zone where over 60% of the world's population live.
Source: Hoffmann (1960). 194

V.12 Effect of estimated change of mean world temperature in the human population-climate interaction diagram.
Source: Weihe (1979). 196

VI. Strategies for Averting a CO_2/Climate Problem

VI.1 Possibilities to influence the CO_2 increase in the atmosphere - an interaction scheme. 200
VI.2 Model projections of atmospheric carbon dioxide variations assuming the entire fossil fuel reserve is depleted as a logistic function of time. All oceanic injections are initiated in 1985. Note fivefold compression of time-scale beyond the year 2300.
Source: Hoffert et al. (1979). 203
VI.3 Per capita primary energy use for selected regions according to renewable energy models.
Source: Sorensen (1981) and supplemented. 222
VI.4 The 30 TW fossil fuel strategy.
Source: Niehaus (1979). 233
VI.5 Simulated CO_2 and temperature change for the 30 TW fossil fuel strategy.
Source: Niehaus (1979). 234
VI.6 CO_2 emission for a variety of energy scenarios.
Computation: 1860-1975. Simulation: 1975-2030. 236
VI.7 CO_2 concentration for a variety of energy scenarios.
Computation: 1860-1957. Observation: 1957-1975.
Simulation: 1975-2030. 237
VI.8 Temperature change for a variety of energy scenarios. Observation: 1880-1975. Simulation: 1975-2030. 237

VII. Opportunities for the Future

VII.1 What's your problem?
Source: Asian Action (1980). 250
VII.2 Comparison of different energy systems for residential heating.
Source: Herrmann (1978). 256

LIST OF TABLES

III. Sociopolitical Aspects of the CO_2/Climate Problem

III.1 Population projections for world, major regions, and selected countries. Source: Global 2000 (1980). 47

III.2 Estimates and rough projections of selected urban agglomerations in developing countries. Source: Global 2000 (1980). 50

III.3 Actual (1975) and possible (2030) urban population share. Source: Colombo and Bernardini (1979). 52

III.4 Total and per capita GNP estimates (1975) and projections (2000) by major regions. Data extracted from: Tables 3 and 4, Global 2000 (1980, Vol. 1). 53

III.5 World resources and reserves of fossil fuels. Source: Ziegler and Holighaus (1979). 55

III.6 World coal resources and reserves (btce), by major coal-producing countries. After: WOCOL (1980). 55

III.7 Development of coal production and export (mtce), by major coal-producing countries, 1975-2030. After: World Energy Conference (1978). 56

III.8 Global primary energy (TWyr/yr) by source for two supply scenarios (1975-2030) and growth factors (in brackets). After: Häfele and Rogner (1980) and Häfele (1980). 60

III.9 Scenarios for population and primary energy use (per capita and total) by regions. After: Häfele et al. (1981); data extracted from 1) Table 14.2; 2) Table 25.1; 3) Table 15.2. ... 62

III.10 Estimates of world forest resources, 1978 and 2000. Source: Global 2000 (1980). 67

IV. Influence of Society on Climate

IV.1 Comparison of carbon dioxide release from production of thermal energy by various fuels (average values). Adapted from: JASON (1980). 78

IV.2 Possible carbon balance, ca. 1978. 92

IV.3 Overview of the most important factors influencing climate. Sources: 1. Almquist (1974a); 2. Bach (1976a); 3. Ellsaesser (1979); 4. Crutzen and Ehhalt (1977); 5. Ehhalt (1976/77); 6. Freyer (1979); 7. Ehhalt and Schmidt (1978); 8. Grassl (1980); 9. Almquist (1974b); 10. Ehhalt (1980); 11. Ramanathan (1980);

12. Isaksen and Stordal (1981); 13. Rasmussen et al. (1981); 14. Wang et al. (1976); 15. Groves and Tuck (1979); 16. Potter et al. (1980); 17. Ehhalt (pers. comm., 1981); 18. Crutzen (pers. comm., 1981); 19. Weiss (1981); 20. Ko and Sze (1982); 21. Zimmerman et al. (1982); 22. Ehhalt et al. (1982); 23. Fraser et al. (1983). 128

IV.4 Estimates of global particle production (diameter ⩽ 5 μm) from nature and man-made sources.
Adapted from: Bach (1976) and Dittberner (1978). 136

IV.5 Causes of large-scale changes in land surface albedo.
Adapted from: Munn and Machta (1979) and supplements. 140

IV.6 Comparison of natural and artificial heat production.
Adapted from: Bartholomäi and Kinzelbach (1979); Bach (1980). 145

V. Impacts of Climate Change on Society

V.1 Changes in expenditure for energy in selected U.S. cities for given changes in temperature (dollars per year).
Adapted from: d'Arge (1979). 167

V.2 The changing regional patterns of world grain trade (mill. tons), 1934-1980.
Adapted from: Coakley and Schneider (1976); Brown (1980, 1981). 169

V.3 Possible effects of climatic change on the average yields of corn, wheat, and rice in the major growing regions expressed in percent of average yield.
Source for corn yield (Benci et al., 1975); for wheat yield (Ramirez et al., 1975); for rice yield (Stansel and Huke, 1975). 172

V.4 Estimated global water needs and water resources.
Adapted from: Flohn (1974) and Lvovich (1977) 177

V.5 Per capita water availability (thousands of m^3/yr) for selected countries, 1971 and 2000.
Adapted from: Global 2000 (1980). 179

V.6 Summary of estimated geographic, demographic and economic impacts of 15- and 25-foot (4.6 and 7.6 m) rises in sea level for the continental United States.
Adapted from: Schneider and Chen (1980). 189

VI. Strategies for Averting a CO_2/Climate Problem

VI.1 Cost of collecting and processing carbon dioxide from stack gas.
Source: Baes et al. (1980). 201

VI.2 Projections of primary energy demand (mtce) of the Federal Republic of Germany and the Institute of Applied Systems Research and Prognosis (ISP) (consumption 1982: 362 mtce).
Source: Pestel et al. (1978) and Die Zeit (1981). 210

VI.3 Evolution of approximate estimates of U.S. primary energy demand in the year 2000 (q/yr $\simeq 10^{15}$ BTU/yr $\simeq$ EJ/yr $\simeq$ 36 mtce/yr $\simeq 3.35 \times 10^{10}$W (1980: 76 q/yr $\simeq$ 2.6 TW).
Sources: Lovins (1980) and Lovins et al. (1981) ... 211

VI.4 Range of recent world energy projections for global primary energy use (TW) in 2000.
All data with * extracted from Bienewitz et al. (1981); for all other data see references. 212

VI.5 Primary energy use (mtce/yr) in three different scenarios for the Federal Republic of Germany.
Extracted from: Krause et al. (1980). 215

VI.6 Results of the four energy paths (mtce) of the Enquête-Kommission of the German Parliament.
Source: Enquête-Kommission (1980). 217

VI.7 Estimated global potential of renewable energy sources for 2030.
Sources: (1) Häfele (1980); (2) Häfele and Rogner (1980); (3) Weingart (1980); (4) Caputo (1980); (5) Matthöfer (1976); Bach and Schwanhäusser (1978); Bach (1979b). 224

VI.8 Global primary energy supply (TWyr/yr) for different energy sources and a variety of energy scenarios, 1975-2030.
Sources: Colombo and Bernardini (1979); Häfele et al. (1981) and Lovins et al. (1981). 228

VI.9 Possible world primary energy demand (excluding feedstocks) with efficient use, strong economic growth, and constant urban fraction.
After: Lovins et al. (1981). 231

VI.10 Projection of synfuels industry growth and effect on atmospheric carbon dioxide.
Adapted from: Rotty (1979). 232

VI.11 Development of CO_2 emission, CO_2 concentration and temperature change from 1975-2030 for the energy scenarios in Table VI.8. 235

LIST OF APPENDICES

II. Climate and Climatic Change

II.1 The use of various types of data for the reconstruction of climate. Source: Pfister et al. (1980). 313

II.2 Review of the Earth's climatic history. After: Bach (1976) and Schönwiese (1979). 314

II.3 Basic equations for a selection of climate model types. After: Holloway and Manabe (1971); Schneider and Gal-Chen (1973); Lemke (1977). 315

III. Sociopolitical Aspects of the CO_2/Climate Problem

III.1 The influence of civilization upon European and Asian Forests. After: Armentano and Hett (1980). 318

IV. Influence of Society on Climate

IV.1 Short historical sketch of CO_2/climate research. Sources: Tyndall (1861); Arrhenius (1896, 1903); Chamberlin (1897); Tolman (1899); Callendar (1938); Flohn (1941); Plass (1956, 1961); Revelle and Suess (1957); Kaplan (1961); Mitchell (1961); Möller (1963); Manabe and Wetherald (1967); Machta (1972); Manabe and Wetherald (1975); Schneider (1975); Oeschger et al. (1975, 1980, 1983); Baes et al. (1977); Dittberner (1978); MacCracken and Potter (cit. Kellogg and Schware, 1981); Hansen et al. (cit. US NAS, 1979); Manabe and Stouffer (1979, 1980); Manabe and Wetherald (1980); Gates (1980); Washington and Ramanathan (1980); Hansen et al. (1981); Michael et al. (1981); Cess and Goldenberg (1981); Wetherald and Manabe (1981); Manabe et al. (1981); Bryan et al. (1982); Thompson and Schneider (1982b); Schlesinger (1982, 1983a); Washington and Meehl (1982, 1983); Gilchrist (1983); Mitchell (1983). 320

IV.2 Calculation of the global atmospheric carbon content. 326

IV.3 Calculation of the surface air temperature as a function of the atmospheric CO_2 level, other trace gases, and model-derived sensitivity parameters. 327

V. Impacts of Climate Change on Society

V.1 Effects of weather and climate on energy demand and supply.
After: USDOE (1980). 329

LIST OF ABBREVATIONS AND CONVERSION FACTORS

Measure Prefixes

E	exa	$= 10^{18}$	d	deci	$= 10^{-1}$
P	peta	$= 10^{15}$	c	centi	$= 10^{-2}$
T	tera	$= 10^{12}$	m	milli	$= 10^{-3}$
G	giga	$= 10^{9}$	μ	micro	$= 10^{-6}$
M	mega	$= 10^{6}$	n	nano	$= 10^{-9}$
k	kilo	$= 10^{3}$	p	pico	$= 10^{-12}$
h	hecto	$= 10^{2}$	f	femto	$= 10^{-15}$
da	deca	$= 10^{1}$	a	atto	$= 10^{-18}$

Units of Energy and Power

1 Terawatt (TW)	$= 10^3$ Gigawatt (GW)
1 Gigawatt (GW)	$= 10^3$ Megawatt (MW)
1 Megawatt (MW)	$= 10^3$ Kilowatt (kW)
1 Kilowatt (kW)	$= 10^3$ Watt (W)
1 Kilowattyear (kWyr)	= 8,760 Kilowatthours (kWh)
1 Kilowattyear (kWyr)	≃ 1 tonne of coal equivalent (tce)
1 Kilowattyear (kWyr)	≃ 0.7 tonne of oil equivalent (toe)
1 Terawattyear (TWyr)	$\simeq 10^9$ tce
1 Terawattyear (TWyr)	≃ 30 quads
1 quad	$= 10^{15}$ Btu (British thermal units) $= 3.34 \cdot 10^{10}$ Wattyears (Wyr)
1 Btu	= 0.293 Watthours (Wh) = 1050 Joules (J)
1 Joule	= 1 Wattsecond (Ws)
1 tonne of coal equivalent	≃ 8,140 kWh = 0.93 kWyr $= 7 \cdot 10^6$ kcal
1 tonne of oil equivalent	≃ 11,650 kWh = 1.33 kWyr $= 10 \cdot 10^6$ kcal
1 barrel (bbl) of oil equivalent	≃ 1,610 kWh = 0.18 kWyr
1 cubic metre (m^3) of gas equivalent	≃ 9.3 kWh

Some Examples

1 kW = the power of ten 100 W light bulbs
1 MW ≃ the power of a Diesel train engine or of a windmill with an 80 m diameter of the rotor blades
1 GW ≃ the power of a large electric power station
1 TW ≃ 1/9 of the present global commercial primary energy use

EXECUTIVE SUMMARY

I Introduction

From research into the climate of the past we know that it is governed by variations with which we shall also have to deal in the future. The exact causes of these changes in climate are not at present known. It is, however, ascertainable that the shift of climate zones, often associated with catastrophic droughts and floods, have led to harvest failures, whose consequences repeatedly prove difficult to overcome.

Both the increasing potential of Man to influence climate as well as the increased impact of changes in climate on Man are factors which were not present in the past. Through his activities Man is now in a position to influence climate not only locally, but also globally in quite drastic and possibly irreversible ways. In addition, the constantly increasing population pressure and the limited environmental and food resources, further increase the vulnerability of society to climatic change. The fear that the already shrunken safety margins will further narrow is not without ground. The struggles for a fair distribution of resources, which will be more pressing in the future, pose a serious threat to world peace.

The main reason for the significant enhancement of the natural greenhouse warming in the lower atmosphere which is expected in the next few decades is the increase in the concentration of atmospheric CO_2, due to both the burning of fossil fuels (coal, oil and gas above all) and the deforestation of large areas of woodland and intensive soil cultivation. This Man-induced "global experiment" is also influenced by other trace elements which absorb radiation, by changes in the characteristics of the Earth's surface, and, to a lesser extent, by the direct release of heat into the atmosphere.

We must now decide whether we wish to expose ourselves unprepared to the potentially grave effects of a change in climate which may take place in the near future, or whether we wish to do something about it while there is yet time. Should we choose the safer option, then the uncertainties in our present knowledge suggest the following safety strategy: We must strive to reduce the gaps in our knowledge through vigorous research, but at the same time we must take full advantage of the already available knowledge and initiate preventive measures.

If the grave consequences of the expected CO_2/climate problem are to be avoided, we must above all reduce the use of fossil fuels and achieve a balance between de- and reforestation. One of the most effective measures is the more efficient use of energy. This can lead to an immediate reduction in the use of fossil fuels and thereby to a reduction in CO_2 emission. Quite apart from the CO_2/climate issue this is also a necessity from the point of economic and energy policies. Such a step would give the necessary breathing-space for the transition to a sustainable future based on CO_2-free renewable energy resources.

Such a preventive strategy makes it clear that the CO_2/climate problem is not our inescapable fate. On the contrary, Man has the capability of influencing what will happen in the future.

II Climate and Climatic Change

By reconstructing the climatic history of the Earth and estimating future climatic changes by numerical modelling we hope to discover how climate could be influenced by CO_2 and what effects this influence might have.

The Earth's climatic history

Instrumental measurements and proxy data (e.g. changes in sea level, tree rings, pollen and ice cores) give information about air temperature, precipitation, soil moisture, etc. From this evidence parts of the history of the Earth's climate can be reconstructed. Throughout the history of the Earth, with the possible exception of the Mesozoic, periods of extreme heat and cold have alternated with each other on a variety of time scales - often abruptly within decades. In the long term, if the climate continues to follow its natural trend, a weak Ice Age will occur within the next 10,000 years or so. In the short term, it is likely that the activities of Man will lead, however, to an increase in temperature. This warming is not as yet rigorously detectable, partly because it is obscured by other natural variations in climate. For example, the oceans have the ability to absorb heat and this delays changes in the temperature of the atmosphere. Between the late-1930s and the mid-1960s the Northern Hemisphere actually experienced a cooling as a result of the climate system's natural variability.

The climate system

Climate, defined as the totality of all meteorological effects over longer time scales, is created by the interaction of the component parts of the climate system, i.e. the atmosphere, hydrosphere (oceans, lakes, and rivers), cryosphere (ice, snow), lithosphere (land and soil) and biosphere (Man, plants and animals). The parts interact by means of physical processes, e.g. radiation, heat transport and momentum. The internal interactions of the system are not linear, and include positive and negative feedbacks which may amplify or dampen variations. The response times and the spatial extents of the various interacting parts vary quite considerably. External influences must also be considered; for example, alterations in the intensity of solar radiation, changes in Earth's orbit, volcanic activity, and the ever stronger influence of Man, in particular his effect on climate through the increase in atmospheric CO_2. It is, therefore, not surprising that such a complex system is under-pinned by very complicated temporal and spatial variations.

The sun is the energy source for all climatic processes. The differing distributions of energy flows in the atmosphere and the heat transported by ocean currents determine the different forms of weather systems and thereby different types of climate. About half of the sun's short wave radiation reaches the Earth's surface and is thence re-emitted as long wave radiation. A portion of this is absorbed and re-radiated back to the Earth by gases which are found naturally in the troposphere and by gases which occur there as a result of human activity (such as carbon dioxide (CO_2), water vapour (H_2O), ozone (O_3), nitrous oxide (N_2O), chlorofluoromethanes and aerosols). Increasing the concentrations of these gases causes a strengthening of the so-called "greenhouse effect" and this is a significant factor in the expected global warming.

Climate: models, prediction and research

A main task in climatic research is to grasp, by means of numerical modelling, the complicated interactions within the climatic system. One of the goals of climate simulation is to detect different anthropogenic influences on climate before they appear in actual climatic data; and also, hopefully, to predict climatic trends over a long period. The mathematical models developed for this purpose vary from simple one-dimensional models to three-dimensional general circulation models which consider atmosphere-ocean interaction. Depending on the type of model, the individual physical processes are dealt with in exact or approximate forms.

For the purposes of the practical application of these models two types of climate-forecasting are distinguished. In the first, the internal, temporal development of the climate system is determined by given external parameters. By this method the development of climatic conditions can be predicted in terms of decades (a time scale relevant to agriculture) or millions of years (a time scale relevant for studies of the Earth's

Ice Ages). The second type of prediction is not concerned directly with the temporal development of the climate system, but, by using sensitivity tests, investigates the reaction of the model climate to artificially introduced changes (such as the elimination of feedback processes or increasing the CO_2 concentration in the atmosphere). The aim of these studies is to understand the mechanisms of the climate system so well that the best possible information can be made available to those decision-makers who are concerned with climatic change.

III Sociopolitical Aspects of the CO_2/Climate Problem

The future development of atmospheric CO_2 content and of other variables which influence climate is dependent upon a number of sociopolitical factors.

Population growth

World population is growing at break-neck speed. Death and fertility rates, as well as net migration figures, point to an increase from about 4 billion in 1975 to more than 6 billion by the year 2000. It is important to note that, despite a slightly lower rate of increase, in absolute terms, world population will grow faster (in 1975 the rate of growth was 75 million per year; in 2000 it will be about 100 million per year). It seems that 90% of the increase will be in the developing countries (DCs). The matter of when, and at what level the population will in future stabilize depends upon a series of unpredictable social, economic and religious parameters. Although we cannot be certain that world population will settle in 50 years at 8 or 12 billion, it is one important factor which determines future demands for energy and land resources and this cannot be without influence upon climate and the environment.

Settlement patterns

In the past, industrialisation has led to rural exodus and to urbanization. According to one United Nations forecast, urbanization in the industrialised countries (ICs) might grow from 70% in 1980 to almost 90% in 2030. In the same period, it might double from 30% to 60% in the DCs. On the other hand, an EEC report doubts that the decrease in the quality of life and the vanishing opportunities for employment in the large cities will lead to a worldwide increase in the level of urbanization. Should urbanization increase, however, an increase in the demand for energy could follow. The future level of the effect of these prognoses on climate and the environment depends on which of them proves to be true.

Economic development

The Gross National Product (GNP) is generally used as an index of economic growth, although it provides only a crude reflection of the economic and social conditions in a country. If, in the future, economic growth is concentrated on the service sector, which is more benign to the environment, climate and environment will benefit. Qualitative economic growth without an automatic growth in energy demand is not only possible, but also much to be desired.

Availability of fossil fuels

The climate of the future will be influenced by the use of fossil fuels (i.e. coal, oil, gas). An important consideration here is the amount of CO_2 released into the atmosphere per unit of time. This depends not only on the size and availability of fossil resources, but also on political decisions on output as well as export and tariff agreements.

The reserves of coal which can be economically extracted by present-day technology equal about 545 x 10^9 tce*. This quantity could be increased to about 1760 x 10^9 tce if techniques of extraction were improved. If we hypothesize that the present level of world demand (10 x 10^9 tce/yr) persists, coal could satisfy the world's energy requirement for the next 170 years. Three countries, the USA, USSR and China possess 82% of coal resources and 57% of coal reserves. There are, however, severe limitations to a significant upturn in the world coal production (about 2.6 x 10^9 tce in 1975) and in the world coal trade (about 200 x 10^6 tce in 1975), such as the high cost, the lengthy mining process and the protection of landscape, environment and climate.

The present-day levels of economically recoverable oil reserves is about 140 x 10^9 tce. Unconventional oil deposits (e.g. tar sands and oil shale) could lead to an increase in the amount of available oil, but they require a capital investment 10-20 times higher than in the case for conventional deposits and their exploitation would cause considerable damage to the environment. The high point of world oil consumption may be exceeded in the middle of the 1990s with a volume of 5 x 10^9 tce. Conventional oil deposits will probably last only 30 years.

Reserves of natural gas, which are economical today, are estimated to be of the order of 96 x 10^9 tce. Unconventional gas reserves can be found in geopressurized zones, as gas hydrates in the Arctic, in the Devonian Shale, and so on. The amount of available gas could, therefore, increase greatly, but the possi-

* 1 tce = 1 tonne coal equivalent (approximately 8140 kWh).

bility of using this gas is limited again by considerations of cost and environmental protection.

There are currently available, by present standards of technology, some 1,000 to 3,000 x 10^9 tce of fossil energy sources. Viewed simply from the point of view of availability and the likely level of use, fossil fuels could lead, within the next century, to the anticipated CO_2/climate problem; but reserves and projected fuel usage are not the whole story.

Future energy use

The degree to which climate may be influenced by Man depends, above all, on the amount and the type of energy sources, on efforts to end the inequities in energy use, and on market penetration times and pricing developments for certain types of energy.

For the purposes of the following discussion, we shall consider two well-known world energy scenarios for 2030 (a high scenario with about 36 TW* and a low one with ca. 22 TW energy demand) developed by the International Institute for Applied Systems Analysis (IIASA). Regarding the CO_2/climate problem, it is important to note that, in both scenarios, the amount of energy supplied by fossil fuels does not decrease much, namely from about 90% in the base-year 1975 to about 70% in 2030. Even more importantly, fossil fuel consumption is predicted to be 2-5 times higher in 2030 as compared with 1975. Such growth is, however, highly unlikely because of considerations such as availability of the huge sums of investment, and the problems of environmental protection, siting and acceptance, to mention but a few. If, however, such increases in fossil fuel consumption were to take place, they could result in a noticeable CO_2-induced climate change.

Moreover, the IIASA scenarios indicate that the inequities in energy use between industrialised and developing countries (and, therefore, also the tension between North and South) are more likely to be dissipated by a decrease in energy use than by an increase.

Observations of market penetration dynamics show that the market share of the different energy sources during the next 30-50 years is already being determined by present-day energy policies. A policy concentrating on coal and on products refined from it would mean an increased output of CO_2 over the next decades, while a strategy of more rational energy use without long market penetration times can lead to a lessening of the CO_2/climate problem.

---

* 1 TW = 1 Terawatt = 10^{12} Watt = 10^9 Kilowatt.

Price increases in the energy sector can damage the social and economic structure of a country. They can also, however, foster ingenuity and lead to better use of energy. The future development of energy will be determined not so much by the technical availability of resources as by the capital and maintenance costs of the conventional forms of energy and their alternatives.

Degradation of forest and soil

The destruction of forest and soil perturbs the carbon cycle which has a severe impact on the entire ecosystem, notably on climate and food production. Depending on whether the terrestrial ecosystems (forests, soils) act as a carbon source (e.g. through deforestation and soil degradation) or as a carbon sink (e.g. through reforestation) the CO_2/climate problem will either become worse or less severe.

IV Influence of Society on Climate

The activities of mankind have set in motion a geophysical experiment of global proportions which will, in a few decades, lead to detectable effects on the Earth's climate. If measures are not taken to correct this trend, irreversible changes to the climate could occur in the span of a human lifetime.

The CO_2/climate problem

The most important anthropogenic factor to affect climate is the increase in CO_2 concentration in the atmosphere. This has been continuously observed for the past 25 years. The causes of this increase are the burning of fossil fuels (coal, oil, gas and so on), gas flaring, cement production, deforestation of large areas and oxidation of soil organic matter. The chain reaction triggered by these operations affects the increase in CO_2 concentration and this increase causes heating of the lower atmosphere (the greenhouse effect). This factor will, in turn, bring about a distinct alteration in the regional and seasonal behaviour of climate; an alteration which, in view of increasing world population, could have catastrophic consequences for our water, food and energy supplies. In order to avoid these problems the influencing factors must be analysed and the results of this analysis used to determine what preventive measures should be introduced.

The natural carbon cycle

If undisturbed, the carbon cycle is in a state of quasi-equilibrium. In order to understand anthropogenic effects on the cycle, this natural state must be clearly understood. Three slow geological cycles and two fast atmospheric-biological-oceanic cycles can be distinguished within the carbon cycle. Only the relatively rapidly exchanging biological and oceanic reservoirs can absorb anthropogenic CO_2. However, these mechanisms - in particular oceanic absorption of CO_2 - work slowly compared to the human scale, so, given that the oceans could eventually absorb the excess carbon dioxide, it would be centuries before the atmospheric CO_2 concentration returned to its original state of equilibrium.

Perturbation of the natural carbon cycle by Man

The assessment of CO_2 production is the basis for the investigation of CO_2 concentration in the atmosphere and of the resulting climatic effects. Carbon dioxide emission from fossil fuel combustion has grown from about 0.1 Gt C in 1860 to about 5.3 Gt C in 1980. This gives an average exponential rate of growth of about 3.4% per year. Since the energy crisis of 1973/4 this has been reduced to about 2% per year. In 1860 the fossil fuel carbon emission was almost entirely from the burning of coal, whereas in 1975 43% was from oil, 39% from coal, 14% from gas, and 2% each was from gas flaring and cement production. The extraction of oil from shale and tar sands as well as synthetic fuel production (liquefaction and gasification) especially in autothermal processes (without the use of CO_2-free catalysts) frees even more CO_2 than does the burning of conventional fossil fuels.

The role of the biosphere in the carbon cycle is one of the most uncertain factors. Human interference in the biosphere is manifested by deforestation in the tropics, by biomass increases in some existing wooded areas in temperate latitudes, by influencing ecosystems in the polar areas and organic soils, by the formation of charcoal, the flux of organic carbon in rivers, stimulation of photosynthesis, and the reduction of CO_2-sequestering because of atmospheric pollution, notably acid rain.

The ocean is an important sink for atmospheric CO_2. The rate of uptake depends on the following factors: the speed of transfer of CO_2 into the surface water, the CO_2 uptake capacity determined by chemical equilibria, the transfer of CO_2 from the surface to the deep sea, and on CO_2 transport in the sea by biological processes. Although eventually all the CO_2 liberated by fossil fuels can be absorbed (provided the carbonates in the sediments become dissolved), in the near future the amount of CO_2 taken in by the oceans would, however, decrease with the increasing atmospheric CO_2 concentration.

Our present state of knowledge gives the following picture of the total budget. It is relatively certain that about 5 Gt C/yr are the result of fossil fuel use of which about 3 Gt C/yr remain in the atmosphere, whilst the oceans absorb about 2 Gt C/yr. The largest uncertainties exist in the biological sources and sinks. Based on the overlapping values of 1–4 Gt C/yr and 0.5–3 Gt C/yr for sources and sinks, respectively, one could deduce a balanced C-budget on purely numerical grounds. But major uncertainties remain and their removal is a necessary precondition for a realistic estimate of future CO_2 and climate developments.

Estimation, measurement and calculation of atmospheric CO_2 changes

To understand possible future developments it is useful to look at what has happened in the past. This we can do by considering the analysis of isotopes in tree rings and ice cores, for example. In Greenland the CO_2 concentration varied from around 200 ppmv* during the peak of the last glaciation (around 15,000 YBP**) to perhaps as high as 500 ppmv during the warm period of the Holocene (6,000 YBP), which suggests a breadth of temperature variation of -2.2 to $+1.8^{\circ}C$. Improved methods of measurement indicate that the value for the warm period is probably too high.

It is estimated that the CO_2 concentration in the atmosphere during the pre-industrial period was probably 260 to 290 ppmv. Given that the present value is around 340 ppmv, this indicates an increase by about 15 to 25% during the past century. CO_2 levels have only been measured regularly since 1958 at Mauna Loa, Hawaii and at the South Pole. At the moment the CO_2 concentration is increasing by 1–2 ppmv per year, and the amount of fossil fuel generated CO_2 which stays in the atmosphere (the airborne fraction) is currently around 55%.

In order to calculate the future development of CO_2 one needs a carbon cycle model (e.g. box models, box diffusion models, ocean-atmosphere-circulation models), into which one feeds assumptions about the temporal variation of the growth rate of fossil fuel use, use of individual fuels, oceanic absorption capacity, and biogenic CO_2 sources and sinks.

Simulations show that if, for example, fossil fuels were exhausted, even at a growth rate of only 1.5% per annum, the CO_2 concentration would still increase to a value of 7 fold higher than the present one and, independent of the rate of consumption, the concentration would remain roughly 6 times higher than at

---

* ppmv = parts of CO_2 by volume per million parts of air.

** YBP = years before present

present for centuries. This sort of level can be avoided by sensible energy use, which permits both the reduction of fossil fuel burning and the greater application of renewable non-CO_2 energy sources.

Climatic effects of a CO_2 increase

A CO_2 increase in the atmosphere affects the radiation processes of the energy budget of the climate system and hence will cause climatic changes. Numerical models and studies of palaeoclimatic and more recent analogues can be used to investigate these mechanisms and their effects. Climate models are, however, the only tools available to assess the effects of a future CO_2 increase. One can use them to study the perturbation of the model climate as a result of a doubling of CO_2 independent of time (equilibrium studies). Alternatively, one can study the response of the model climate to a specified rate of increase of CO_2 over a given period of time (transient studies).

Equilibrium response experiments

Sensitivity studies using the most sophisticated models available show that a doubling of the CO_2 concentration from 300 to 600 ppmv would raise the mean Northern Hemisphere temperature of the lower troposphere by about 3°C and that when the absorptive capacity of the ocean is taken into account this increase would still be of the order of 2°C. The increase in mean temperature in polar areas is especially marked, around 8°C. In the stratosphere temperatures decrease, with little latitudinal differences in the magnitude of the decrease.

For a four-fold increase of CO_2 to 1200 ppmv, the drift ice could disappear completely during the summer in both the Northern and Southern Hemispheres. The decisive effect is a world-wide change in the distribution of precipitation with corresponding effects on the water budget and on agriculture. A doubling of the CO_2 concentration could intensify the hydrological cycle by as much as 7%. Even more importantly, there would be various different regional and seasonal effects. For example, on the basis of these model calculations, soil moisture levels in the mid-west of the USA would increase in the winter, whereas during the summer the ground would be much drier than is presently the case. As this area is the breadbasket of the US, there could be grave consequences for the global food supply.

Transient response studies and comparison of model results

Temperature measurements in the Northern Hemisphere over the period from 1880 to the 1940s record a mean increase of the order of 0.5°C. Measurements since the 1940s show that after a cooling to around 1970 there has been a further small increase. The 180-year temperature cycle suggested by isotopic measurements in Greenland ice cores would have resulted in a temperature decrease since the 1940s. Since this cycle also implies that the natural trend of temperatures would be upward by the end of the millennium, it is indicated that a continuing increase of CO_2 concentration coupled with this would cause a global warming such as has not been experienced for at least the past 1,000 years.

Contrary to previously held opinion, the CO_2 effect (the signal) should first be observed in climatic data, not for northern areas in winter, but for middle latitudes in the summer. Because of the thermal inertia of the oceans this signal's appearance will be delayed for several decades and it will probably only become identifiable in the natural temperature record in the 1990s.

A critical comparison of the various model results shows that almost all climate models suggest that a doubling of the CO_2 concentration would cause an increase of 3 ± 1.5°C in average global surface temperature. Simulations which give much smaller values can be shown to have ignored important feedback mechanisms.

One must also consider that specially constructed time-dependent models react differently to external disturbances than do models which assume an equilibrium response. The important matter of whether we can reliably estimate the climatic effects of a continuing increase in CO_2 before the climate system itself presents us with the result by carrying out its own geophysical experiment, must for the time being remain unanswered.

Climatic effects of other trace gases

Besides CO_2 there is a certain number of other trace gases produced by Man, whose effects must be added to the global greenhouse effect. Among these are: nitrous oxide, chlorofluoromethanes, ozone, methane, carbon monoxide, ammonia, other nitrogen oxides and sulphur dioxide. It is important to note that while CO_2 is currently believed to account for some 60% of the main causes of anthropogenic global warming of the lower atmosphere, the other trace gases, because some of them tend to have longer residence times and because they have a higher growth rate, may in the future take on a greater importance compared to CO_2. It is therefore necessary that a realistic overall judgement of the effects on climate must include **all** trace gases. Concentrating on CO_2 alone will certainly produce an underestimate.

Climatic effects of aerosols

It is difficult to estimate the total effect of aerosols on climate because it is not only dependent on the surface albedo and the distribution of different types of aerosols over the continents and oceans but also on the varying vertical profiles of aerosols and the change in the mean solar zenith angle. The present state of knowledge indicates that the total influence of aerosols will probably lead to a small degree of regional warming. In some regions aerosols seem to increase atmospheric stability and therefore lead to less convective precipitation which could affect the water budget and agriculture.

Climatic effects of land use changes

Changes in land use due to excessive grazing, destruction of tropical rain forest and desertification, urbanization and industrialization, energy projects (e.g. solar farms and bioplantations), and drainage and irrigation schemes all have consequences for the climatic system. The associated changes in the albedo and in the aerodynamic and hydrological characteristics of the Earth's surface can affect the global heat and water budgets and therefore also the climate.

Climatic effects of waste heat

Every time energy is generated, waste heat is set free. Although at the moment high emissions of waste heat can produce marked effects in climate at the local and even regional levels, it is rather unlikely that, in the future, world energy development will give rise to global effects.

Combined greenhouse effect and critical threshold values

To obtain a realistic grasp of the total greenhouse effect it is important to look at both the "real" CO_2 concentration (i.e. without the other trace gases) and the "virtual" CO_2 concentration (i.e. CO_2 plus the other trace gases). Because of this an equivalent level of the "real" CO_2 concentration would lead to an increase in temperature at an earlier point in time.

Comparison with various warm climatic periods in the past permits a meaningful estimation of the critical threshold values. If we want our climate to remain stable the equivalent CO_2 concentration should not exceed 400-450 ppmv. This first critical threshold range implies a warming of between 1 and 1.5°C - something which has not occurred since the Middle Ages (around 1000 AD). A CO_2 concentration of 600-700 ppmv, which includes the effects of the other trace gases and would therefore correspond to a substantially higher "virtual" CO_2 concentration, could produce a catastrophic change in climate. Such a level would lead to a 4-5°C warming in mean global temperature, such as was the norm during the early Tertiary period (5-3 million years ago). The palaeoclimatic evidence shows that such temperatures result not only in an ice-free Arctic ocean, but also in a shift of climatic zones which would have wide-ranging implications for world food supply and which might eventually lead to irreversible changes in climate.

Unless preventive measures are taken, this critical value will be reached some time during the next century. There is, therefore, the danger that, the longer such precautionary measures are delayed, the more drastic they will have to be. The CO_2/climate problem is therefore already very pressing.

V Impacts of Climate Change on Society

The probability of a global change in climate set in motion by an increase in atmospheric CO_2 and other factors, underlines the advisability of assessing now the possible wide-ranging consequences this would have for human societies.

Nature and methods of impact analyses

In order to assess the consequences of increasing CO_2 we use climate impact studies which have been specifically developed to investigate the interaction between climate and society. These studies use systems analysis, which, with the help of internally consistent scenarios, illuminate the complex set of relationships between climate and society. The insights gained by this method are an essential precondition for a pertinent evaluation of future risks and of the necessity and suitability of precautionary measures. As yet, we lack a suitable instrument which can adequately describe the chain of cause and effect which runs as follows: CO_2 increase → climatic change → socio-economic impacts → response of society.

Ecosystems

There is now scarcely any ecosystem which has not been influenced by the activities of Man, among the most threatened being the tropical rain forests (through the search for cultivable land, fire clearance and the use of wood), the steppes and semi-arid areas (by overgrazing and removal of wood for fires), and the forests of the middle and northerly latitudes (by acid rain). A CO_2-induced warming goes hand in hand with a quite different distribution of the temperature and precipitation patterns. In marginal areas, e.g. in semi-arid zones, trees are already highly susceptible to small decreases in precipitation.

The much greater increase in temperatures which is to be expected in polar latitudes would especially affect areas of permafrost and tundra. Here, a warming could lead to the dessication of swamps and peat bogs and to the release of their carbon matter into the atmosphere as CO_2, which would increase the trend towards warming even more. This amplification, a classical example of positive feedback, must be taken very seriously, for, apart from the increase it would cause in the rate of snow and ice-melt, it would additionally affect the whole global climatic system.

Energy use

It is only since energy has become an expensive commodity that we have realized what a decisive role climate plays in the field of energy production. Climate is both a threat and a resource. Climatic extremes cause an increase in the demand for energy through increased heating in very cold winters and increased use of air-conditioning units in very hot summers. Much energy is also required to shield us from other aspects of weather and climate such as storms, snow and ice. On the other hand such climate elements as solar radiation, wind, precipitation and so on, supply us with inexhaustible resources from the solar, wind, biological and oceanic energy systems as well as hydroelectric power.

In the eastern half of the USA alone, cold winters regularly cause several billion dollars expenditure on additional energy requirements, while in California many hundreds of millions of dollars must be spent in combatting periods of drought. Cost estimates for the USA show that an increase in temperature of 2^oC (the assumed value for a doubling of CO_2) will, against all expectations, lead to an increase in energy costs. This is connected with the transfer of requirements from the cheaper primary energy sources, gas and oil, to the dearer, secondary energy source electricity which powers air-conditioning units.

Food security

One of the greatest problems currently facing Mankind is the provision of the growing world population with sufficient food. As there are already 500 million people who go hungry and as in the next 20 years the world population will increase by a further

2,000 million, the question whether the world food situation will be further endangered by a CO_2-induced climatic change is of more than academic interest.

Modelling the dependence of harvest yields upon climate shows that with a temperature increase of 2°C, which is expected in the middle of the next century, US corn (i.e. maize) production might fall by about 26%. It generally holds for the USA's granary that cooler and moister conditions increase the yield whereas warmer and drier conditions decrease it. Climate models show that we may expect drier conditions in the USA's corn belt. Wheat is especially susceptible to changes in temperature. A rise of 2°C in temperature could lead to a fall in yield of about 10% in the USA's wheat belt, while in Kazakstan, USSR, a temperature increase of 1°C in combination with a 10% decrease in precipitation could conceivably lead to a 20% loss of production. On the edge of the tropics, however, a CO_2-induced temperature increase could work positively, for climatic modelling shows that the hydrologic cycle would be intensified in these regions. The expected increase in temperature of 0.5–1°C could lead to an increase of rice production of 15%.

Warm and moist conditions result in increased damage to crops by pests and in a higher incidence of plant-disease, both of which can cause poor harvests. An integrated pest management (e.g. methods of biological and climatological control, increased resistance to pests, preservation of the world's genetic diversity, better methods of cultivation) could offset the risk of a climatically-induced fall in yield due to parasites.

Under artificial, optimal conditions of temperature and humidity, higher levels of CO_2 can lead to an increase in photosynthetic activity and plant dry-matter production. This is somewhat different under natural conditions since the availability of water and other nutrients essential to growth apparently play a more decisive role. It is therefore improbable that a climatically-induced reduction in yield due to an increase in CO_2 would be balanced by a direct increase in the amount of growth of the biomass.

Even a higher level of available technology does not seem capable of reducing the vulnerability to climatic impacts. The periods of extreme heat and of drought in 1980 alone caused damage to US agriculture estimated at $19 billion.

Water resources

All life depends on the water cycle. Drought and flood alike cause harvest failure. If the water table falls, the water temperature increases as does the level of toxic matter and, finally, fish suffocate. Problems of drinking water, hygiene and health increase. A lack of water also implies a reduction in the capacity for energy generation (too little cooling water, rivers too warm, empty reservoirs), a reduction in fuel extraction (oil-shale, refining of coal), and a disruption to the transport of goods and to canal traffic.

CO_2-induced climatic change could, in causing the disappearance of the Arctic sea ice during the summer, cause a shifting of the Earth's climatic zones which could in turn lead to a critical state of affairs in the matter of water supply, especially in the industrialized countries of the temperate zones, but also in the Mediterranean area, in the Near East and in California. Suggestions for increasing the amount of water available have included desalination of sea water, cloud seeding, the transporting of icebergs from the Antarctic or the diversion of rivers. Large-scale desalination and the transport of icebergs are too costly in terms of energy, and the production of rain by artificial means has not met with statistically significant success. The diversion southwards of great rivers in the USSR and Canada in order to bring water to large agricultural areas could cause an increase in the salinity of the Arctic ocean and thereby delay the formation of the Arctic pack ice, which would further influence the global energy budget, atmospheric circulation and the climate.

Fisheries

Solar radiation, atmospheric pressure and wind patterns, as well as the circulation system of the oceans have a great influence on fish stocks. The survival of the fish larvae, the migration to the spawning grounds and the time to reach pubescence all depend heavily on these elements of climate. Herring prefer cold ocean currents while cod prefer warmer waters. The anchovy catch off the Peruvian coast is good when oceanic upwelling brings water rich in nutrients near to the surface and bad when the upwelling weakens. A CO_2-induced climatic change could increase the magnitude of such perturbations. Aquaculture is a way of trying to achieve fisheries which are independent of climatic fluctuations, but how far economic considerations will allow this method to achieve a reasonable volume of production remains to be seen.

Population and settlement

In human history climate has always played a decisive role in the growth and decline of civilizations. For example, the sudden decline of the Mycenaean culture in Greece between 1300–1200 BC has been connected with regional precipitation anomalies. A mild climate (1–2°C higher than the present day mean) and luxurious vegetation fostered the settlement of Iceland by the Vikings around the end of the first millennium. The fall of the ancient native Indian culture in the USA Mid-West may have been hastened by a fall in temperature (the Little Ice Age) and a reduction in precipitation. This example is of special significance, because this region, in which the drought lasted for more than 200 years, is today the USA's main producer of spring wheat, corn and soy beans. Between 1300 and 1500 AD in central, western and northern Europe, about 20–60% of all villages were deserted. The Little Ice Age was also an important contributory factor here.

A warmer climate could raise the sea level by thermal expansion. However, the melting of the Arctic ice masses and a resultant increase in sea level are those phenomena which the public most often relates to a CO_2-induced warming. But, sea level cannot be influenced in this way, because the Arctic pack ice is floating. Even the great ice sheets of Greenland and the East Antarctic are relatively insensitive to a warming of the expected extent.

The relationship is, however, different as far as the smaller, western Antarctic ice sheet is concerned. Here the expected rise in temperature could cause about 2–2.5 million km^3 of ice to surge into the sea, melt and increase the sea level by 5–6 m. Estimates show that this could flood 2% of the land area of the USA and, in the case of the present distribution of the population, render 12 million people (6% of the total population) homeless. As another example, in the Federal Republic of Germany, Lower Saxony, Bremen, Hamburg and Schleswig-Holstein could be inundated with a loss of 16% of land and the evacuation of about 2 million people (17% of the total population).

Although the melting of the Arctic sea ice would not increase mean sea level, an ice-free Arctic ocean could lead to shifts of the climatic zones. The situation would be especially critical were the arid areas to spread into southern central Europe and into the central portion of North America – a development which could have serious consequences for the world's water and food supplies. On the basis of the geophysical processes involved, it has been argued that the rate of melting of ice due to a CO_2-induced warming would be faster for the Arctic sea ice than for the Antarctic ice sheets.

Health, illness, welfare and leisure

Human health is heavily affected for good or ill by prevailing weather conditions. The climatic conditions propitious to human physiology are narrowly defined. Specific causes of disease also have their optimal climates. It is generally true that warm, humid conditions favour the spread of disease.

The expected warming could strengthen summer heatwaves. The summer 1980 heatwave which affected the USA caused the deaths of over 1,300 people and damage to the value of about $20 billion.

Leisure habits are also acutely susceptible to prevailing weather, climate and their contingent anomalies. Climate impact studies could save this lucrative industry very large sums of money.

VI Strategies for Averting a CO_2/Climate Problem

A precautionary strategy, based on low-risk climate, energy and land use policies gives decision-makers more room to manoeuvre.

Range of strategies

The effect of increasing CO_2 on the climate is not, at present, unequivocally detectable in instrumental data. This makes it easy to put aside the thought of a possible CO_2/climate problem. Our dilemma is that today's energy policies, relying as they do primarily on fossil fuel energy sources, are already now determining the damage to the climate system which will occur in the next few decades and in the more distant future. If we were to wait for detectable, global effects, it would be too late to take preventive action. Such action must, if it is to be effective, be taken now. Room to manoeuvre is, however, not inconsistent with a flexible approach which allows wrong decisions to be revised and new knowledge to enter and to help shape decision-making.

Technical fixes

Technical means have, until recently, been the only ones considered pertinent to the reduction of the CO_2/climate danger. Amongst these are removal of CO_2 from flue gases, or from the atmosphere and the oceans; storage of CO_2 in the oceans and in caverns; partly also the use of CO_2 in coal-based chemical industries; and lastly, efforts to compensate for increasing CO_2 by modifying the albedo. The problem is one of volume and cost. The present-day volume of CO_2 production is about 20,000 million tonnes per year, compared to about 300 million tonnes of dust per year and about 150 million tonnes of SO_2 per year. The unanimous opinion of all experts who have looked closely at such technical fixes to the CO_2 problem is that they require too much energy and are therefore unaffordable.

Biological methods

Plants and soils can be considered candidates for biogenic absorption of CO_2. A carbon bank, similar in function to a blood bank, but composed of fast-growing trees and water plants, has been suggested. An American species of Sycamore absorbs

750 t C/km^2/yr. These CO_2 sinks would only be effective for a few decades, i.e. for as long as the plants are actually growing. Such a strategy would, however, win valuable time, enabling long-term measures for the reduction of CO_2 to be brought in. In addition, reforestation and the preservation of humus is essential from an ecological point of view alone.

Energy conservation through rational energy use

As a result of the energy crises, most countries began to revise their energy policies and to concentrate on more rational uses of energy and thereby also on its conservation. In as far as measures for energy conservation preserve and even replace the energy sources, more efficient energy use effectively offers an additional, domestic energy source which reduces the requirements for imports. Conservation allows a reduced use of fossil fuels, thereby resulting in a reduced impact of CO_2 on climate. Some sample studies on this topic are given below.

The study of the Öko-Institut (1980) comes to the conclusion that present-day (1973 taken as base year) Federal German energy use technology could be improved by a factor of 2. This could be achieved by the improved technical efficiency in the conversion, distribution and use of energy, as well as through better ways of putting energy to work. Thus in the year 2030, primary energy use could be 40% less than in 1973. For a variety of energy scenarios it was shown that CO_2 emissions could be significantly reduced by 29% to 53%.

In a report of the Enquete Kommission of the German Federal Parliament (1980) four paths for future energy use have been defined. Paths 1 and 2 follow the supply-oriented conventional approach. In paths 3 and 4, on the other hand, demand is the starting point, and it is considered how demand can be reduced by more rational energy use. For the year 2030, the four assumed cases give the following alterations in CO_2 emissions (compared with 1978): path 1, an increase of about 35%, as opposed to decreases in the cases of paths 2, 3 and 4 of about 9%, 15% and 36%, respectively.

An energy study for Great Britain (1979) shows that, if energy saving technologies are introduced, it is possible to achieve economic growth while maintaining zero growth in energy use. The analyses show that, over the period from 1976-2025, it would be possible, in a low-energy scenario, to achieve a reduction of CO_2 emissions by about 24%, while a high energy scenario would give a reduction of only 10%.

The Swedish Parliament, in 1981, passed an energy law whose aim was the halving of oil imports and more efficient energy use within the space of ten years. According to the measures contained in this law, by 1990, in comparison to 1975, CO_2 emission would be reduced by 3.9% through oil-substitution and by 7.5% due to more efficient energy use.

The Harvard Study (1979) comes to the conclusion that about 40% of the energy used in the USA in 1973 could have been saved through rational energy use. By 2000 renewable energy sources could account for about 20% of total primary energy use in the USA. If such a policy were introduced in the world's most important industrialized nation, it would constitute a decisive signal to other countries and could thereby reduce other environmental problems as well as that of the effect of CO_2 on climate.

In a further US study, the CONAES report (1980), it is argued that the growth in energy use can be slowed down without endangering economic well-being. If, as suggested, the synthetic fuels industry increases sharply, this could rather accelerate the present rate of increase of CO_2 in the atmosphere.

Energy productivity is low in the industrialized countries. In developing countries, despite their much lower consumption, it is even worse (e.g. because of cooking on open fires, obsolete machines). It is precisely when the level of use is very low, and especially when domestic resources and private capital are lacking, that more efficient energy becomes the most important energy source.

Renewable energy sources

The chances of fulfilling a significant portion of energy requirements with renewable energy sources remain slim for a strong exponential growth rate, but they increase rapidly when energy requirements are reduced by efficient energy use. The CO_2/climate problem can be avoided by the following two-pronged strategy: first, energy use should be significantly reduced, and, second, the much reduced energy requirement could then be met using CO_2-free renewable energy sources. There are also many other reasons why we should conserve exhaustible and irreplaceable fuels such as crude oil, natural gas, coal and uranium, and cover our energy needs as quickly as possible by renewable energy sources. A sustainable and ecologically sound solution is only possible if based on renewable energy sources.

There already exist energy programmes for the USA, California, Quebec, Saskatchewan, Sweden, Scandinavia, France, Sardinia and India, which accept the complete replacement of non-renewable energy sources by renewable ones between the years 2015 and 2050. The maximum global potential of such resources has been assessed at about 20 TW and 26 TW by IIASA and a UN Conference, respectively. A former researcher at IIASA gives the extreme limit as being between 78 and 283 TW (present world energy use is about 9 TW), though this could only be achieved by massive development in desert areas which receive a lot of sunshine. The problem of using renewable energy sources is, therefore, not one of quantity, but rather one of choosing the most economically rewarding possibilities.

The effects of various energy strategies on CO_2 and temperature changes

According to the energy policy principle "move away from oil", the synthetic fuel industry (i.e. coal gasification and coal liquefaction) has been fostered. Analyses of scenarios have come to the surprising conclusion that the level of CO_2 will double between 1978 and 2050, independent of whether only the USA or the whole world builds up the coal-refining industry. This leads one to suppose that it is not so much the sort of fossil energy source as the volume of fossil energy used which is decisive in the matter of high CO_2 concentrations. How far allothermic processes, where the refining process takes place with the help of non-fossil fuels, affect CO_2 concentrations is not known at present because of the lack of research in this area.

In the strategy relying on fossil fuels, it was assumed that the global use of energy would asymptotically approach a value of 30 TW by the end of the 21st century. Modelling exercises show that this would result in a roughly 500% increase in CO_2 emission rate to about 24 billion tC/yr, and about a 300% increase in CO_2 concentration (to over 1,000 ppmv) with a consequent mean global temperature change of about 4°C above the 1967 base value.

The high IIASA scenario assumes an increase of 435% in the energy requirement for the year 2030 as compared with the base year 1975 (8 TW) and the low scenario an increase of 273%. In these scenarios the high (roughly 92%) proportion of fossil fuel use is reduced only to 69% by the year 2030. In the case of zero growth scenario - a misleading term - a doubling of the energy requirement to 16 TW is assumed. The efficiency scenario, on the other hand, has the following starting point: by first raising productivity, energy services can be provided with a significantly reduced energy input. The resulting, sharply reduced requirement for fossil fuels can then quickly be replaced by renewable and CO_2-free energy sources. In this case the demand for primary energy sources would be reduced by 14% in 2000 and 37% in 2030 compared with 1975. At the same time, the high fossil fuel proportion of 92% in 1975 would be lowered to 71% in 2000 and to 18% in 2030.

From what we know of the global carbon cycle and from climate modelling, IIASA's high scenario in 2030 would give CO_2 emissions 342% higher than in 1975 and a CO_2 concentration 50% higher than in 1975, resulting in a 4 fold increase in excess temperature over the 1975 value from 0.4 to 1.6°C. Against this the efficiency scenario indicates that the level of CO_2 emission in 2030 would be an eighth of that in 1975 so that the concentration would increase by only 11%. The increase in temperature is then, at 0.8°C, only half as great as in the case of the high IIASA scenario.

The purely numerical values of these simulations must be interpreted with the usual reservation because of the many remaining uncertainties - this is, of course, true of all projections. It appears, however, that only the steps envisaged in the efficiency scenario will lead to a perceptible weakening of the trend towards a temperature increase. This would win us something very important, namely time for an orderly transition to a sustainable energy future on the basis of renewable energy sources which do not uncontrollably alter the environment and the climate system.

Associated measures and aids to decision making

Alongside these direct, precautionary measures there is a multitude of other activities which have goals which are unrelated to the reduction of CO_2, but which, because they make our environment less susceptible to a change in climate, have a positive side effect. Amongst these are soil conservation, the fight against desertification, the application of innovative agricultural technology, the establishment of food reserves, improvements in water management, building sea defences and the development of aid programmes.

Appropriate decisions depend on a good information system. In order that these decisions can be made in good time, we need a warning and prevention system. It is not sufficient simply to register climatic events, or merely to calculate possible CO_2-induced climatic changes. Such knowledge must be brought into the decision-making process and must have an influence there. Only with all of the steps - information collection, dissemination of information, use of information and introduction of precautionary measures - is the circle of actions complete.

VII Opportunities for the Future

What realistic possibilities can lead us out of the climatic danger brought about by CO_2 and other factors? We must learn to think differently and to realise that the climate, as a part of our environment, is common property which must be treated with respect. The key to avoiding the CO_2/climate problem lies in laying new emphasis on a rational energy and land use policy.

The old credence in progress - the new credibility

Technical developments influence the environment and climate. The reaction to this is frequently only to ascertain and to measure the effect after it has been noticed, without, at the same time, taking precautionary measures to hinder its escalation. Helplessness in the face of complex technological developments causes fear and a lack of trust. The lost credibility can only be restored and the fear of life in the face of a future inimical to it can only be reduced, when, in its place, we adopt a policy of mutual trust and active cooperation towards a more humane way of life.

Man: master of his own fate?

The sudden collapse of the Maya culture about 1,000 years ago and the chronic drought and famine in the Sahel are often adduced as examples of what happens when the carrying capacity and hence its own ecological foundations are destroyed. Are these and the possible CO_2-induced climatic changes fated events to which Man is helplessly doomed? The CO_2/climate experiment which we have set in motion is already developing its own momentum within the environment. But for a short time, about the next 15–20 years, there is a good chance of slowing down the chain of events and of bringing about a change in their direction. Unlike the Maya we have a greater potential, not only to analyse our problems, but also to estimate their consequences and thus, by timely (i.e. now) application of precautionary measures to avert the threatening catastrophe. We still have the chance of choosing between different roads to the future.

Prospects: extrapolating old trends

One approach to the future could be the continuation of present energy policies. Many managers in business as well as in the field of energy cannot conceive of a future without quantitative growth. On the other hand, it is clear to systems analysts that in a closed system, such as that of the Earth, boundless growth at a constant rate is absolutely impossible. Clearly, a new philosophy, namely that of qualitative growth is needed, for the decisive question is not whether there should be growth or not, but rather how, and in which direction it should proceed.

The inequality of opportunity between industrialized (IC) and developing (DC) countries presents one of the greatest areas of potential conflict for the future. Many forecasts advance the idea of a large growth of energy use in DCs as well as in ICs, as if only in this way the inequity between them could be reduced. This theory offers no explanation of how, in the past, during times of rapid economic growth, rapidly increasing energy use and low cost energy, the gulf between ICs and DCs grew ever greater. Recent scenario analyses show, however, the exact opposite; that is, when energy use grows more slowly there is a shrinking of the differences between ICs and DCs and, therefore, improved opportunities for both types of country. A more efficient use of energy, and the consequent reduction in energy requirements (especially in ICs) can, in preserving both climate and evironment, lessen the tensions between north and south. Alternatively, if the present trends continue, one must reckon with a quicker dwindling of resources, causing price rises and more acute disputes over distribution. This, in turn, will cause greater stress on world supplies of food, raw materials and energy, thus increasing the potential of world-wide conflict.

Prospects: shifting emphases

A more reasonable approach to the future demands a new perspective. This is necessary because, until now, official forecasts of economic and energy growth have been very wide off the mark, and the former strict rule that economic growth and increased energy use were linked has been exposed as a myth. A decline in the absolute amounts of energy can be achieved while maintaining standards of living equal to today's or even higher. This decline is, furthermore, essential in the interest of a healthy world economy (since a falling rather than a rising level of use of fossil fuels is the most economical way of supplying energy), and thus essential for a reduction of the potential for global conflict.

Much could be achieved simply by applying our present technical knowledge in a rational way. The overall efficiency of conventional power stations could be increased from about 36% to 85% simply by cogeneration, i.e. by combined heat and power generation. By this expedient not only could a considerable amount of energy be saved, but emission of gases (particularly CO_2, SO_2, NO_x, etc.), aerosols and waste heat would be significantly reduced, too.

By building decentralised local power stations for heating, the losses incurred in transporting energy and the investment needed for power lines and grid systems could be substantially reduced. Over and above this, such power stations, when they comprise a series of generators which could be shut down or brought into service according to demand, operate at maximum capacity and therefore very efficiently.

With fluidized-bed combustion, not only is efficiency increased by 10%, but emission of sulphur dioxide and nitrogen oxides, and hence the acid rain problem, is also drastically reduced. If such technology is used in conjunction with combined heat and power generation, a considerable amount of energy is saved. This accords well with an energy concept which seeks a reduced reliance on imported oil and a more intensive and environmentally safer use of existing coal deposits.

Power companies have, until now, shown no great interest in the use of waste heat. The building of decentralised heating stations and the development of district heating grids would constitute a meaningful job programme which could be begun immediately. Besides conserving energy, this would guarantee secure employment for many thousands of people for years to come. No country can, in the future, afford to reject the use of waste heat, i.e. a domestically owned energy source immune to international energy and price cartels.

In the field of heat pumps there already exist proven systems suitable for use in industry and business. These would make possible considerable savings in energy use. As space heating constitutes a large part of the energy requirements of industrialized countries, efficient insulation is an effective and many-sided means of reducing consumption. Sodium-vapour, quartz-halogen and other lamps which are available now use 75% less current than a conventional light bulb while lasting 4-5 times longer. Innovative, but existing techniques can reduce the fuel consumption of motor cars from the current level of 8-10 litre/100 km to about 2 litre/100 km. Co-ordinated transport and town and country planning policies, coupled with a switch to a hydrogen based economy could reduce still further the use of fossil fuels and, therefore, the risk to our climate and environment. Microelectronics can also help to save energy.

Exhaustible fossilised carbon is too valuable simply to burn it. Up to the present about four million different chemical processes have been built on this corner-stone of life. The field of application of carbon-based chemistry extends from medicine to the construction industry.

A new perspective demands, above all, the breaking down of institutional barriers. For instance, in the Federal Republic of Germany, a basic cause of much grievance is the 1935 energy tariff law, which is still in force today. Because of the present tariff structure, many heavy users and even squanderers of energy are subsidized by a special tariff, whereas those who are careful in their use of energy are penalized. The extensive construction of large coal-fired and atomic power stations is a massive hindrance to the chances of more rational energy use based on combined heat and power generation and local heating networks. In the interests of a reasonable energy supply system it is absolutely necessary, not only to set the right price signals, but also to abolish the existing market distortions. The Federal German Enquete Kommission has suggested a detailed catalogue of measures to lower such barriers.

Ways out of the CO_2/climate threat

The CO_2/climate problem is not our inevitable fate, one with which we are helplessly confronted. We have at hand a variety of options which we may use to determine the desired future course of events. Above all, it is our energy policy which sets the points for the future development of our climate.

This insight has spawned much rethinking which itself rests on empirically and methodologically convincing energy studies carried out for about two dozen different countries. These all show that more energy services can be provided with less total energy simply by using it more productively. This encouraging result paves the way for a healthy energy economy which is the key factor in the quest of avoiding a CO_2/climate problem.

At the essence of such a low-risk energy and climate policy is the more efficient use of present energy sources. By following this course, total energy demand is reduced, thus conserving the irreplaceable non-renewable sources and increasing the relative importance of the CO_2-free, renewable energy sources – sources which initially can only be introduced in a limited way. This two-pronged strategy has a decisive advantage over all other possibilities. It can actively introduce more efficient energy use into energy policy without a long lead time and with clearly visible success, while simultaneously leaving sufficient time for the well-planned introduction of renewable energy sources.

This concept must be supplemented by a land use policy as characterised by the balance between de- and reforestation and by the goal of soil conservation. Such a policy brings many benefits. It not only reduces the CO_2/climate problem, but also contributes decisively to the security of the food supply by combating soil erosion, preventing floods and improving the water budget.

These strategies would lead us from the present throughput economy to a cyclic economy, thereby ensuring a stable social order. The advantages of such a policy are many, from lessened dependence on external resources, reduced deficits in the balance of payments, and job security, to the reduction of technological risks and associated problems of acceptance of new technology as well as putting a brake on the destruction of the environment.

This problem for humanity demands increased attention now. The pressing nature of the CO_2/climate threat should give an added impulse to the introduction of precautionary measures by which this danger can be reduced. The chances of success are by no means small, because a policy involving the least economic cost also involves the least risk to the climate. Nonetheless, this rational economic and energy policy will have to fight against considerable opposition by influential lobbies. In this respect we must call upon every citizen to help.

It is not enough to know,
application is required.
It is not enough to wish,
action is required.

Johann Wolfgang von Goethe

I. INTRODUCTION

The climatic history of the Earth is marked by a great number of **natural climatic variations.** In the last two million years of geological time roughly 20 glacials and interglacials have succeeded each other. We also have convincing evidence from the varves and tree rings which point to quite abrupt changes in climate on time scales relevant to human life (Flohn, 1978).

A well known example of sudden cooling occurred in the Lüneburg Heath area. Here Müller (1974) found that, 300,000 years ago, a forest which indicated a warm climate disappeared and was replaced within the space of 30 years by an Arctic birch wood. Abrupt warmings are also known. The research of Coope (1977) in central England has shown that a sub-Arctic beetle fauna was forced to retreat by a fauna acclimatized to warmer conditions in only 50 years. This indicates an increase of 8–9°C in the July temperature. What has happened in the past is also possible in the future.

Recently another factor capable of perturbing the delicate balance of forces which determine the Earth's climate has come into play. **Human activity** has reached magnitudes which can affect climate locally and, to some degree, regionally. There are strong grounds for suspecting that the "**geophysical experiment**" (Revelle and Suess, 1957) has already reached a global extent, and that, within a few decades, it might lead to detectable changes in climate which could be irreversible on the human time scale.

The main cause of this is the increasing concentration of CO_2 in the atmosphere due to the burning of fossil fuels (oil, coal and gas) and to the destruction of large areas of forest leading to an enhanced oxidation of soil organic matter. Through these processes CO_2 is given off into the atmosphere more quickly than it can be absorbed by natural sinks. The chain reaction of this experiment leads on from increasing CO_2 to an increased absorption of long wave radiation by the lower atmosphere (the **greenhouse effect**) and to a general warming. This, in turn, leads to a great variety of regional and seasonal variations in climatic events which, given the Earth's increasing population, could lead to shortages in food and energy supplies. Over and above this, other gases, aerosols, changes in land use and, to some degree, waste heat have an influence on climate which should not be underestimated. The total effect of all these factors could seriously endanger the global climate. The CO_2/climate problem* is, therefore, a crucial matter for the coexistence of people and the survival of Mankind.

It is important to make it very clear at the beginning of this book that the statements made so far and the following presentations are based on the present state of knowledge which is fraught with many **uncertainties.** It is, for example, possible for us to convincingly demonstrate past climatic changes and to date them fairly precisely, even though we do not know the exact causes of these changes. We also know that climatic change in different periods has occurred abruptly and in different ways from one region to another, but here too we cannot precisely forecast when and where this will recur in the future. We cannot yet detect the influence of Man on global climate - the "signal" - because its present, limited scale lies hidden in the "noise" of natural climatic variability. Lags in response due to the thermal inertia of the oceans may delay its recognition for decades.

It is, therefore, feared that by the time we can unambiguously detect an anthropogenic effect on climate, the experiment we have set in motion will have gone too far for us to do anything to halt it. Finally, we continually are made aware that climatic variability and anomalies may drastically and fatefully affect social and political events in particular regions. We are, however, unable at present to make confident quantitative statements about the effects of future climate changes on the highly differentiated areas of society.

The fact that, at present, considerable gaps in our knowledge exist, must not lead to the false conclusion that there is no **cause for concern.** Quite the opposite is true, for if one is in a state of uncertainty, it necessarily follows that things may go either way. In such a situation it is advisable to follow a safety strategy: while no effort should be spared to lessen the gaps in our knowledge, the information to hand should already be used to introduce **precautionary measures.**

This approach is also suggested in the formal **Declaration of the World Climate Conference**, which took place in Geneva in 1979 under the auspices of the World Meteorological Organization and other international bodies (WMO, 1979). The following **appeal to all nations** went out from the 400 or so experts from various disciplines:

> "Having regard of the all-pervading influence of climate on human society and on many fields of human activity and endeavour, the Conference finds that it is now urgently

---

* From hereon this abbreviation will also include all the other effects.

necessary for the nations of the world:

- to take advantage of man's present knowledge of climate;
- to take steps to improve significantly that knowledge;
- to foresee and to present man-made changes in climate that might be adverse to the well-being of humanity."

The content and structure of this book reflect this approach. The foundations of climatic and sociopolitical knowledge will be used to gain a better understanding of the CO_2/climate problem in Chapters II and III, respectively. In Chapter IV, I shall deal with the relative order of magnitude of the different anthropogenic factors which influence climate and the climatic effects which they cause, with constant regard to the caution which should be exercised given the imperfect state of our knowledge on this matter. As the results do not rule out a serious influence on climate, Chapter V, as the next logical step, will include an assessment of possible effects on the impacts of climatic change on society. The consequences for society, which are not yet foreseeable but potentially very serious, and economic considerations, lead naturally to the discussion in Chapter VI of the precautionary measures required to avert the CO_2/climate threat. Finally, it will be shown in Chapter VII that to reduce the CO_2/climate risk or even avert the CO_2/ climate problem demands a **new perspective**, in which **different emphases** must be placed on energy and economic problems. It thus becomes clear that the CO_2/climate problem is not our inevitable fate, but that Mankind can partake in shaping his future.

Theories are only hypotheses supported by more or less numerous facts. The more supporting facts, the better the theory.

Claude Bernard,
French Physiologist

II. CLIMATE AND CLIMATIC CHANGE

In recent years, the frequent occurrence of catastrophic drought and floods have provoked interest in the climate and its possible variations. We have learned, as a result, that despite remarkable technological progress we are still, to a large degree, dependent on climate. Above all, the security of food, energy, and water supplies, and, not least, well-being and health are influenced by the prevailing weather and its variability. Basically, all countries and regions are susceptible to unforeseeable climatic anomalies. No country is yet in a position to protect itself sufficiently from climatic influence by means of an effective early warning system. This is because our present understanding of the mechanisms and causes of climatic variations is not good enough to permit us to make reliable forecasts. Reconstruction of past climates and climate modelling are our most important tools in the endeavour to describe and understand possible future changes in the climate.

II.1 History of the Earth's climate

An important prerequisite for the understanding and prediction of climatic change is the reconstruction of the Earth's climatic history. The changes undergone by our climate can be investigated by numerous methods. These are brought together in Appendix II.1. Apart from the short period of **instrumental measurements** (since the 17th century) we must almost exclusively turn to indirect information about climate, so-called **proxy data.** Examples of this are varves and changes in sea and lake levels (period covered 10^3–10^4yr), tree rings (10^3–10^4yr), fossil pollen (10^4–10^5yr), ice cores (10^4–10^5yr), ice cover (10^4–10^6yr) and cores from the ocean floor (10^5–10^7yr). We may also use evidence from past glacial activity (such as old moraines), loess deposits, the direction and form of sand dunes, and the colour of fossilized soils to give us important information on earlier

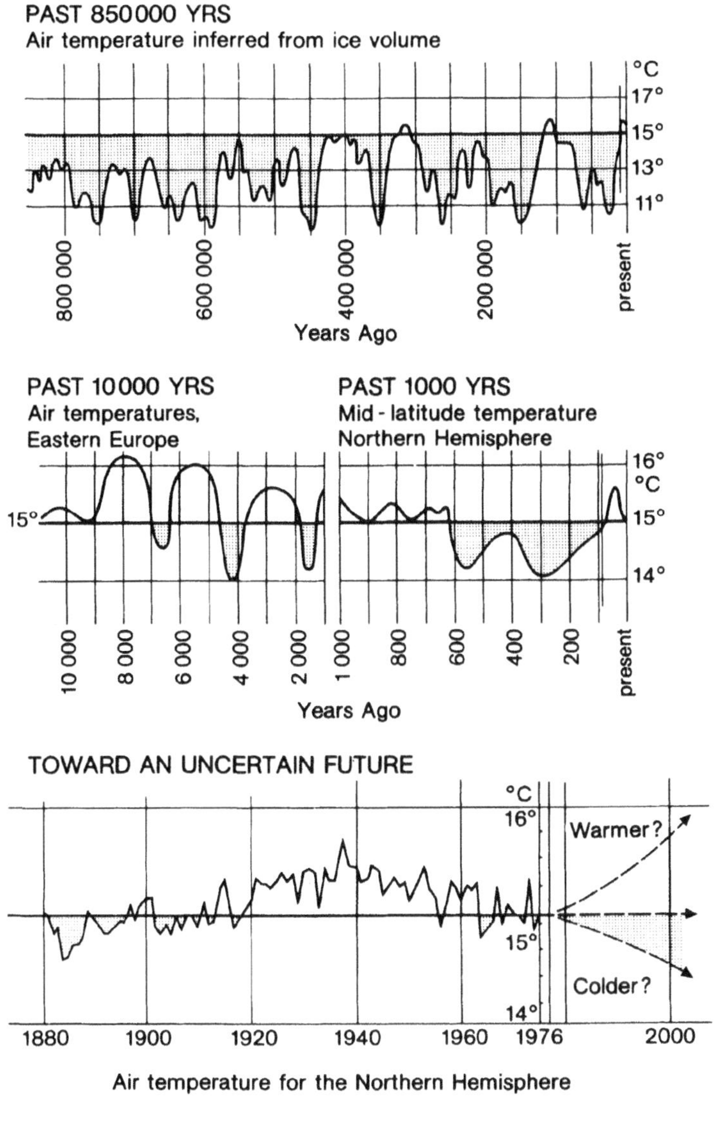

Fig. II.1 : History of mean air temperature as inferred from a variety of palaeoclimatic indicators.
After: Mitchell Jr. (1977).

climatic conditions. A synthesis and intercomparison of many types of climatic evidence reveals information about **climatic elements** such as air temperature, precipitation, evaporation, soil moisture, runoff, ice volume, sea surface temperature and salinity, etc. (Kutzbach, 1975; Haxel, 1976; Barry et al., 1979; Pfister, 1980.)

The **climatic history** of the Earth can be reconstructed using this **climatic evidence.** This history is presented in Appendix II.2 in summary form and in Fig. II.1 for the last 850,000 years.

With the exception of the Mesozoic, glacials and interglacials followed each other in all the other ages of the Earth. In the Pleistocene alone we know of a series of about 20 glaciations during the last 2 MY BP*. We are dealing then with real periodic processes which involve changes from glacial to interglacial in cycles which last from 100,000 to as little as 20,000 years. Within these cycles there exist even smaller fluctuations, such as the "Climatic Optimum" (also known as the Altithermal or Hypsithermal) which occurred between 8,000-5,000 YBP and which will be of use in Chapter V in the analysis of climatic scenarios, the Little Ice Age, ca. 1500-1700 AD, and the recent warm period from about 1880-1940. It should be noted that at the beginning and end of the last glaciation and in the periods between ice ages abrupt coolings of almost glacial intensity have taken place within periods of only 100 years (Flohn, 1978, 1979; Müller, 1979).

The much asked question of whether we are now entering an ice age or a period of warming can be answered as follows. Viewed on a geological time scale we are now, following the natural trend, leaving an interglacial which began about 12,000 YBP (Holocene) and, in the course of the next 10,000 years can expect to enter another glacial period - albeit a fairly mild one. In the short term there is a strong possibility that human activity will bring about a warming such as is shown in Fig. II.1. The natural, long term cooling trend is apparently still predominant at present. Moreover, the ability of the oceans to act as heat sinks seems clearly capable of delaying the warming for a period of one or two decades, so that it might only become detectable in climatic data about the turn of the century (Hoffert et al., 1980). There is, however, good reason to believe that anthropogenic influences have already reached an order of magnitude which could noticeably alter world climate in the space of a few decades. A better understanding of the complex processes involved demands a firm understanding of the climatic system and its mechanisms.

---

* MY BP = million years before present.

II.2 Climate system

The **weather** experienced by Man is characterized by **meteorological variables** such as temperature, precipitation and wind. Weather is the totality of **meteorological conditions** of the atmosphere over a particular area and at a particular point in time. The **climate** of a particular area is the average of these meteorological conditions observed over a period long enough to fulfill the conditions necessary for statistical stationarity. The **Atmosphere** of the Earth is not an isolated physical system, rather it interacts with other spheres, such as the **Hydrosphere** (the oceans, lakes, rivers and groundwater); the **Cryosphere** (land and sea ice, ice in lakes and rivers, mountain glaciation, and snow cover); the **Lithosphere** (the land masses, mountains and ocean basins, rocks and soil) and the **Biosphere** (the plant and animal kingdoms, including Man). An overview of these spheres appears in Fig. II.2 and is called the **climate system.**

The individual sub-systems are characterised by their own specific qualities and **meteorological conditions** such as the heat capacity and temperature of air, water, ice and land; the distribution of density and pressure in the atmosphere and oceans and the resulting wind and ocean currents; the distribution of humidity in the atmosphere, and the amount of water contained in the soil or stored as sea ice and snow.

The individual parts of the climate system interact further through **physical processes**, such as radiation, evaporation, precipitation, heat transport and momentum. The total picture is complicated because these **interactions** are non-linear, and, as a result, some processes can be amplified or dampened by positive or negative **feedback processes.** A further complication arises from the varying **response times** and the differing **spatial scales** of these processes (see Fig. II.2). For example, an air molecule may remain in the troposphere for between 4-8 days, while the same molecule might remain in the stratosphere for a period of months to years. An ice crystal can remain in a portion of drift ice for 1-5 years, and a water vapour molecule's residence time can be from a few days in the upper soil layers to 10^4 years in groundwater (Flohn, 1978). Finally, apart from the interactions in the **internal system, external influences** are also brought to bear on the climate system. Amongst these are solar radiation, variations in Earth's orbit, volcanic eruptions, and also human activity with its resulting changes in the characteristics of the Earth's surface, in the heat budget and water cycle and in the cycle of trace elements in the atmosphere (see the outer boxes in Fig. II.2).

It is not surprising that an interacting feedback system with such diverse temporal and spatial dimensions is continually subject to variations. It is also understandable, as will be shown later, that prediction of individual processes within the climate system is faced with enormous difficulties.

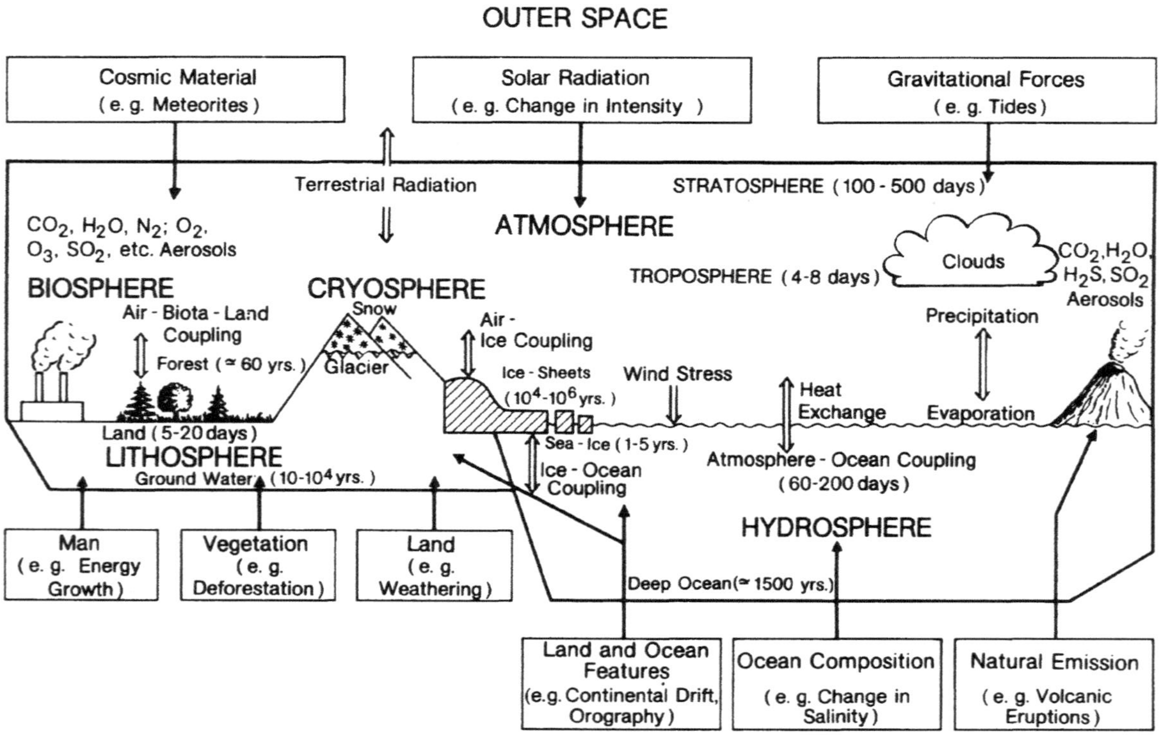

Fig. II.2 : Schematic illustration of the climate system.
After: WMO (1975), Schönwiese (1979), Flohn (1980).

II.2.1 Radiation budget of the earth-atmosphere system

The basic energy source of all climatic processes on Earth is the Sun. The distribution of solar energy over the Earth is dependent on the latitude, on the physical processes occurring between the different parts of the climate system and on the proportion of the Earth covered by land and sea. The variable distribution of energy and water vapour in the lower atmosphere, as well as the processes of heat transport by ocean currents, determine the various states of weather systems and different climates.

Analysis of the radiation budget should deepen our understanding of the complex interactions within the climate system. The diagrammatic description of the radiation budget in Fig. II.3 shows the connections and percentages of individual radiation and energy flows. Short-wave solar radiation is partly absorbed and partly scattered by the stratosphere and troposphere before it reaches the Earth's surface. That which reaches the Earth's surface is partly absorbed and partly reflected. The portion absorbed by the Earth's surface is subsequently transferred back to the atmosphere, partly as long-wave radiation (including infra-red or IR radiation), and partly as direct sensible heat or indirectly as latent heat (the heat absorbed in evaporation is only liberated when condensation occurs). A small part of the re-radiated long-wave radiation escapes through transparent "windows" into space while the greater proportion of such radiation is absorbed by the atmosphere to be released into space or re-radiated back to the Earth's surface. This process of absorption and re-emission of radiation which warms the lower troposphere and the Earth's surface has become known as the **greenhouse effect**, although the heating of air in greenhouses occurs largely by the control of vertical heat flow and only partly by the restriction of the loss of heat radiation. Fig. II.3 shows the energy balance for each sphere in the form of an input/output analysis: the important greenhouse effect is represented separately under "clouds".

II.2.2 Anthropogenic influence on the natural radiation budget

It is of special interest to examine the ways in which human activities can influence the natural radiation budget and thereby also the climate. Among these are the influence of the composition of trace substances in the atmosphere, the alteration of the Earth's surface characteristics, and the increase of direct heat release. All combustion processes and most agricultural activities increase the concentration of atmospheric trace gases. Deforestation, overgrazing, large irrigation projects, and extensive urban and industrial agglomerations alter the characteristics of the Earth's surface. In addition, all processes of combustion contribute to the heating of the troposphere.

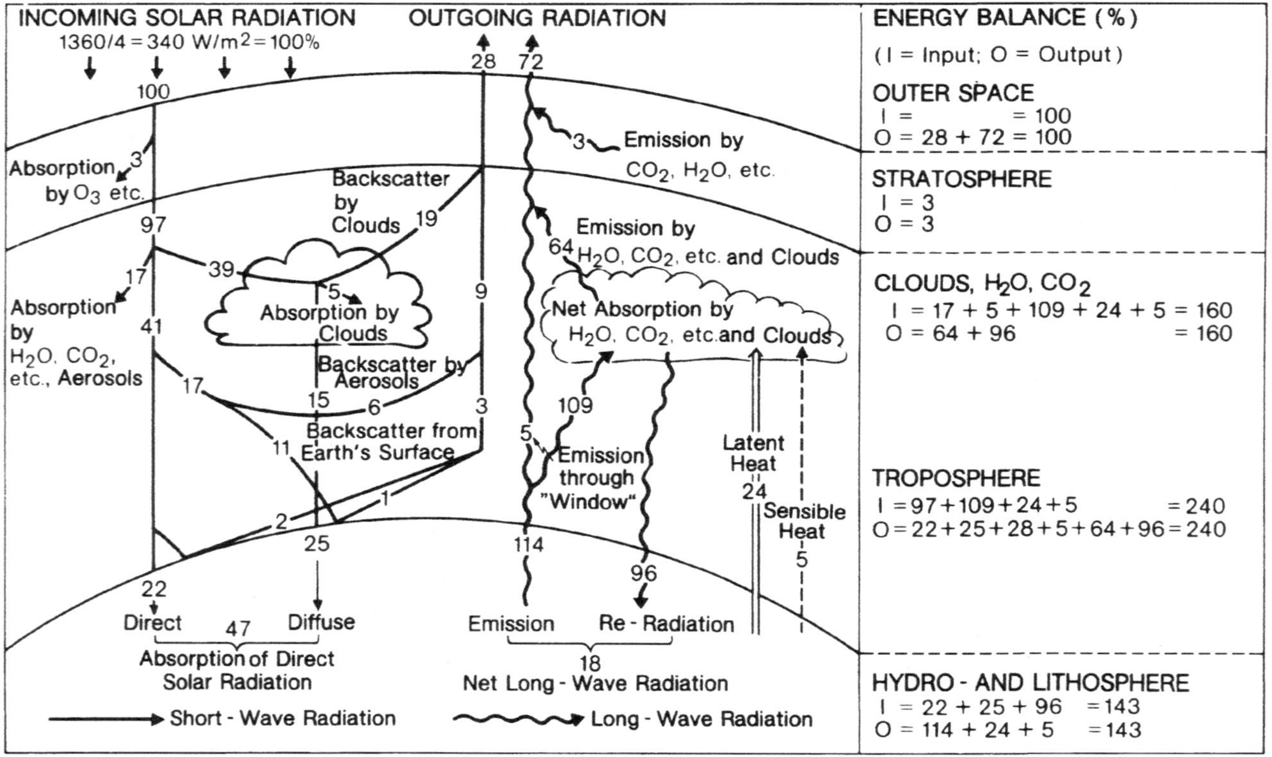

Fig. II.3 : The average radiation and heat balance of the earth – atmosphere system, relative to 100 units of incoming solar radiation.
After: Rotty (1975) and Gates (1979).

The increasing emissions of carbon dioxide (CO_2), nitrous oxide (N_2O), water vapour (H_2O), methane (CH_4) and spray gases (chlorofluoromethanes, also known by the trade name of freons) strengthen the natural greenhouse effect and therefore contribute to the warming of the air-mass near the ground. As the troposphere provides no sink for freons and only a minor one for N_2O, these gases reach the stratosphere, where, by a series of chemical reactions, they break down the ozone (O_3). Because less solar energy is absorbed by O_3, more energy reaches the Earth's surface, which in turn results in a slightly cooler stratosphere and a slightly warmer Earth's surface (Fig. II.3, see also Chapter IV.2).

Even more complicated is the influence of aerosols and clouds. Aerosols influence the natural radiation budget directly through absorption and scattering, and indirectly by providing condensation nuclei for cloud formation. These effects can occur on the long-wave and short-wave (absorption, scattering) parts of the spectrum, as well as in either the troposphere or stratosphere. According to their particular optical characteristics, aerosols can either warm or cool the atmosphere and therefore also the planet Earth (see Chapter IV.3). The influence of clouds on the radiation budget is likewise very complicated, and there are still great difficulties in incorporating this in climate models. Back-scattering from clouds and aerosols and reflection from the Earth's surface together give the value of the Earth's reflective capability or albedo, which is here given as 28% (Fig. II.3).

A change in albedo at the Earth's surface changes the relationship between surface absorption and reflection and therefore contributes either to an increase or to a decrease in temperature. As small changes in albedo could already result in large temperature changes, the continual monitoring of changes in land use by satellites becomes even more important (see Chapter IV.4). All combustion processes continuously add both sensible and latent heat to the radiation budget, but in very different proportions depending on the region concerned (see Chapter IV.5). In conclusion, it must be stated that all these processes can influence the radiation budget of the Earth and can also, therefore, influence the climate.

II.2.3 Causes of climatic change and possibilities of early detection

The causes of climatic change are complex and there are many theories and many possible mechanisms (Mason, 1977; Schonwiese, 1979). Basically, the nature of such changes is both statistical and physical. The latter comprise extraterrestrial and terrestrial processes which can be divided into internal and external causes (Kutzbach, 1974). Among **internal causes** are non-linear feedback mechanisms which work within the component systems - atmosphere, hydrosphere, cryosphere, lithosphere and biosphere (WMO, 1975). The **external causes** include the mechanisms which determine the Earth's orbit (e.g. eccentricity, obliquity, precession), possible changes in the solar constant, continental drift, mountain-building and volcanic activity. The external causes also include anthropogenic influences such as changes in the gas composition and aerosol content of the atmosphere and changes in land use, as well as changes in the heat budget and water cycle.

In addition to changes of a physical nature, climate is also subject to changes of a statistical nature, since climate is defined as an ensemble average or an average over a certain period of time. The changes detectable by statistical means are partly caused by daily fluctuations of weather which cannot be predicted for periods of climatological interest (i.e. several weeks and longer). These changes are therefore accepted as the inherent variability of the climate system, or as **climate noise** (US NAS, 1975). The main problem of anthropogenically caused climate change is to distinguish the **anthropogenic signal** (e.g. due to the increase of atmospheric CO_2) from climatic noise and to estimate its order of magnitude (Chervin, Washington and Schneider, 1976). The problems which hinder such an exercise are threefold: few long series of observations are available; construction of spatial averages and the non-representative distribution of stations obscure the true climatic trend; and, lastly, the expected global increase in temperature due to increasing atmospheric CO_2 will not reach an order of magnitude which makes it possible to distinguish it from natural variations in temperature until about the year 2000. Because of the long residence time of CO_2 in the atmosphere and the long lead times necessary for the introduction of lower-risk energy sources, it would be perilous to wait for Nature's own announcement of a human-induced climatic change around the year 2000; it would then be too late for preventive measures.

A variety of methods is available for the estimation of the likely degree of climatic change and its effects in the future. For example, the methods of **palaeoclimatology** (see Appendix II.1) allow us to look at analogous periods of the Earth's history when the Earth was a few degrees warmer than is the case today, such as the already mentioned Altithermal period 8,000-5,000 YBP (Kellogg, 1977/8). In this sort of **scenario analysis** one must be aware that the factors producing the warming vary considerably between the two periods, and that any interpretation of the analogy must therefore be cautious (Flohn, 1980; see also Chapter V.1). The main role of palaeoclimatology is to provide the empirical data with which the theories formulated in climatic models can be tested. In addition, one can analyse zonal instead of global means of temperature variation (Williams and van Loon,

1976). Since it is known that the higher latitudes, in particular the polar areas, react more strongly to climate-influencing factors than the equatorial zones, the climatic data obtained there can act as a sort of warning. Etkins and Epstein (1982) suggest that the most important factors to monitor when looking for signs of climatic change are changes in the polar ice masses, mean global sea level and the speed of the Earth's rotation (see also Chapter IV.1.7.2). Lastly, mention must be made of the **climate models** (see also Chapter IV.1.7.1), whose goal is not only the estimation of individual anthropogenic influences on climate before they can be detected in climatic data, but also, hopefully, the forecasting of the behaviour of the whole climate system over longer periods.

II.3 Climate models and prediction

One of the main tasks of climatic research is to establish by modelling experiments, the complicated range of interactions which play a part in the climatic system as it is described in Fig. II.2. The skill of the climatologist is to develop a series of climate models which incorporate the most important climatic processes and which give realistic results with a reasonable amount of computation. Since no model can be an exact copy of the climate system, the models must be improved by continuous comparison of the calculated values with the observed data. The necessary, world-wide data collection is a long-term undertaking so that it will certainly be some decades before a reliable verification of modelling results is possible.

The models developed by climatic research vary in their complexity from simple zero-dimensional climate models to the three-dimensional general circulation models including atmosphere-ocean feedback, which can accommodate a large number of calculations. Climate models are our most powerful tools for dealing in a consistent and quantitative way with the highly interactive processes that determine the complex climate system. The following overview describes the major types of climate model with which, amongst others, CO_2 simulations can be carried out (Schneider and Dickinson, 1974; Fraedrich, Grassl, Hantel, Herterich, Reiser and Renner, 1980; Dickinson, 1982; US NAS, 1982; Schlesinger, 1983; Manabe, 1983; Mitchell, 1983; Hansen et al., 1983; Washington and Meehl, 1983).

II.3.1 Simple climate models

The simplest climate models are, geometrically, zero-dimensional (0-D); i.e. the spatial resolution of the Earth is reduced to one point. The basis of these models is a global energy balance. This sort of model is useful above all for the discussion of methodological problems; for example, the possible existence of several different conditions of equilibrium and of instabilities. These are usually more obvious in zero-dimensional models than in the more complex ones.

Using these relatively unsophisticated models, which are, incidentally, the only ones which do not require computers, sensitivity tests may also be carried out; for example, investigations of emissivity or the planetary albedo, which allow inferences to be made, amongst other things, about the CO_2 content of the atmosphere. In general, however, 0-D climate models are of little use for providing quantitative information about the effects of, for example, a doubling of atmospheric CO_2.

II.3.2 Energy balance models

The most frequently applied climate models are energy balance models. As the name indicates, of all the equations governing climate, they consider only those involving energy. Appendix II.3 shows, as an example, the energy equation of a vertically integrated one-dimensional (1-D) energy balance model. As this operates with averaged values it is termed a **statistical-dynamical model** or a **quasi-stationary equilibrium model**. The averaging time employed varies from one month to a season or to a year and the spatial averaging is carried out zonally or meridionally. Here one can already detect one weakness of energy balance models, for they necessarily **parameterize** effects on the synoptic scale; that is to say, that small scale processes which are not described explicitly are expressed in averaged values (e.g. meridional transport by large scale temperature gradients).

Energy balance models can be used successfully in the **simulation** of **past** and **present climates.** The radiation parameters seem to reflect all significant processes, probably because dynamic processes become more significant as the period of time under consideration increases. The simulation of palaeoclimates is very useful, because the verification of the models with the aid of palaeoclimatic data is necessary before a confident assessment of future processes can be made.

Sensitivity studies such as the investigation of internal processes (feedback mechanisms) or external influences (e.g. **increased CO_2**), mean that energy balance models have further important applications. Because they are more transparent and require fewer calculations than general circulation models, and are thus very well suited to the study of the component parts of the climate system, energy balance models are sure of retaining their place in climatic research.

II.3.3 Zonally averaged models

The conspicuous **zonality** of atmospheric circulation is the basis of these models. Because of this zoning between the equator and the poles it is correct to speak of the climatic zones used in the classification of climate. In an east-west direction the meteorological fields do not change much, while there is a large change in the north-south direction. Although a zonally averaged climate model has no longitudinal dependence, it does realistically represent the broad structure of the field. Since the equations involve one less dimension there is a commensurate reduction in the number of calculations necessary.

A conspicuous feature of these models is that the rapid changes of the weather system are not explicitly calculated. Rather, the statistical effects of the processes for calculating eddy fluxes (turbulence) are used. This displaces the closure problem to the higher-order moments (for example, covariances). The hope that statistical-dynamical models which require less computation could be used in place of the more detailed general circulation models has not yet been realised. It should not be forgotten that important phenomena of the climatic system such as the Indian monsoon are hidden by zonal averaging. It is nonetheless important to experiment with zonally averaged models, since it is possible, given certain externally specified parameters, to assess the likelihood of certain climatic developments without excessive computing requirements.

II.3.4 Stochastic models

The basis of these models is the fact that observed climatic variations arise within the climate system. Unaveraged short-scale weather events influence the slower elements of climate by a series of **random effects (stochastic processes)**. Examples of these are the ever changing high and low pressure systems, while the oceans and ice cover represent the sluggish elements. Weather variables, taken as stochastic elements (from which the climatic variables are driven) are specified by a Markov Process (i.e. a first order autoregressive process) because of their statistical attributes (see Appendix II.3 for an example of a 1-D stochastic climate model).

Modelling experience has shown that the incorporation of weather processes results in good simulations of observed climatic variations in stochastic models. Even the level of variation in different climatic periods is simulated in a quantitatively correct way.

II.3.5 General circulation models

These **dynamical, nonlinear models** calculate the time evolution of circulation processes of the climate system with optimal spatial resolution. The processes of the climate system are described by the governing equations such as the horizontal momentum equation, the mass continuity equation, and the thermodynamic energy equation. Together with the hydrostatic equation, the equation of state and appropriate boundary conditions, these equations form a closed system (see Appendix II.3 for an example of the basic equations of a 3-D general circulation model (GCM)). Some models use an additional prognostic equation, namely the surface pressure tendency equation. This set of equations is used to determine the time rates of change of magnitude of the **prognostic variables** whose principal ones include temperature, surface pressure, horizontal velocity, water vapour concentration and soil moisture, the latter two being related to the hydrological cycle. Furthermore, GCMs have many **diagnostic variables,** of which clouds, surface albedo and vertical velocity are the more important ones.

An important feature of GCMs is the three-dimensional description of the atmosphere and the high resolution of atmospheric processes. Depending on computer capacity, the atmosphere is divided into a horizontal grid of an approximate 300-400 km mesh-size and into 2 to about 10 discrete vertical layers. With the nonlinear partial differential equations a step-wise integration is performed to determine, for the various layers, the horizontal variations of the predicted quantities either at discrete grid points over the earth (**grid point models**), or by a finite number of prescribed mathematical functions (**spectral models**).

Several of the subgrid-scale physical processes important to climate are not resolved due to the limited spatial resolution of GCMs. They must therefore be **parameterized** i.e. they are either statistically or empirically related to the scale of those variables which are resolved. Some of the parameterized subgrid-scale processes include transfer of solar and terrestrial radiation, turbulent transfer of heat, moisture and momentum between the Earth's surface and the atmosphere and within the atmosphere, and condensation of water vapour. Moreover, the simulation of climatic change with a GCM requires the prescription of certain parameters and boundary conditions such as the solar constant and orbital parameters, land-sea distribution and topography, total atmospheric mass and composition, surface albedo, etc.

The Earth's climate cannot be successfully simulated without coupling the components of the atmosphere with those of the ocean, the ice and the biosphere. The ocean has a significant influence on climate because of its high storage capacity, and this influence must be simulated as realistically as possible. Special attention is given to the computation of sea surface temperature, since it is determined both by ocean circulation and the energy exchange between ocean and atmosphere.

In the early experiments the sea surface temperatures were prescribed to equal their observed values (the ocean acts as an infinite heat sink), and later the ocean was modelled as a flat **"swamp"** (with zero heat capacity, no ocean currents). In more recent experiments the ocean consisted either of one upper mixed layer (**slab model**), or of two vertically homogeneous ocean layers of variable depth. Work is in progress with the even more realistic **ocean general circulation models** which can explicitly compute the ocean circulation. Also the sea ice models require further improvement. Eventually, **atmosphere-, ocean-** and **sea ice models** must be coupled with a **biosphere model** which can accommodate all significant changes in land use. (For further details see Chapter IV 1.7.1.)

General circulation models are the most advanced climate models. Compared with the complexity of the real climate, models, however, must still be seen as primitive. Because they require such a high computational capacity, only a few research centres in the world have worked with such models. Although much work remains to be done, general circulation models have been successfully used in so-called **sensitivity studies**, i.e. in estimating the effect of internal forcing (by changing the feedback processes), or by investigating the influence of external factors on climate. Examples of external forcing such as by increasing CO_2, changes in land-use and excess heat will be given in Chapters IV.1, IV.4 and IV.5, respectively.

II.3.6 Climatic predictability

The limit of weather forecasts is currently a few days, and the absolute **predictability** of daily weather fluctuations will probably never extend beyond a few weeks. As these fluctuations show no underlying periodicities, it will therefore be impossible to predict, on the daily time scale, changes in weather for a season ahead, much less for future decades or centuries. Only the mean characteristics of the climate system are predictable.

Lorenz (1975) has distinguished two different sorts of climate forecasting in the practical application of climate models. In **forecasts** of the **first kind**, the internal, temporal development of the climate system is determined by set external parameters. This stability consideration of climate depends on the assumption that the autovariations of the climate system and their interactions can cause climatic change in the absence of external forcing. The following phenomena are investigated: **transitivity** (i.e. for constant external influences there is only one stable **climatic condition** and the climate, in the event of a perturbation, always reverts to the former state of equilibrium); **intransitivity** (i.e. there is no stable climatic condition, so a perturbation may possibly cause the advent of a new, arbitrary climatic condition); **almost-intransitivity** (i.e. the existence of a climate system with several quasi-stable climatic conditions and sudden changes from one to the other). With this type of forecasting climatic factors could be predicted for decades, or for millions of years which would be interesting both from the point of view of agriculture and of the study of ice ages.

Climate forecasts of the second kind simulate the equilibrium state of the climate when the boundary conditions are held constant. These **sensitivity studies** investigate the reaction of the internal climate model system to artificially induced changes in the internal system (such as the removal of feedback mechanisms) or, in the external system (such as an increase in the concentration of atmospheric CO_2).

II.4 Climate research programme

Until recently the WMO (World Meteorological Organization) concerned itself chiefly with the exchange of meteorological data and with the improvement of weather forecasting. Only since the World Climate Conference in Geneva in 1979 (WMO, 1979) has the question of the future development of climate been given greater prominence, as is shown by the working out of a concrete World Climate Programme (WMO, 1980). Other international organizations, the United Nations Environment Programme (UNEP), and the European Economic Community (EEC), also participate in aspects of the World Climate Programme.

Complementing the World Climate Programme are the national climate research programmes being developed in various countries. For example, the climatic research programme of the Federal Republic of Germany contains, since May 1980, the following main aims:

- Research into the climate system
- Collection, processing and dissemination of data
- Research into the sources and sinks of atmospheric trace elements which affect climate
- Investigations of the socioeconomic impacts of climatic change
- Research into the interactions between climate and the biosphere.

These programmes can only be managed in a coordinated and integrated way, such as is described in Fig. II.4. Individual contributions to this research come from disciplines which are sometimes interrelated and sometimes widely different, e.g. Biology, Chemistry, Climatology, Ecology, Economics, Environmental Research, Geography, Geology, Geophysics, Hydrology, Meteorology, Oceanography, Physics, Sociology, and others.

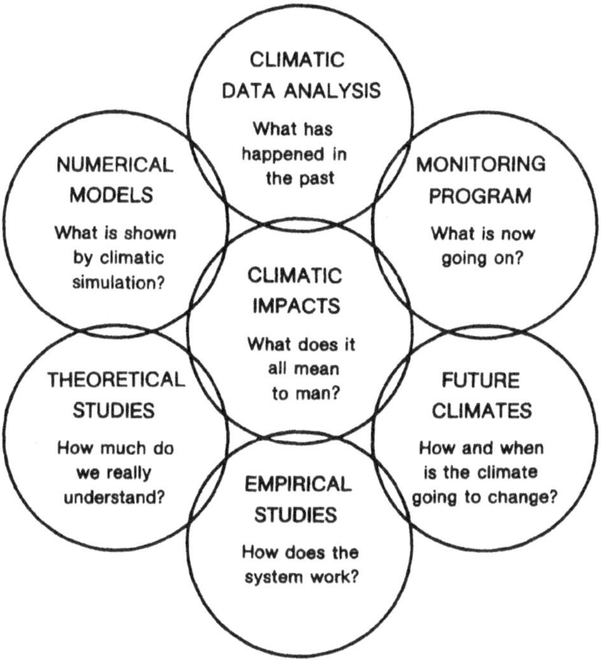

Fig. II.4 : Interdependence of the major components of a climate research programme.
Source: US NAS (1975).

The main aim of integrated climatic research is to achieve a degree of understanding of the climate system which enables the anthropogenic influences on climate and its future development to be assessed with a sufficient degree of accuracy. With the help of such knowledge, decision-makers can take more rational precautionary measures.

We did not inherit the Earth from our fathers, we are borrowing it from our children.

Lester R. Brown
President of the
Worldwatch Institute

III. SOCIOPOLITICAL ASPECTS OF THE CO_2/CLIMATE PROBLEM

The future development of CO_2 and its influence on climate depends on a series of sociopolitical factors, the most important of which are population growth, changes in settlement patterns, economic development, the use of the fossil fuel potential, future energy use, forestry, and agricultural policies.

III.1 Population growth

The human population is increasing exponentially. How far this growth can continue depends on one limiting factor: **the Earth's carrying capacity.** This capacity is in turn dependent on such variables as the production of food, raw materials and energy, on political conditions and, not least, on the influence of environment and climate. The level at which world population will eventually stabilise is of great importance for the avoidance of an unacceptable climatic change.

III.1.1 Past growth

World population is increasing at a breath-taking rate. About 1810 the world's population was 1 billion. The 2 billion level was reached in 1924 after 114 years, 3 billion in 1961 after 37 years, and 4 billion in 1975 after only 14 years (Keyfitz, 1977). In the last 30 years the population of developing countries (DCs) has almost doubled from 1.7 to about 3.3 billion, while in the industrialised countries (ICs) population has increased by about 47% from 750 million to 1.1 billion (Mauldin, 1980).

III.1.2 Future growth

Two groups of experts have investigated the future development of population growth: The Bureau of the Census and the Agency for

International Development on behalf of the U.S. President's Council on Environmental Quality (Global 2000, 1980), and the Rockefeller Foundation on behalf of the U.N. (Mauldin, 1980). The following results are based on these investigations.

Table III.1 : Population projections for world, major regions, and selected countries.

	1975	2000	Per cent Increase by 2000	Average Annual Per cent Increase	Per cent of World Population in 2000
	millions				
World	4,090	6,351	55	1.8	100
Developed Regions	1,131	1,323	17	0.6	21
Developing Regions	2,959	5,028	70	2.1	79
Major Regions					
Africa	399	814	104	2.9	13
Asia and Oceania	2,274	3,630	60	1.9	57
Latin America	325	637	96	2.7	10
USSR and Eastern Europe	384	460	20	0.7	7
North America, Western Europe, Japan, Australia, and New Zealand	708	809	14	0.5	13
Selected countries and regions					
People's Republic of China	935	1,329	42	1.4	21
India	618	1,021	65	2.0	16
Indonesia	135	226	68	2.1	4
Bangladesh	79	159	100	2.8	2
Pakistan	71	149	111	3.0	2
Philippines	43	73	71	2.1	1
Thailand	42	75	77	2.3	1
South Korea	37	57	55	1.7	1
Egypt	37	65	77	2.3	1
Nigeria	63	135	114	3.0	2
Brazil	109	226	108	2.9	4
Mexico	60	131	119	3.1	2
United States	214	248	16	0.6	4
USSR	254	309	21	0.8	5
Japan	112	133	19	0.7	2
Eastern Europe	130	152	17	0.6	2
Western Europe	344	378	10	0.4	6

Source: Global 2000 (1980).

According to the projections in Table III.1, world population will increase from 4 billion in 1975 by a further 2 billion to 6 billion by the year 2000 - an increase of 57%. The projections were made using the cohort-component methodology, i.e. every age-group (cohort) is divided into males and females for each individual country and investigated with regard to the following demographic components: mortality, fertility and net migration. On the basis of this detailed research it can be said with a fair amount of certainty that, by the year 2000, world population will not be less than 6 billion people.

More important still is the realisation that the rate of population growth will not noticeably slow down. By the year 2000 the current annual rate of growth will only have decreased by 0.1% to 1.7% for the world as a whole and to 2.0% for DCs (Table III.1). The fact must not be overlooked, however, that the absolute total population will increase faster in the future. While mankind increased its numbers by 75 million in 1975, in the year 2000 this figure could be 100 million. Almost 90% of this increase will be in DCs - this is an important problem for the future.

A large number of factors will determine when, and at what

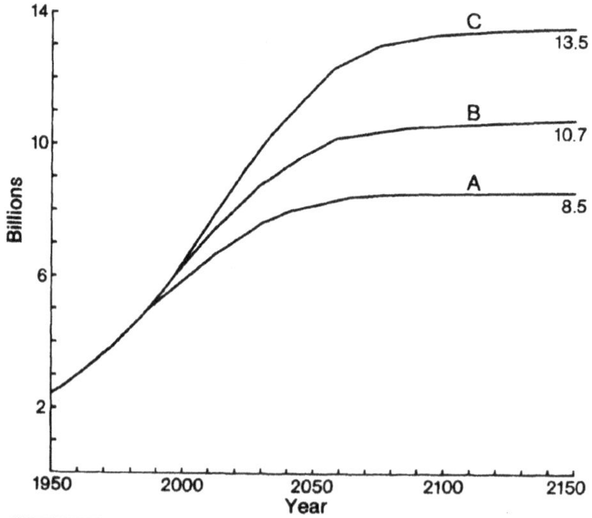

If the world attained replacement fertility in	World population (bill.) would then stand at	and world population (bill.) could be expected eventually to stabilize at
(A) 2000-2005	5.9 (2000)	8.5
(B) 2020-2025	8.0 (2020)	10.7
(C) 2040-2045	10.8 (2040)	13.5

Fig. III.1 : Momentum of world population growth.
Source: Mauldin (1980).
(reprinted with permission from *Science*, Vol. 209, p. 156, ©1980 AAAS)

level, world population stabilises. An important criterion is the inherent **momentum of growth**, which depends on the contemporary age-structure (the percentage of young women attaining child-bearing age is especially high in developing countries), as well as on present and future mortality and fertility rates. Even if world population should reach an equilibrium (i.e. a rate of live births of not more than 2.3 per woman (Klauder, 1980)), the momentum of growth shown in Fig. III.1 would still lead to a world population of 8.5 billion. Due to age structure and socio-economic parameters, a rapid reduction in fertility rates, especially in developing countries, is highly improbable, so that population stabilisation at 13.5 billion cannot be ruled out.

The further into the future these prognoses attempt to go, the greater becomes the range of uncertainty. It is uncertain whether world population will stabilise at 8 or 12 billion. What is certain is that future energy requirements and the intensive use of agricultural land will depend on the level of this growth and that this will not be without consequences for the environment and for climate.

III.2 Settlement patterns

In the past, increasing industrialisation has led to depopulation of the countryside and an increase in urbanisation. While world population increased by a factor of 2.6 between 1800 and 1850, the number of people living in large towns (100,000 or more inhabitants) increased by a factor of 35 (Global 2000, 1980).

As Fig. III.2 shows, in 1900 about 20% of the world's popu-

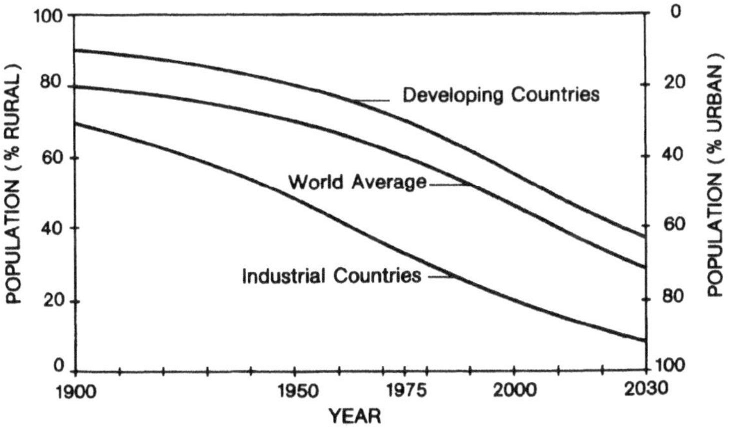

Fig. III.2 : Rural and urban population distribution. After: U.N. (1975) and Sassin (1980a).

Table III.2 : Estimates and rough projections of selected urban agglomerations in developing countries.

	1960	1970	1975	2000
		Millions of persons		
Calcutta	5.5	6.9	8.1	19.7
Mexico City	4.9	8.6	10.9	31.6
Greater Bombay	4.1	5.8	7.1	19.1
Greater Cairo	3.7	5.7	6.9	16.4
Jakarta	2.7	4.3	5.6	16.9
Seoul	2.4	5.4	7.3	18.7
Delhi	2.3	3.5	4.5	13.2
Manila	2.2	3.5	4.4	12.7
Tehran	1.9	3.4	4.4	13.8
Karachi	1.8	3.3	4.5	15.9
Bogota	1.7	2.6	3.4	9.5
Lagos	0.8	1.4	2.1	9.4

Source: Global 2000 (1980).

lation lived in towns. By 1980 this had increased to 40%. According to the prognoses of the U.N. (1975) the degree of urbanisation could reach 50% by the year 2000 and almost 70% by 2030. A social structure with such a high degree of urbanisation would have a disproportionately high energy requirement with resulting damage to the environment and climate.

III.2.1 Urbanisation in industrialised countries

Urbanisation in ICs is a result of industrialisation and was encouraged by the availability of cheap energy (Laconte, 1980). The surrounding countryside provided the pool of potential migrants. As Fig. III.2 shows, the level of urbanisation in ICs was 30% in 1900, reaching almost 70% by 1980. According to the forecast of the U.N. (1975) the figure will be almost 90% by 2030. It is doubtful whether such a sharp rate of increase is realistic for ICs, due to the widely observed stagnation of urban growth.

III.2.2 Urbanisation in developing countries

The degree of urbanisation in the DCs was 10% in 1900 and had reached about 30% in 1980. According to U.N. (1975) estimates, this level will increase rapidly, almost doubling by 2030 (Fig. III.2). This rapid rate of increase is based on the assumption that agriculture and industry are becoming increasingly capital-

and energy-intensive and correspondingly less labour-intensive, so that the depopulation of the countryside increases and the slums and poverty-stricken areas of cities in the third world will grow ever larger in the future. These developments and past trends were used to project the development of cities in DCs, shown in Table III.2. According to this, by the year 2000 a city such as Mexico City could have grown to an ungovernable monstrosity of about 32 million inhabitants - a population about four times as large as that of present-day New York City. In view of this prospect, it is no wonder that the administrators and town-planners of the future Megacities, Megalopoli and Oecumenopoli had only gloomy views when they came together in 1980 in Rome for the U.N. Conference on population and the future of the city (Peterson, 1980).

As indicated above, urbanisation is connected with a sharp increase in energy requirements. If, in 2030, about 70% of the expected 8 billion people live in cities (see Fig. III.2), then, given an average population density of 1,000 people per square kilometre and a mean energy use density of 10 watts per square metre, the amount of energy used at this level of urbanisation would be 56 TW* (Bach and Matthews, 1980; Sassin, 1980a, b). In a study commissioned by the EEC, Colombo and Bernardini (1979) come to quite different conclusions from those contained in the U.N's published estimates. They contend that, due to the continuing deterioration in the quality of life in cities and a broadening of employment opportunities in the countryside, the level of urbanisation will not increase further. Accordingly, the urban population shares would change in individual regions between 1975 and 2030, but the fraction of the total population living in urban areas would remain constant at about 30% (Table III.3). If this level of urbanization were sustained the 2030 energy requirement, on the basis of this information alone, would fall from 56 to 24 TW.

III.3 Economic development

The Gross National Product (GNP) says little about either the distribution of income among social classes in different countries or about changes in the structure of this distribution. Although it reflects little of the prevailing economic and social conditions of a country, it is, nonetheless, generally used as an index of economic growth. In this section it will serve to illustrate the great economic gulf which separates ICs from DCs which is, in its turn, adduced as an argument for strong energy growth.

---

* See the conversion factors for energy and power at the beginning of the book.

Table III.3 : Actual (1975) and possible (2030) urban population share.

Region	1975 Mill.	1975 Per cent of Total	2030 Mill.	2030 Per cent of Total
1. North America (USA and Canada)	149	63	150	51
2. Soviet Union and Eastern Europe	171	47	216	45
3. Western Europe, Japan, Australia, New Zealand, South Africa and Israel	259	46	340	44
4. Latin America	131	41	279	35
5. South and Southeast Asia, Africa (except North and South Africa)	274	19	782	22
6. Middle East and North Africa (Egypt, Algeria, Lybia)	40	30	131	37
7. China and Centrally Planned Asia	215	24	479	28
World	1239	31	2377	29

Source: Colombo and Bernardini (1979).

As Table III.4 shows, in 1975 about 80% of global income can be attributed to industrialised countries and only 20% to developing countries, although the latter account for 72% of world population (see Table III.1). The average annual GNP growth rates assumed by Global 2000 (1980) for the years between 1975 and 2000, show lower rates of growth for ICs than for DCs of 3.4% and 4.5% respectively. This differential in growth rates, over the short period of 25 years, cannot significantly reduce the disparities of the poorest regions. Such a reduction could only be achieved, especially given the increasing level of population in developing countries, by a disproportionate economic recovery, which is, however, improbable (Leontief, 1980; Bach, 1980).

The differences are even graver when one considers the per-capita development of the GNP (Table III.4). Although the average annual growth rate between 1975 and 2000 is, in most regions, of roughly the same order (2.2–2.9%), consideration of the absolute figures makes it clear that the gulf between rich and poor areas will grow even greater when assuming a doubling of the per capita GNP of the western industrial nations to 11,117 dollars.

It remains to be seen whether the growth rates accepted for these projections will prove to be realistic. The present growth

Table III.4 : Total and per capita GNP estimates (1975) and projections (2000) by major regions.

Regions	Total GNP (Billions of constant 1975 U.S. dollars)					Per Capita GNP (Constant 1975 U.S. dollars)		
	1975	%	2000	%	Annual Growth Rate (%) 1975-2000	1975	2000	Annual Growth Rate (%) 1975-2000
World	6025	100	14677	100	3.6	1473	2311	1.8
Developed regions	4892	81	11224	76	3.4	4325	8485	2.7
Developing regions	1133	19	3452	24	4.5	382	587	1.7
Major Regions								
North America, Western Europe, Japan, Australia, and New Zealand	3844	64	8996	61	3.4	5431	11117	2.9
USSR and Eastern Europe	996	17	2060	14	2.9	2591	4472	2.2
Asia and Oceania	697	11	2023	14	4.3	306	557	2.4
Latin America	326	5	1092	8	4.9	1005	1715	2.2
Africa	162	3	505	3	4.6	405	620	1.7

Note: The GNP estimates and projected growth rates are based on data presented in the 1976 World Bank Atlas and on SIMLINK (SIMulated trade LINKages), an econometric model developed by the World Bank.

Data extracted from: Tables 3 and 4, Global 2000 (1980, Vol. 1).

rates in ICs are lower. A boom in the late 1980s or 1990s is not impossible, but rather unlikely.

The question of how far the continuing growth of the GNP could affect the climate and the environment, depends above all on the sectors in which it occurs. In ICs, economic growth can be increasingly diverted into the service industries which cause relatively little harm to the environment. In DCs, however, economic activities such as the exploitation of resources, development of raw-material industries and the intensification of agriculture, which can all damage the environment, should be more prominent. If economic growth in the DCs remains at a low level, and if one does not accept as axiomatic a coupling between economic growth and increased energy use (Müller and Stoy, 1978), then controlled economic growth with no unacceptable impact on the environment is not only possible, but also something to be encouraged.

III.4 Availability of fossil fuels

The climate of the future will be decisively influenced by the increase of atmospheric CO_2, which is, in turn, largely dependent on the use of fossil fuels such as coal, oil and gas. The amount of CO_2 which actually enters the atmosphere depends not only on the amount of fossil fuel resources and their level of availability, but also on the accessibility of these deposits and on political decisions about production. These criteria will be used below to estimate possible developments in the use of fossil fuels.

III.4.1 Coal

As Table III.5 shows, the present level of available, geologically detectable, coal deposits (**resources**) is about 10,000 billion tonnes coal equivalent (tce)*. Of this total, only about 545 billion tce are, given the present level of technology, economically recoverable (**reserves**). With progressive improvements in extraction techniques and the relative increase in the prices of our energy resources, the level of recoverable coal reserves could be raised to 1,760 billion tce. Given the present world energy use of about 10 billion tce/yr, the coal supply would be sufficient to cover the total world energy demand for the next 176 years.

World coal deposits (resources and reserves) are concentrated in only a few countries (Table III.6). Only 10 countries (including two developing countries) possess 98% of world coal

---

* See the conversion factors at the beginning of the book.

Table III.5 : World resources and reserves of fossil fuels.

Type	Total Resources		Presumed Technically Recoverable Reserves		Present Economically Recoverable Reserves	
	btce*	%	btce	%	btce	%
Hard Coal	7,900	63.5	1,425	42.9	420	47.4
Lignite	1,900	15.3	333	10.0	125	14.1
Total Coal	9,800	78.8	1,758	52.9	545	61.5
Crude Oil	1,044	8.4	418	12.7	141	15.9
Natural Gas	313	2.5	313	9.4	96	10.8
Tar Sand	490	3.9	392	11.7	57	6.5
Oil Shale	705	5.7	353	10.6	47	5.3
Hydro-Carbons	2,552	20.5	1,476	44.4	341	38.5
Peat	90	0.7	90	0.7	unknown	
Total	12,442	100.0	3,324	100.0	886	100.0

* btce = Billion tons of coal equivalent

Source: Ziegler and Holighaus (1979).

Table III.6 : World coal resources and reserves (btce), by major coal-producing countries.

Country	Geological Resources	Technically and Economically Recoverable Reserves
Soviet Union	4,860	109
USA	2,570	166
China	1,438	98
Australia	600	32
Canada	323	4
Federal Republic of Germany	246	34
United Kingdom	190	45
Poland	139	59
India	81	12
South Africa	72	43
Other Countries	229	55
Total World	10,748	657

After: WOCOL (1980).

deposits. Contrary to the case for oil, the three countries which possess the lion's share of coal deposits (82% of the coal resources and 57% of the coal reserves), USA, USSR and China (People's Republic), are also the heaviest coal users. The world coal trade is, therefore, relatively small (Table III.7). In 1975 it was at a level of about 200 million tce and, according to the estimates of a group of experts at the World Energy Conference, could rise to about 790 million tce/yr by 2020 (World Energy Conference, 1978; Anderheggen, 1980). In the future, Canada, South Africa and the USSR could join the current export leaders, the USA, Poland and Australia. The greater part of coal imports go to Japan and western Europe. A sharp increase in world coal trade is highly improbable, due to sharply increasing use in the major coal-producing countries, to the expected, large price increase, to protectionism and to the lack of cheaper strategies of application in the DCs. Furthermore, given the OPEC experience, it is doubtful whether the world will readily accept dependence on an additional coal cartel.

In 1975 the world coal production was about 2.6 billion tce (Table III.7) and, according to various estimates could probably more than treble by 2020 to a level of about 8.8 billion tce (World Energy Conference, 1978; Griffith and Clarke, 1979). Whether such sharp rates of increase in production can be achieved is, because of the expected problems, doubtful. The permafrost of Siberia (Bottke, 1979) and the lack of water in

Table III.7 : Development of coal production and export (mtce), by major coal-producing countries, 1975 - 2030.

	1975		1985		2000		2020	
Country	Prodn.	Export	Prodn.	Export	Prodn.	Export	Prodn.	Export
Soviet Union	614	26	851	37	1,100	50	1,800	60
USA	581	60	842	68	1,340	90	2,400	145
China	349	3	725	7	1,200	30	1,800	50
Poland	181	39	258	45	300	50	320	50
United Kingdom	129	2	137	10	173	10	200	10
F.R. Germany	126	23	129	25	145	30	155	30
India	73	-	135	7	235	13	500	32
Australia	69	29	150	60	300	180	400	240
South Africa	69	3	119	23	233	55	300	60
Canada	23	12	35	15	115	40	200	65
Other Countries	360	2	483	6	619	34	751	46
Total World	2,593	199	3,884	303	5,780	582	8,846	788

After: World Energy Conference (1978).

semi-arid zones of the mid-western United States and, not least, confrontations with laws protecting the countryside and the environment might limit coal production and processing (Bach, 1979a).

III.4.2 Oil

Oil reserves which can be economically exploited by present-day methods total approximately 140 billion tce (Table III.5). Improved methods of production could increase this figure to 400 billion tce. It is probable that only a fraction of the estimated resources of about 1,000 billion tce can be economically extracted due to rapid increases in costs and to difficulties of production in offshore oil fields.

It is estimated that the amount of oil which can at present be economically extracted from tar sands and oil shale is 50 billion tce (Table III.5). Of the heavy oil sediments in the Orinoco region of Venezuela, containing about 200 billion tce and the largest known deposit in the world, probably only 10% are economically recoverable (Kemmer, 1981). There are strict limits on the production of oil from unconventional sources, since capital investment is up to 10-20 times greater than for conventional oil production (Voss, 1980). Besides this factor, one must consider the level of environmental damage, which would probably be unacceptable (Bach, 1981).

Of the economically exploitable world reserves of conventionally produced oil, about 62% are located in the Near East and North Africa, 12% in the USSR and eastern Europe, 7% in the rest of Africa and 5% to 6% in the Americas and western Europe (Bischoff and Gocht, 1979; Gerwin, 1980). Present production equals about 4.5 billion tce. The OPEC countries provide 60% of the oil used in the western world (Voss, 1980). The continuing political instability in the Near East and heavily increasing production and distribution costs will probably not allow a significant increase in production, so that the high point of global production will probably be reached in the middle of the 1990s with a production level of about 5 billion tce. Given the maximum level of production, world oil supplies would last for no more than 30 years.

III.4.3 Natural gas

In the course of oil production large amounts of natural gas are frequently encountered. These are either flared off or pumped back into the oil fields to maintain pressure. Fields of pure natural gas also occur and these are exploited all over the world. Table III.5 shows that present-day economically recoverable reserves are estimated at about 96 billion tce. Total available resources comprise about 300 billion tce. Recently, the possibility of obtaining natural gas from unconventional sources has been discussed. Such sources include: natural gas in geopressurised zones, gas hydrates in the Arctic, natural gas in the Devonian Shales and the production of methanes by gasification of subterranean coal fields (Bach, 1981). These additional resources are estimated at 1,000 billion tce.

Economically recoverable reserves of gas are concentrated, essentially in two regions, the USSR and eastern Europe with 35.6%, and the Near East and North Africa with 33.4% (Gerwin, 1980). A considerable way behind come Canada and the USA with 12.2%, then western Europe with 8%. It is assumed that the present world demand for gas will have doubled to a level of 4 billion tce by 2000. This doubling will require increasing development of natural gas pipelines and an extension of the transportation of liquid gas in tankers. With this sort of increased production, conventional gas reserves will last only another 20 or 30 years. The relocation of raw material industries to the production countries could increase the efficiency of gas utilisation, especially by avoiding the wasteful gas flaring. If, over and above this, the exploitation of unconventional natural gas resources proves to be economical, the amount of gas available could double.

In conclusion, at the current state of knowledge, the total fossil fuels which are probably economically recoverable are of the order of 900 to 3,300 billion tce (see Table III.5). It therefore follows that, even given an annual rate of increase of 2%, the total world energy demand could be met for the next 200-300 years by fossil fuels alone. Looked at purely from the point of view of availability, the use of fossil fuels could lead to the feared CO_2/climate problem by the middle of the next century. Whether this in fact happens depends on the future development of world energy policy.

III.5 Future energy use

The possible influence of future energy use on climate depends, to a large extent, upon the following factors: the amount and mix of energy, inequities in energy use, market penetration times, energy cost developments, and, especially, the efficiency with which energy is used and the deployment of CO_2-free sustainable resources, the latter two being discussed in detail in Chapter VI.

III.5.1 Energy amount and mix

The technique of **scenario analysis** is nowadays widely used in the estimation of future energy demand, since the simple projection of past trends into the future - much favoured until now - has proved erroneous in most cases. These energy scenarios are energy paths which must be thought of, not as predictions, but as planning aids. They are determined by criteria such as population growth, changes in societal structure, economic growth and availability of resources, which have all been discussed above. Usually, two scenarios, one high and one low, are developed for the future, and the increasing spread between the two paths reflects the increasing uncertainty about what will happen in the future. Such scenarios, calculated with the help of different models, provide **energy strategies** which demonstrate, either for individual countries or for regions, both the quantity and the type of the energy sources required to cover the energy demand. The calculations usually deal with a period of around 50 years - a period which, as will be shown in a later section, can be taken as a typical market penetration time. The purpose of such scenario analyses is to give an overview of the possibilities and to determine in a normative way, the amount of room for manoeuvering.

While there are a great number of energy scenario analyses on the national level, the number of global-scale analyses currently available is relatively small. Amongst the best known **world models** are: the cybernetic models with dynamic feedback processes sponsored by the Club of Rome and known as World 2 (Forrester, 1971) and World 3 (Meadows et al., 1972); the multi-level model of Mesarovic and Pestel (1974); the U.N. world-model, a model of the world economy based on input-output analysis (Leontief, 1977); a world energy model constructed by the Institute for Energy Analysis in Oak Ridge, USA which is based on demographic, economic, social and political factors (Rotty and Marland, 1980); the integrated world model of the U.S. Government (Global 2000, 1980); and a world energy model, based on conventional approaches to economics and population growth, which concentrates upon the efficient use of energy and the supply of energy services at the least possible costs (Lovins et al., 1981). Lastly, mention should be made of the world energy analysis carried out by the International Institute for Applied Systems Analysis at Laxenburg near Vienna as part of a large-scale energy project (Häfele, 1980; Häfele and Rogner, 1980; Rogner and Sassin, 1980). This study has been published both in a popular German edition (Gerwin, 1980) and in a comprehensive English version (Häfele et al., 1981). The following remarks are, to a great extent, based on this study. (For additional world energy studies see Chapter VI.4.1.)

Table III.8 shows how, according to the IIASA scenarios, the world energy demand might develop up to the year 2030 and what fraction of primary energy sources could be used to satisfy it. According to this, the year 2000 will see a doubling of present energy demand (to about 16 TW), and, by 2030, an energy requirement which is three to four times higher, dependent upon which scenario one uses.

Important to the CO_2/climate issue is that the IIASA scenarios maintain the dominant role of fossil fuels, reducing them only slightly from 90% in the base year (1975) to about 83% in the year 2000 and about 70% in 2030. Still more important in this connection are the absolute values, which show 2 to 5 fold

Table III.8 : Global primary energy (TWyr/yr) by source for two supply scenarios (1975–2030) and growth factors (in brackets).

Primary Source	Base Year	High Scenario				Low Scenario			
	1975	2000		2030		2000		2030	
Oil	3.62	5.89	(1.6)	6.83	(1.9)	4.75	(1.3)	5.02	(1.4)
Nat. Gas	1.51	3.11	(2.1)	5.97	(4.0)	2.53	(1.7)	3.47	(2.3)
Coal	2.26	4.94	(2.2)	11.98	(5.3)	3.92	(1.7)	6.45	(2.9)
LWR[1]	0.12	1.70	(14.2)	3.21	(26.8)	1.27	(10.6)	1.89	(15.8)
FBR[2]	0	0.04		4.88	(122)	0.02		3.28	(164)
Hydro	0.50	0.83	(1.7)	1.46	(2.9)	0.83	(1.7)	1.46	(2.9)
Solar[3]	0	0.10		0.49	(4.9)	0.09		0.30	(3.3)
Other[4]	0.21	0.22	(1.0)	0.81	(3.9)	0.17	(0.8)	0.52	(2.5)
Total[5]	8.21	16.84	(2.1)	35.65	(4.3)	13.59	(1.6)	22.39	(2.7)

[1] LWR = Light water reactor; [2] FBR = Fast breeder reactor; [3] includes mostly soft solar (e.g. individual rooftop collectors) and also small amounts of centralized solar electricity; [4] "Other" includes biogas, geothermal, and commercial wood use; [5] Columns may not sum to totals due to rounding.

After: Häfele and Rogner (1980) and Häfele (1980).

increases for oil, natural gas and coal (see the growth factors in brackets in Table III.8). Even given much more enhanced production of fossil fuels and an expeditious transfer from conventional fossil fuels to highly capital-intensive unconventional fossil fuels, analyses of the availability of fossil fuels (c.f. III.4) indicate that such growth rates, at least for oil and coal, are improbable. If, however, the actual fossil fuel consumptions were in accord with the IIASA scenarios, then the CO_2/climate problems discussed in Chapters IV and VI could occur.

Basically, there are two possible ways of defusing the CO_2/climate problem: to achieve, by means of more efficient energy use, a smaller total energy demand and consequently a lower usage of fossil fuels, and to replace fossil fuels by those energy resources which emit little or no CO_2 into the atmosphere. The assertion that energy production can be increased 10 to 27 times by the use of fission reactors, or 122 to 164 times by the use of breeder reactors in only 30 years is more than questionable (see Table III.8). On the other hand, it is hard to see why energy production from biomass (under the heading "Other") should achieve only one tenth of the growth rate of conventional atomic energy. It is even less plausible that growth rates for breeder reactors have been assumed which are 25 (50) times higher than the growth rate for solar energy in the high (low) IIASA scenario. What is certain is that, in this field, questions of acceptance and economic viability will decisively influence future developments. Investigations of the possible effects on climate of different schemes of fossil fuel use and better energy use will be discussed in Chapter VI.

III.5.2 Inequities in energy use

Both the total amount of energy used and the energy used per capita highlight significant differences between the energy used in different regions and in individual countries. For example, only 6% of the world population (region 1) used more than 11 kW/cap, whereas 70% (regions 4 to 7) of the world population used 1 kW/cap or less (Table III.9).

A particularly informative comparison can be made between region 1 (North America) which is the richest region, and region 5 (South and South East Asia, as well as Africa) which is the poorest. In the base year 1975, region 5 had a population six times greater than that of region 1 but used about 2 TW less energy in total. The per capita energy consumption in region 5 was only one fiftieth of that in region 1. Even given the growth in energy use assumed by the IIASA scenarios for the year 2030 to reach a level of 22 or 36 TW, the disparity between region 1 and region 5 remains (for both high and low scenarios) at the 2 TW level observed in 1975. In region 5 in 2030, the per capita energy use will also be well below the world average - a result largely due to the rapid energy growth in the industrialised countries in regions 1-3 and the rapid population growth in the developing countries. Even though North America had already a per capita energy consumption level of 11.2 kW/cap in 1975, more than double that of the Federal Republic of Germany (about 5 kW/cap), the high IIASA scenario envisages a further growth to over 19 kW/cap by 2030, thereby widening the gap between the rich and the poor countries.

Considering the high rate of population growth, especially in the developing countries, and the price increases in the energy sector which may, on average, also be expected to continue in the future, then the energy-gap which appears in the IIASA scenarios becomes quite plausible. Even if account is taken of the non-commercial use of primary energy sources, which runs at about 1 TW in the developing countries (Fritz, 1980), and if one is of the opinion that developing countries can make do with less

Table III.9 : Scenarios for population and primary energy use (per capita and total) by regions.

Region	Population (millions)		PRIMARY ENERGY USE: 1975 Base Year		Scenarios for the year 2030: High		Low		Zero-Growth*	
	1975[1]	2030[1]	kW/cap[2]	TW[3]	kW/cap[2]	TW[3]	kW/cap[2]	TW[3]	kW/cap[2]	TW
1. North America (USA and Canada)	237	315	11.2	2.65	19.1	6.02	13.9	4.37	8.0	2.52
2. Soviet Union and E. Europe	363	480	5.1	1.84	15.3	7.33	10.4	5.00	6.2	2.98
3. W. Europe, Japan, Australia, New Zealand, South Africa, and Israel	560	767	4.0	2.26	9.3	7.14	5.9	4.54	3.2	2.45
4. Latin America	319	797	1.1	0.34	4.6	3.68	2.9	2.31	2.8	2.23
5. South and Southeast Asia, Africa (except North and South)	1,422	3,550	0.2	0.33	1.3	4.65	0.7	2.48	0.7	2.48
6. Middle East and North Africa (Egypt, Algeria and Libya)	133	353	0.9	0.13	6.7	2.38	3.5	1.23	3.6	1.27
7. China and Centrally Planned Asia	912	1,714	0.5	0.46	2.6	4.46	1.3	2.29	1.2	2.06
World	3,946	7,976	2.0	8.21	4.5	35.7	2.8	22.2	2.0	15.99

* Zero growth because the per capita energy consumption of 2 kW does not change; a doubling of population results, however, in a doubling of energy use to 16 TW.

After: Häfele et al. (1981), data extracted from
[1] *Table 14.2;* [2] *Table 25.1;* [3] *Table 15.2.*

energy because of climatic factors, there is still a huge gulf, which must be reduced as soon as possible, if tension between North and South is to be eased (Brandt et al., 1980).

How can these disparities be most quickly removed - by increased or decreased energy growth? As Table III.9 clearly shows, the chance of lessening the energy-gap between industrialised and developing countries is much greater for the so-called zero growth scenario than for those which envisage higher levels of energy use. For example, in 1975 the per capita level of energy use in region 1 was fifty times higher than that in region 5. In all scenarios this difference is lessened, but it is most markedly lessened in the zero growth scenario. According to this scenario, region 1 would have only 11 times the energy consumption of region 5 by 2030. This result shows that North-South tension can be most effectively reduced by a less wasteful use of energy - a strategy which also preserves and protects our environment and climate.

III.5.3 Market penetration times

A potential influence on climate does not only depend on the shares of the total energy market that the various primary energy carriers have, but also on the speed with which they penetrate the energy market. Fig. III.3 shows the market shares of the energy carriers as a function of time. The market share is plotted logarithmically as the ratio of the market share of the new energy carrier (F), to the market shares of the old energy carriers (1-F) (Marchetti, 1978). The set of curves is based on a simple logistic function, which obviously gives a good description of the actual development of the individual primary energy shares (see the irregular lines). If the point of market introduction (i.e. the point where the energy curve reaches 1% of the market share) and the slope of the curve of the new energy type are known, then, according to Marchetti (1977a, b), this logistic function can be used to make long-term projections of the market development. Of course, this claim has met with scepticism, since it implies that the development of new energy market shares can be determined by causes that have their roots in the distant past. On the basis of about 300 investigations for the various energy carriers in many countries, Marchetti and Nakicenovic (1979) attempt to underpin their theory of **market penetration dynamics**.

As Fig. III.3 shows, wood was the dominant energy carrier before the beginning of the Industrial Revolution with a market share of about 70%. Because it is easier to store and transport, coal replaced wood as the main energy carrier in 1880. Coal, which made the Industrial Revolution possible, reached a unique maximum with a market share of 80% already in 1910. This maximum cannot be reached by the subsequent energy carriers, oil and natural gas, because of the limited availability of resources and the large price increases.

The curves in Fig. III.3 show that globally a new energy carrier has, on average, required about 100 years until it acquired 50% of the market share. In highly industrialised countries, such as the USA, market penetration usually lasted about 50 years (Bach, 1979b). Under special circumstances, as for example during the reconstruction of the Federal Republic of Germany after the Second World War, the build up of the energy supply took place in less than 25 years as a result of availability, of know-how and skilled labour (Bach and Matthews, 1980). This leads to the conclusion that the market shares of the energy carriers that can be expected between 2010 and 2050 are apparently already determined by the current energy policy. The future magnitude of the CO_2/climate effects will be significantly affected by current energy policy that prescribes a global revival of coal with increased CO_2-emissions as a result of synthetic fuels production or that prescribes an intensive development of alternative energy carriers that release little or no CO_2 into the atmosphere. An energy policy that supports more efficient energy use contributes, on the other hand, to a reduction of the fossil fuel demand immediately, without long market penetration times (see Chapter VI).

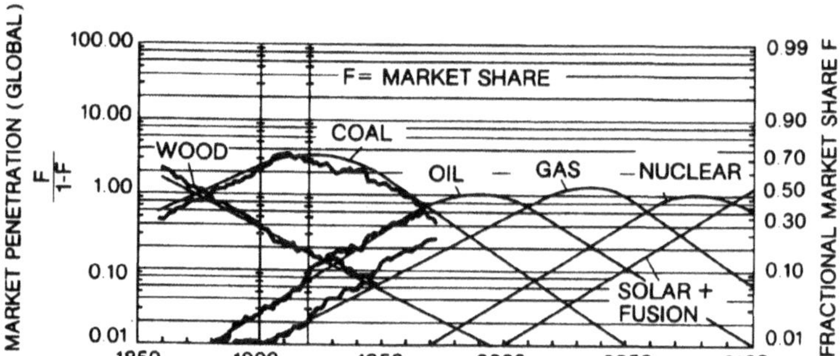

Fig. III.3 : World primary energy substitution.
Source: Marchetti and Nakicenović (1979).

III.5.4 Capital requirements

The development of the energy supply will also be affected by the future energy price policies. Before the energy crisis of 1973/74, the energy prices (after the subtraction of the inflation rate) fell, while the total primary energy consumption simultaneously rose. There are sufficient indications that the lower prices supported the increase of the energy consumption

(Doblin, 1980). With the large energy price increase after 1973/74 the growth of the primary energy consumption slowed down and began to lag behind the Gross National Product and industrial production.

Supply and demand determine the future development of the energy price level. The supply depends mainly on three factors: the market form (price agreements by cartels like OPEC), the future expectations and economic plans of the producing countries (investment needs, build up of infrastructure, defence expenditure) and the costs of production (capital investments and running costs). It is generally assumed that in the short- and medium-term the most important influence is the form of the market, while the other two factors will become more important as the time of exhaustion of the resources is approached. In the case of demand, aspects that play an important role include general standard of living, buying power, the significance of energy in the goods that are to be produced, and the availability of alternative energies and their price in comparison to the conventional energy carriers. All of these factors and others (such as, for example, bank policies and international exchange rates) have complex effects on the future energy price development, but should contribute, overall, to further energy price increases.

Price increases in the energy sector can have quite different effects. On the one hand they can lead to drastic changes or adjustments in economic and social structures. On the other hand, they can also accelerate attempts at innovation and force an improved energy efficiency. Price increases can only be met effectively by a more rational energy use and more rapid transition to available unconventional energy sources which are cheaper than present energy supply systems (see Chapter VI).

Estimates of future price increases and their effects on energy prices depend on future political developments that are not predictable, and on the necessary investment expenses that can be estimated with the aid of coupled energy-economics models (e.g. Grenon and Lapillone, 1976; Beaujean and Charpentier, 1978; Lapillone, 1978; Spinrad, 1980).

The total investments consist of the direct investments (e.g. costs of opening, production, conversion and transportation; costs for the development of new technologies and plants) and indirect investments (e.g. costs for land, water, raw materials, manpower and environmental protection). For the energy scenarios presented in Table III.8, the following total investments cumulated over the period 1980–2030 have been estimated: 46.5 trillion dollars (10^{12} dollars) for the high scenario and 29.7 trillion dollars for the low scenario (Häfele et al., 1981). For comparison, the total global gross national product was 4 trillion dollars in 1980 and the GNP of the Federal Republic of Germany was about 0.8 trillion dollars.

The studies show that the investments required up to 2030 in the developing countries must increase twice as much as those in the industrialised countries. The share of the energy development costs in the developing countries must increase from 2% of the GNP in 1975 to 6.1%-7.4% in 2030, while in the industrialised countries an increase from 2.3% in 1975 to 3.3%-3.8% in 2030 would be sufficient (Häfele et al., 1981). As the estimates of the growth of GNP in Table III.4 show, this would be very difficult, especially in the developing countries. The future energy development will presumably be determined not so much by the technical availability of resources (see Tables III.5-7) but much more by the financing possibilities within the capital market. It must be noted in conclusion that the statements about the level of necessary investments and the possibilities for financing are subject to great uncertainties.

III.6 Degradation of forest and soil

Destruction of forests and soil has serious effects on the entire ecosystem, especially on climate and food production. The removal of entire forests leads to soil erosion, floods, landslides and avalanches, as well as silting up of rivers, reservoirs and irrigation plants. In contrast, a forest has a positive effect on the quality and uniformity of the groundwater. Within the ecosystem, forest and soil are in a complex balance, being linked by soil types, soil characteristics and the distribution of nutrients in the soil, and by climatic conditions. This balance can be disturbed by human activities. Deforestation and the dying of forests through air pollution (in particular through acid rain and toxic metals) are the most visible signs of such disturbance.

The perturbation of the carbon budget is of particular interest within the context of this book. Depending on whether the terrestrial ecosystems (forests, soils) act as **carbon sources** (e.g. deforestation, soil destruction) or **carbon sinks** (e.g. reforestation), the CO_2 problem will be either amplified or reduced.

III.6.1 Deforestation

The history of Mankind and the history of deforestation are closely related. Appendix III.1 shows a brief overview of the influences of civilisation on the European and Asian forests during the past 10,000 years. In about 1000 BC there were already attempts in China to put a stop to the increasing destruction of forests. Only in recent times are deforestation and reforestation apparently being kept in balance through large-scale reforestation programmes in middle latitudes.

In spite of this, the global deforestation is proceeding

rapidly, especially as a result of high deforestation rates in the tropical forests. About 1950 1/4 of the land surface was still forest-covered, today it is only 1/5. In about 2000 it will presumably be only 1/6 and for the year 2020 it is hoped that the forest cover will stabilise at about 1/7 of the land area (Global 2000, 1980). As Table III.10 shows, experts suggest that the present global forest cover of about 2,600 million hectares will be reduced by about 18% by 2000. The subtotals indicate that it is assumed there will be a constant forest resource in the developed countries up to 2000 and a decrease of 40% in the developing countries. The assumption of such a large deforestation in the tropics appears plausible on the basis of the rapidly growing population and the related demand for agricultural and grazing land for indigenous consumption, or for export products and fuelwood. The increasing demand for tropical timber in the developed world also plays an important role.

Table III.10 : Estimates of world forest resources, 1978 and 2000.

Region	Closed Forest (millions of hectares)		Growing Stock (billions cu m overbark)	
	1978	2000	1978	2000
USSR	785	775	79	77
Europe	140	150	15	13
North America	470	464	58	55
Japan, Australia, New Zealand	69	68	4	4
Subtotal (Developed Countries)	1,464	1,457	156	149
Latin America	550	329	94	54
Africa	188	150	39	31
Asia and Pacific Developing Countries	361	181	38	19
Subtotal (Developing Countries)	1,099	660	171	104
Total (World)	2,563	2,117	327	253
			Growing Stock per capita (cu m biomass)	
Developed Countries			142	114
Developing Countries			57	21
Global			76	40

Source: Global 2000 (1980).

How likely is it that wood will be used even more as an energy carrier in the future and thereby become an additional carbon source for the atmosphere? As Fig. III.3 shows, wood was still the dominant energy carrier in 1860 with a market share of almost 70%. Today the share of wood in the total energy production has sunk to less than 1% in western Europe, but in the developing countries it plays an important role, as, for example, in Brazil, where it has a share of about 20% (in some poorer developing countries it is much higher) (Bischoff and Gocht, 1979). If energy prices increase in the future, as can be assumed in respect to the overall trend, the developing countries will have no choice but to make use of their growing stock, in order to meet the increasing demand for fuelwood.

According to Table III.10, the growing stock was about 170 billion m^3 in 1978, which has an energy content of about 23 billion tce (about 4% of the global coal reserves). Also in the developed countries as a result of high energy prices, **silviculture**, i.e. the production of biomass in **energy plantations**, is becoming noticeably more attractive, especially for the production of liquid fuels for the reduction of dependence on oil imports (Armentano and Hett, 1980). The extent to which conflicts of interest will arise between food and energy production, as the world population increases, is an important **ethical question** (Brown, 1980).

III.6.2 Loss of soil

After deforestation has taken place, erosion and loss of humus are increased, so that the carbon bound in the soils is released into the atmosphere. The carbon content of the soil is reduced by about 40% as a result of removing the original plant cover and subsequent cultivation (Delcourt and Harris, 1980). Of course, these losses could be practically reduced to zero using ecologically more suited methods. Under natural conditions with low runoff, organic soils bind large amounts of carbon, but return it to the atmosphere when they are drained. According to the best estimates possible at the moment, the carbon loss from organic soils to the atmosphere currently amounts to 0.03–0.37 billion tonnes. If 50% remains in the atmosphere, this means a contribution of 0.6–8% of the annual atmospheric CO_2 increase (Armentano, 1980; see also Chapter IV.1.4.3).

Even with the American soil conservation methods, that are held to be good, there was a loss of almost 3 billion tonnes of agricultural land as a result of soil erosion in 1975, which corresponds to about 22 tonnes top soil per hectare (U.S. Department of Agriculture, 1977). The soil losses are especially large in the developing countries. In Nepal – which might be an example for many developing countries – rivers wash away about 240 million m^3 of soil annually (Brown, 1981). As the need to intensify food production increases in the future, still more and mostly marginal land will certainly be ploughed up. Thus, increasing soil losses and growing carbon emissions can be expected.

The statements made in this Chapter have shown the ways that could lead to a CO_2/climate problem when plausible assumptions are made for the future. The following Chapter looks at the specific factors through which Mankind influences the climate.

The responsibility for the CO_2 problem is ours - we should accept it and act in a way that recognises our role as trustees of the earth for future generations.

Gus Speth,
former Chairman of the Council
on Environmental Quality, U.S.A.

IV. INFLUENCE OF SOCIETY ON CLIMATE

IV.1 Climatic effects of CO_2

IV.1.1 The CO_2/climate problem

Carbon dioxide (CO_2), a colourless and tasteless gas, is breathed out by humans and animals as a metabolic product and is absorbed by plants through photosynthesis for building up biomass. Given the present global concentration of about 0.040% or 340 ppmv (i.e. 340 parts CO_2 per million parts of air by volume), the CO_2 represents no danger to the health of mankind. However, as a result of its characteristic of absorbing heat energy, it could contribute to the warming of the lower layers of the atmosphere (greenhouse effect) and thus give rise to climatic changes.

The CO_2 problem is certainly not new, as the compilation of the CO_2/climate research in Appendix IV.1 shows. However, it has been underestimated and ignored for a long time. Revelle and Suess (1957) were the first to point out with some urgency that mankind's activities were setting in motion a **geophysical experiment on a global scale**, which, according to present-day expert opinion, could lead to detectable climate changes in a few decades. The most important causal factor is the rapid increase of the atmospheric CO_2 content, observed now for 24 years, resulting from the combustion of fossil fuels (coal, oil, gas, etc.), flaring of natural gas, cement production, as well as deforestation and the related soil oxidation. The resulting chain reaction then leads to quite different regional and seasonal changes in climate via the CO_2 increase and the **greenhouse effect**. With the increasing world population this must lead to increased supply problems, especially in the food, water and energy supply systems. The CO_2 problem, therefore, becomes a central question for the coexistence of humans and the survival of mankind.

Why is there a worldwide interest in the CO_2 issue? What

makes it such a central question of survival? The CO_2 problem is an inescapable global and international phenomenon. While individual countries, especially those in climatically marginal areas, can be affected differently, no country is immune, not even the technologically highly-developed industrialised countries. The measures of any individual country cannot bring much, only a coordinated action promises to be successful. However, since the strongest effects will first be seen in the climatically marginal areas, i.e. in the developing countries, and will be less noticeable in the middle latitudes of the industrialised countries, there is a danger that the necessary joint activities will not be started in due time. In addition, the CO_2 problem is closely interwoven with a large number of human problems that are not less controversial, such as, for example, with the future population, economic, and energy development, the unequal distribution of resources and wealth, a major cause of the north-south conflict, and environmental impacts, notably those through acid rain, heavy metals and nuclear waste, all of which can threaten our life-support system. These various aspects of our complex environment are characterised by large **uncertainties.** This has given rise to many speculations, which lead some to fear an immediate catastrophe, while for others the belief has been reinforced that there is no CO_2 problem at all. Both points of view are wrong and dangerous. The first view leads to rush actions, while the second could mean that the appropriate time for the introduction of **preventive measures** is missed.

The **CO_2 problem** is **not fate,** but rather the result of errors in human actions. It is, therefore, necessary to increase the efforts to reduce gaps in our knowledge and at the same time to introduce the appropriate preventive measures. The following sections and chapters reflect the spectrum of both the required research and the necessary measures. We start by looking at the natural carbon cycle and investigate how it is influenced by humans. Only after the role of the sources and sinks has been satisfactorily understood is there any hope that the future CO_2 development can be reliably estimated. This is the prerequisite for a realistic estimate of the future climate effects with the aid of climate models.

While CO_2 appears to be the main problem, we must not forget, as often occurs in the estimation of the climatic effects, that other climatic factors must be taken into account. In sections IV.2-6 the effects of other important infrared-absorbing gases, aerosols, land use changes and waste heat are therefore considered and taken into account in the estimation of the total effect. The resulting impacts on the environment and society are treated in Chapter V and measures for reducing or even avoiding these effects in Chapters VI and VII.

IV.1.2 The natural carbon cycle

In its natural, unperturbed state, it is assumed that the carbon cycle is in equilibrium or quasi-equilibrium. An understanding of the natural state is a prerequisite for the evaluation of major disturbances. In the carbon cycle system (Fig. IV.1) we distinguish three slow **geological cycles** and two fast **atmospheric-biological-oceanic cycles** (Junge, 1978; Hampicke and Bach, 1979). The slow carbon cycles include the organic carbon decay, carbonate weathering and the weathering of primary silicates as well as volcanism (magma). The atmosphere, the living and dead organic matter, both on the land and in the ocean, are part of the fast atmospheric-biological carbon cycle. The fast atmosphere-ocean carbon cycle does not only include the atmosphere-ocean exchange at the sea surface, but also the mixing processes within the surface water including the thermocline. The driving forces of the carbon turnover in the slow cycles are weathering, sedimentation and tectonic events; in contrast, in the fast cycles the formation and decomposition of biomass and the exchange between the atmosphere and ocean play the decisive roles.

In the slow cycles, very large amounts of carbon - about 100 x 10^6 billion tC or 4% of the mass of the earth's sediments (Fig. IV.1) - turn over at a slow rate of about 0.5 billion tC/yr. The turnover time amounts to 200 million years (100 x 10^6 billion tC/0.5 billion tC), so that on average the carbon atoms have gone through the geological cycle only about three times since the beginning of the Palaeozoic.

A comparatively small amount of carbon takes part in the fast cycles, amounting to only about 0.04% of the slow cycles. Of this, the largest part (90-95%) is found in the deep sea and with an average residence time of about 1,000 years is fixed for a long term on the human time scale (Fig. IV.1). The most active reservoirs are the atmosphere (710 billion t*), the ocean mixed layer (680 billion t) and the living plants (650 billion t), that together contain only about 2,000 billion t or 0.002% of the total carbon of the earth's crust. The fluxes between these three reservoirs are, however, extremely intensive. A carbon atom passes from the atmosphere into the ocean surface water on average every 7 years (710/100) or every 12 years (710/60) into the terrestrial plants.

Finally, let us consider how the natural carbon cycle system responds to perturbations of its equilibrium, for example, through an intensive CO_2 emission into the atmosphere as a result of human activities. As a result of the consumption of coal, oil and gas, presently more than 5 billion tC/yr flow out of this geological reservoir that was formed millions of years ago, in

---

* See Appendix IV.2 for the calculation of this value.

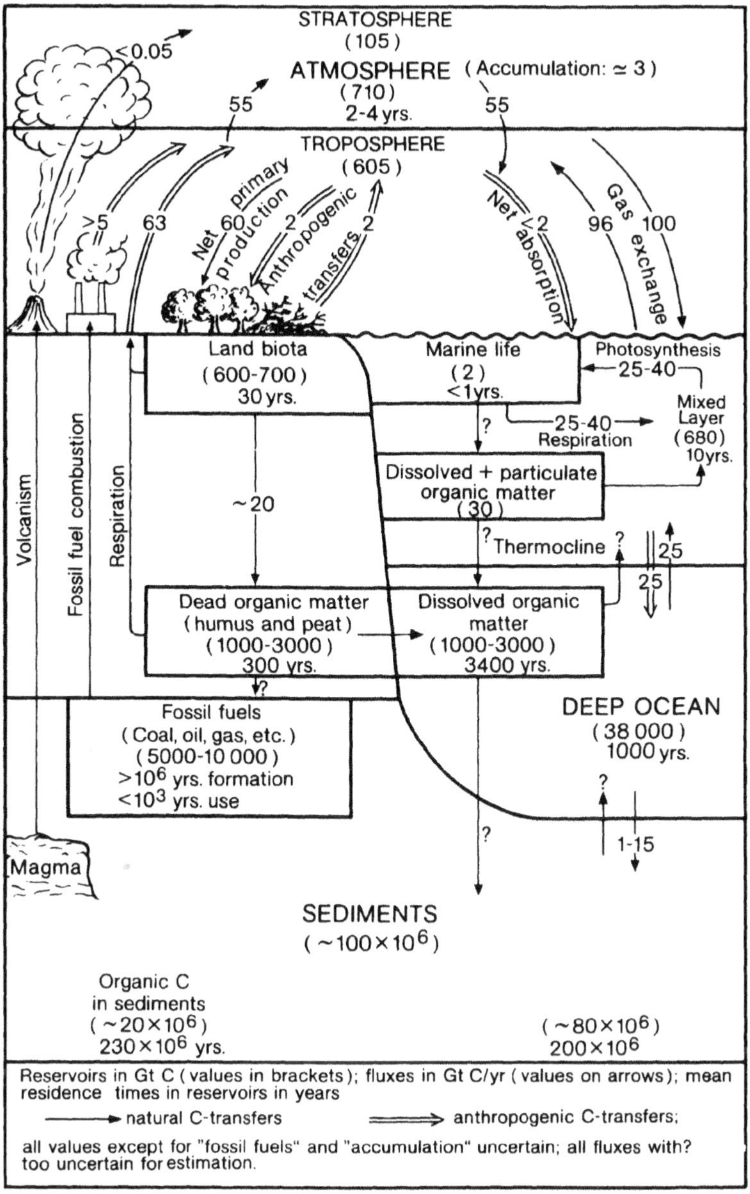

Fig. IV.1 : The global carbon cycle (about 1978).
Sources: Kester and Pytkowicz (1977); Woodwell (1978); Bolin et al. (1979); Hampicke and Bach (1979); Lieth et al. (1980); US DOE (1980); Hampicke (pers. comm., 1981).

addition to the natural geological flows of 0.5 billion tC into the atmosphere, thus accelerating the natural geological cycle by more than a factor of 10. In addition, there are the C-fluxes from the living phytomass through deforestation and from the humus through soil oxidation, whose order of magnitude and flow direction are still largely unclear. These additional carbon amounts will accumulate more and more in the atmosphere, since they are only taken up very slowly (over thousands of years) by the deep ocean. Thus, the estimated preindustrial carbon content of the atmosphere of about 610 billion t has increased by about 100 billion t to the present observed value of about 710 billion tC (US DOE, 1980). The fact that the natural control mechanisms work very slowly in comparison to the human time scale, leads to the important conclusion, that it will take thousands of years before a disturbance of the atmospheric CO_2 content, due to human activities, will reach its original state of equilibrium. If, however, carbonate solution does not occur, then also, in the case of complete mixing of the ocean, the original atmospheric CO_2 level will not be reached again.

IV.1.3 Evidence of carbon cycle perturbation

IV.1.3.1 Observed CO_2 increase in the atmosphere

The observed CO_2 increase shows clearly that the carbon content of the atmosphere has been perturbed. Reliable measurements of the atmospheric CO_2 concentration have been made since 1958 at Mauna Loa (Hawaii, 3,400 m above sea level) and at the South Pole (2,800 m above sea level) (Keeling et al., 1976a, b). Apparently, neither large-scale variations in the atmospheric circulation, nor local sugar cane fires distort the data, and intense outgassing from the summit caldera can be readily eliminated from the record so that Mauna Loa is an excellent site for monitoring background atmospheric CO_2 (Sadler, Hori and Ramage, 1982). As Fig. IV.2 shows, between 1959 and 1978 the CO_2 concentration at Mauna Loa increased from about 315 ppmv to about 335 ppmv, i.e. about 6%. Over the same period the amount of CO_2 released by fossil fuel combustion was about 36 ppmv (Rotty, 1981). If we make the simplifying assumption, that, as a result of the relatively good mixing, the CO_2 content above Mauna Loa is representative of the entire atmosphere, then over the entire 20-year period the fraction of CO_2 remaining in the atmosphere was about 55% (the so-called apparent **airborne fraction** AF = 20 ppmv/36 ppmv). For prediction purposes a so-called effective airborne fraction is defined, as discussed in IV.1.6.

There are strong seasonal variations of the atmospheric CO_2 content as shown in Fig. IV.2. This is related to the fast photosynthesis-respiration cycle of terrestrial plants. In the

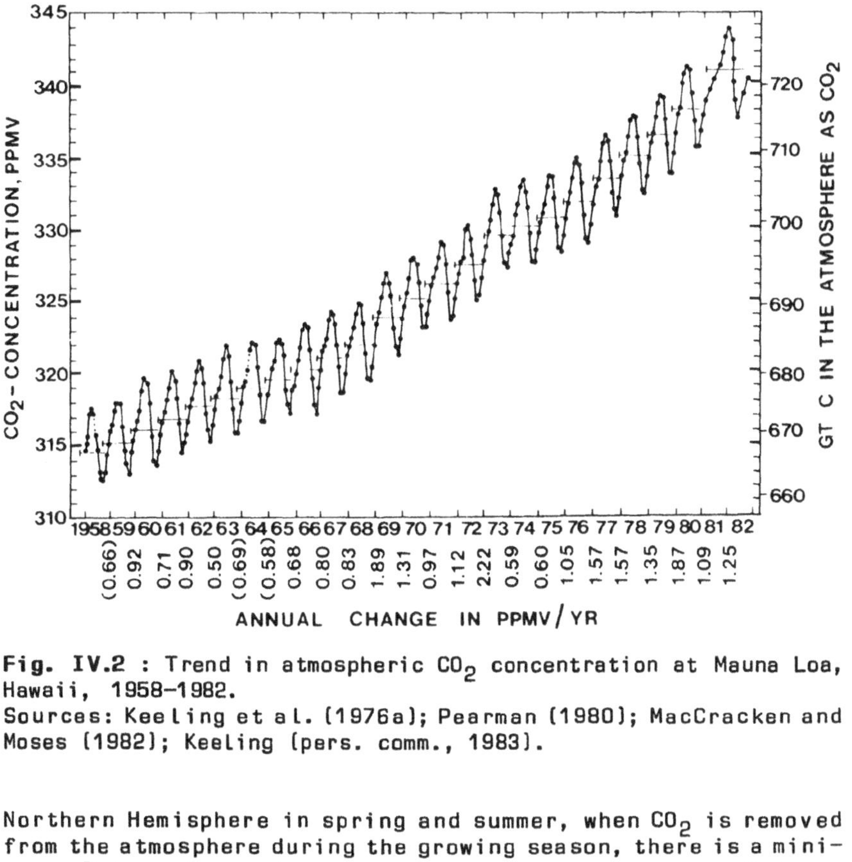

Fig. IV.2 : Trend in atmospheric CO_2 concentration at Mauna Loa, Hawaii, 1958–1982.
Sources: Keeling et al. (1976a); Pearman (1980); MacCracken and Moses (1982); Keeling (pers. comm., 1983).

Northern Hemisphere in spring and summer, when CO_2 is removed from the atmosphere during the growing season, there is a minimum. Decay of the biomass in autumn gives a maximum. Primary production and hence CO_2 uptake occurs only during daylight hours, while CO_2 is returned to the atmosphere by oxidation of organic detritus from plants and soil carbon during both day and night. Therefore, in order to obtain a net gain of biospheric carbon during the growing season, a large fraction of the net primary production (NPP) would be required just to compensate these losses. Thus the evidence found by Keeling (1983) from the Mauna Loa record that the seasonal activity of land plants has increased at an average rate of 0.66 %/yr, or by 13% over the past 20 years, is somewhat surprising, since the fractional increase in atmospheric CO_2 over the same time period is only 8% (see above). Brown, Lugo and Gertner (1982) suggest that CO_2 stimulation of the growth of land plants (see also V.4.4), a net increase in forests of the temperate and boreal zone, and, to a lesser extent, seasonal contributions from fossil fuel use, may account for the enhanced CO_2 uptake. Kohlmaier and Revelle (pers. comm.; Kohlmaier et al., 1983) have estimated that the above relative amplitude increase of 0.66%/yr at Mauna Loa corresponds to a relative increase in NPP of 0.56%/yr, of which 0.36%/yr may be due to CO_2 stimulation and 0.20%/yr to reforestation. There may also be opposing effects, such as through the acid rain ingredients (SO_2/H_2SO_4, NO_x/HNO_3 and tropospheric O_3), which have not been considered.

Moreover, Fig. IV.2 shows that there are strong variations in the annual rate of CO_2 increase that are probably related to the large variations of the sea surface temperatures in the South Pacific, a phenomenon called the Southern Oscillation (Bacastow, 1976, 1979; Newell et al., 1978; Angell, 1981). Sadler et al. (1982) have pointed out, however, that it is not yet clear what, if any, the relationship is and that so far no cause-effect linkages have been shown.

The average growth rate of the atmospheric CO_2 content was about 0.4%/yr over the past 20 years showing a tendency to increase, although the growth rate of fossil fuel combustion has decreased over the past 10 years (see also IV.1.4.1). All of these findings, shrouded with much uncertainty, are important enough to warrant more detailed analyses because of their significance, especially for future food and fibre production.

IV.1.3.2 CO_2 monitoring

Within the framework of the International Geophysical Year the first systematic CO_2 measurement programme was started in 1957 on Mauna Loa on the island of Hawaii and at the South Pole. Two decades later the UN began to build up a global environment measurement system (GEMS), for which the World Meteorological Organization (WMO) with its World Weather Watch (WWW) provide the weather- and climate-relevant data (Wallen, 1980). A Background Air Pollution Monitoring Network (BAPMoN) supported by the WMO and the United Nations Environment Programme (UNEP) observes the influence of humans on the atmosphere on a regional, continental and global basis. The existing or planned CO_2 monitoring stations shown in Fig. IV.3 participate in this programme.

The use of standardised recording methods and the selection of the measurement sites is important, since in the vicinity of biogenic (woods) and anthropogenic (towns, industry) sources, there can be considerable variations of the CO_2 values depending on the meteorological state of the atmosphere.

Over a period of decades the atmosphere is a well-mixed reservoir. The observation of the long-term CO_2 trends can therefore be made with a background monitoring network of about 2-3 stations (Pearman, 1980). For the assessment of many details of the carbon cycle so few monitoring stations would, however, no longer be representative (WCP, 1981a; Keeling, 1983). Airline and ship data show that the mean annual concentration of CO_2 is

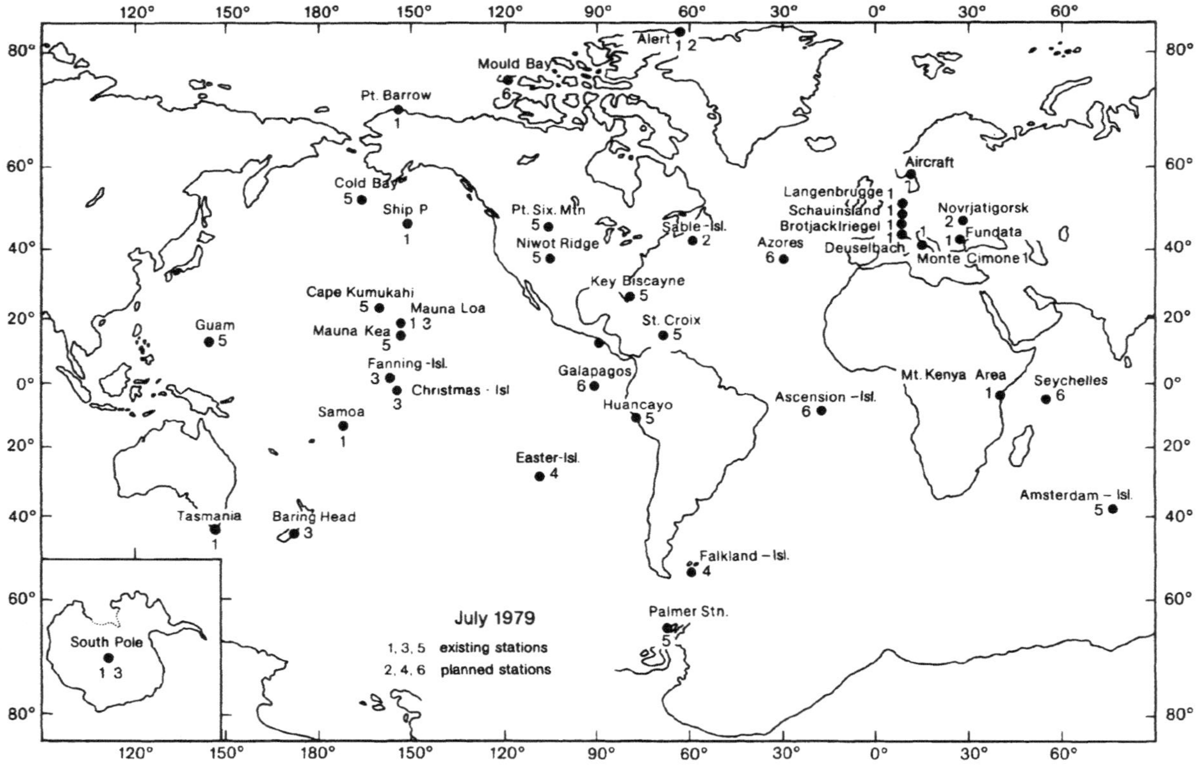

Fig. IV.3 : Stations monitoring atmospheric CO_2 trends, July 1979.
Source: US DOE (1980).

about 1 ppmv higher in the Northern Hemisphere than in the Southern Hemisphere, as a result of the higher industrialisation in the former. Further, meridional CO_2 differences between the pole and equator, or vertical CO_2 differences in the troposphere of up to 10 ppmv, can occur in certain months. Thus, if one wants to detect perturbations of the carbon cycle, the planned extension of the CO_2 measurement network shown in Fig. IV.3 is necessary.

IV.1.4 Causes of perturbation

The estimation of the future CO_2 emissions is the prerequisite for the projection of CO_2 concentrations and thus is the first step in the evaluation of the impacts of a CO_2 increase on climate and society. To do this, an estimate of the future fossil fuel consumption and shares of the individual energy carriers is necessary. An additional uncertainty arises from the fact that the future behaviour of the biosphere and ocean as sources or sinks for CO_2 must be estimated. The difficulties and uncertainties in producing a meaningful estimate appear to be insurmountable. In such a situation, the most reasonable course of action is to make the present state of knowledge the starting point for increased efforts to remove the existing gaps in our knowledge.

IV.1.4.1 Role of the conventional fossil fuels

The CO_2 output of a fossil fuel depends on its hydrogen: carbon ratio (JASON, 1980). As the H:C ratio increases, the energy produced per unit CO_2 output increases. Table IV.1 shows that

Table IV.1 : Comparison of carbon dioxide release from production of thermal energy by various fuels (average values).

Fuel	Carbon dioxide released in kgC/MWh	Ratio of carbon released by fuel to that released by methane
Natural fuels		
Natural gas (methane)	48	1.00
Oil	72	1.50
Coal	89	1.85
Shale oil	102	2.12
Biomass (renewable)	4	0.08
Synthetic fuels		
Gas	144	3.00
Oil	134	2.79
Methanol from biomass	6	0.13

Adapted from: JASON (1980).

oil has a factor of 1.5, coal a factor of 1.8, oil shale more than a factor of 2 and recent biomass (wood) only a fraction of the CO_2 emission from natural gas.

With the aid of the CO_2 emission factors (kg C/MWh) also shown in Table IV.1 and the UN consumption statistics, the amount of carbon released by coal, oil and gas combustion, by flaring of natural gas and by cement production can be derived for the individual countries. The calculations are based on the values for fuel production and not fuel consumption, because a more reliable estimate of the total amount of oxidised carbon can be gained from production values. As Fig. IV.4 shows, the CO_2 emission due to combustion of fossil fuels has increased from about 0.1 Gt C* in 1860 to about 5.3 Gt C in 1980 (Rotty, 1980, 1981). During this 120-year period the exponential growth rate for the carbon emissions amounted to about 3.4%/yr. If one ignores the breaks in growth during the two world wars and the world economic depression of 1930, one finds the generally quoted growth rate of 4.3%/yr. The detailed breakdown in Fig. IV.4 shows that the very sudden increase of energy prices, resulting from the energy crisis of 1973/74, apparently was able to slow down the fast growth of oil consumption from 7.1%/yr to 1.7%/yr and that of gas from 8%/yr to 2.8%/yr. Coal, with a relatively small annual growth rate of 1.9%, was much less influenced by these developments. While in the mid-1960s oil became the largest contributor to atmospheric CO_2, coal may eventually become again the largest source of CO_2, in particular, if more of the unconventional fossil fuel sources are put to use (see IV.1.4.2). The growth rate for the total CO_2 emission has decreased to the present 1.8%/yr with the prospect of a further reduction due to the increasing resource scarcity and the expected long-term price increases. Because of the inertia of the system and perhaps countervailing processes, however, the ongoing reduction in fossil fuel CO_2 emissions in Fig. IV.4 is not yet discernible in the observed CO_2 concentrations shown in Fig. IV.2.

In addition to the trend analysis, the changes of the individual fuel shares and the development of the total amount of carbon that has been introduced into the atmosphere are of interest. In 1860 the carbon emissions came almost entirely from coal combustion. Over the course of time this picture has changed drastically, such that 43% came from oil, 39% from coal, 14% from gas, and 2% each from flaring of gas and cement production in 1975. The cumulated amount of carbon introduced into the atmosphere since 1860 was about 165 Gt in 1980, or about 1/4 of the total carbon content of the atmosphere. Of this amount, 52 Gt or 31% were produced in the 10 years of 1970-1980. The accuracy of the values given for the period 1860-1949 is $\pm$20% and $\pm$13% for

---

* 1 Gt C (Gigatonne = 10^9 or billion tonnes) corresponds to 3.67 Gt CO_2.

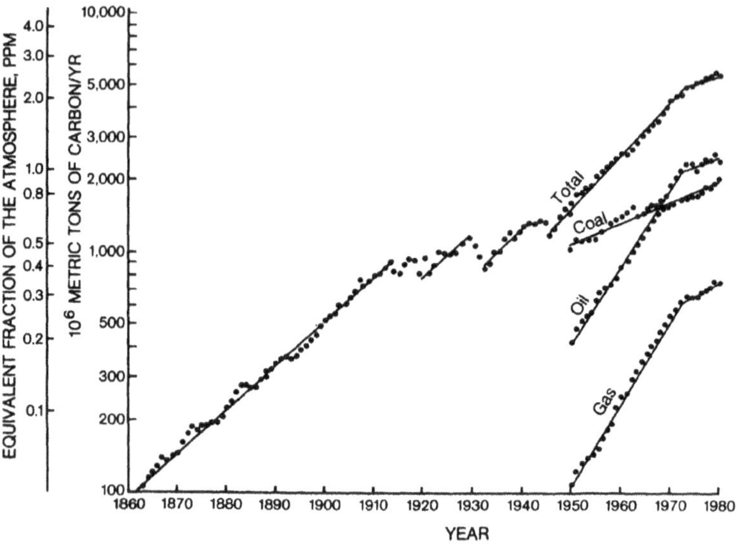

Fig. IV.4 : Global CO_2 production from fossil fuels, cement and gas flaring, 1860–1980.
Sources: Keeling (1973 for 1860–1949) and Rotty (1980, 1981 for 1950–1980).

the time after 1949 (Keeling, 1973).

The asymmetries in the observed CO_2 concentration, discussed in section 1.3.2 and Appendix IV.2, are partially explained by the latitudinally-dependent distribution of the CO_2 emissions in Fig. IV.5. The CO_2 emissions from the combustion of fossil fuels and cement production are almost entirely released from a narrow latitude belt of 30–60°N. Given the long-term price developments, the only half-hearted efforts to begin a serious north-south dialogue, and the fossil resource distribution (see III.4), only a small redistribution of this emission pattern can be expected. The latitude-dependent CO_2 distribution is not only of interest for the modelling of the future regionally differing climate impacts, but also for the required compensation payments.

IV.1.4.2 Role of the unconventional fossil fuels

When US President Carter presented his Energy Program in 1979, from a total investment sum of $144 billion, $88 billion were slated for the development of a synthetic fuels industry, i.e. for the refining of carbon-containing materials. Many scientists at that time warned about the consequences of such a strong build-up of a synthetic fuels industry (U.S. Senate, 1979). Since coal liquefaction and gasification require additional energy, the question arises by how much the CO_2 emission increases in comparison to the direct combustion of gas, oil or coal.

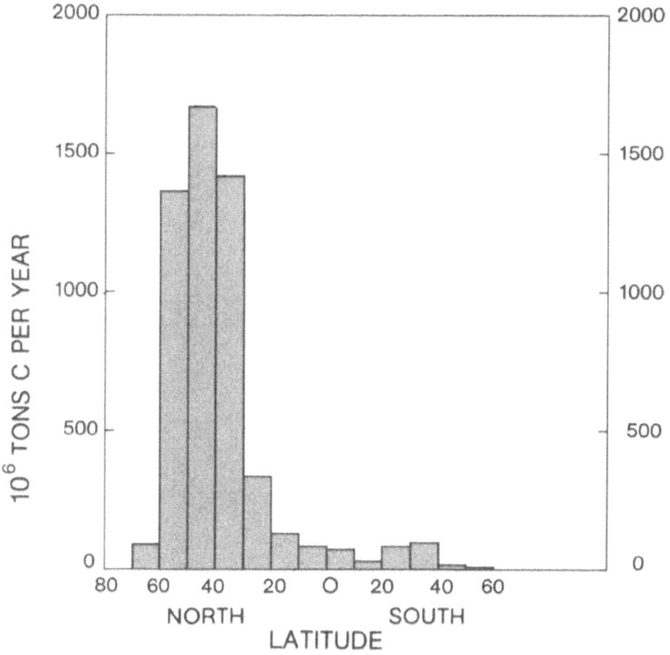

Fig. IV.5 : The 1979 CO_2 emissions from fossil fuels and cement manufacture of 5340 x 10^6 tons C by latitudinal bands. Source: Rotty (1981).

For **coal liquefaction**, there are basically three processes: hydration, extraction and pyrolysis (Franck and Knop, 1980). Coal containing large amounts of ash and sulphur is converted into a fuel containing a low amount of ash and sulphur - but at the cost of the total energetic efficiency with a simultaneous increase of CO_2 emissions. As Table IV.1 shows, the CO_2 emission increases by a factor of 2.8 in comparison to the direct combustion of natural gas.

Coal gasification products can be divided into generator gas or low Btu gas, syngas and reduction gas, town gas and coking gas, and into high Btu gas and synthetic natural gas (Franck and Knop, 1980). According to Table IV.1, the CO_2 emission is increased by a factor of 3 in comparison to the direct use of natural gas.

The methods described above are **autothermal processes**, i.e. methods without introduction of external energy. If, however, the refining process is carried out with the aid of a non-fossil energy source (**allothermal process**), the CO_2 emission is still increased 6% above that of direct coal combustion, because carbon is needed for the splitting of water for the production of hydrogen (Bach, 1981). Only in the case of the use of a non-fossil energy source and hydrogen production without carbon (e.g. through electrolysis or thermo-chemical splitting of water) can the CO_2 emission be reduced by about 40% compared to direct coal combustion.

The carbonisation of **oil shales** to shale oils also releases large amounts of CO_2 (Sundquist and Miller, 1980). According to Table IV.1 the emission is a factor of 2.1 more than from natural gas or about 20% more than from direct coal combustion. **Fluidized-bed combustion** emits less nitrogen oxide because of the lower temperatures, but the CO_2 emission is the same as in conventional coal power plants. In the case of **magneto-hydrodynamic generators** that work with hot, ionised gases, a reduction of the CO_2 emission per energy unit produced can be expected because of the increased efficiency (JASON, 1980). Finally, Table IV.1 shows that both the direct combustion of **biomass** (e.g. wood, energy plantations) and of methanol from biomass lead to only a small net CO_2 emission in comparison to fossil fuels, mainly resulting from diesel oil driven machinery.

IV.1.4.3 Role of the biosphere

As we have seen in Fig. IV.1, the land ecosystems, with the land plants and humus and peat, contain about 3,000 Gt carbon or about 4 times as much as in the atmosphere. A change of this reservoir by only $1^{o}/oo$/yr (which would be hardly detectable) would lead to a net exchange with the atmosphere of about 3 Gt carbon, or about 50% of the present CO_2 emission from fossil fuels. This example shows clearly the critical role of the land ecosystems in the global carbon cycle.

The importance of the land ecosystems is underlined by the work of Moore et al. (1980), who have reconstructed the annual carbon flux into the atmosphere from 1860 to 1970 with the aid of historical records. Fig. IV.6a shows that the net carbon transfers, as a result of agricultural use, were greater than those from deforestation during the entire time period. According to Fig. IV.6b, up until 1960 the annual carbon emissions from land ecosystems (including plants and soil oxidation) were greater than the carbon emissions from fossil fuels. In this compilation, however, the so-called **pioneer effect** (i.e. deforestation and claiming of land without documentation) has been omitted, so that the biogenic carbon transfers in Figs. IV6a and b have to be

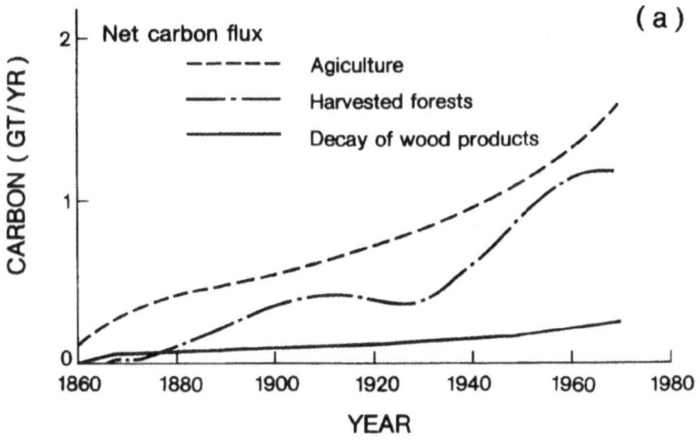

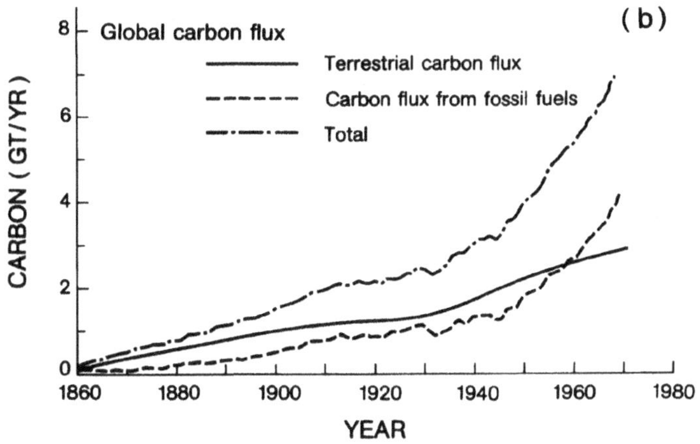

Fig. IV.6: a) Carbon release from forest clearing and agricultural activity, 1860–1970; b) Global carbon flux, showing terrestrial and fossil fuel components, 1860–1970.
Source: Moore et al. (1980).

taken as lower limits.

One of the most controversial problems that requires urgent clarification is the question whether the land vegetation is together with the oceans a sink for CO_2, as geophysicists have assumed till now on the basis of their investigations, or a source, as claimed by ecologists on the basis of deforestation of tropical forests (Bach and Breuer, 1980; Breuer, 1981). A reasonably sensible estimation of the future atmospheric CO_2 content (see IV.1.7) can only be made if the biogenic transfers, that make up a considerable part of the total carbon cycle, can be reliably quantified. In the following, we consider the extensive range of the presently known or assumed sources and sinks and a possible balancing sheet of the total carbon budget.

The tropical forests - The influence of mankind on the carbon cycle of tropical ecosystems can be assessed using three independent methods: estimation of the extension of cultivated areas, extrapolation of remote sensing data, and evaluation of the historical development (Hampicke and Bach, 1979; Woodwell, 1980).

The **population growth** in the tropics of about 2.4%/yr has been shown to result in an increase in the **cultivated area** of about 2%/yr, or some 120,000-210,000 km^2 at the cost of tropical forests. The conversion of natural ecosystems into agricultural systems almost always results in a net carbon flux into the atmosphere. According to our estimates, the annual net transfer amounts to 1.3-2.3 Gt C, or about 27-46% of the annual carbon emissions from fossil fuels (Hampicke and Bach, 1979).

A comparison of **air photographs** from the 1950s, with **satellite photographs** from the 1970s shows, for example, that the forest area of Thailand was reduced by at least 2% during this time period. LANDSAT data of the Brazilian Institute for Forest Development suggest that the deforested area in the tropical rainforest of the Amazon, that was still 28,595 km^2 in 1973-75, had already reached 48,576 km^2 in 1976-78 (Freyer, pers. comm., 1981). From this and similar investigations, with extrapolation for comparable areas, we obtain a net carbon transfer into the atmosphere of about 1.6 Gt/yr, or 33% of the fossil emissions. It appears that data gathered by aircraft, satellites and near-surface measurements can be used to provide a quantitative assessment of source/sink distributions of CO_2 in the biosphere (Desjardins et al., 1982).

When the net loss of phytomass takes place in a relatively short time period, the CO_2 content of the atmosphere can be strongly affected. The temporal distribution can be reconstructed approximately using **historical vegetation statistics.** Using this method, an effective net carbon transfer into the atmosphere of 1.8-4.0 Gt/yr is computed for 1977 (Hampicke and Bach, 1979).

From these evaluations we obtain the result that the tropics, for the study year 1977, represented a net carbon source for the atmosphere on the order of 1.3-4 Gt/yr (about 27-80% of the fossil fuel emissions). One may attach some significance to these values, since they are based on a variety of methods and independent assumptions. There are, however, large differences of opinion. At a symposium in Puerto Rico in 1979, it was claimed, on the basis of reforestation programmes on this Caribbean island, that the tropical rainforests are presently more a sink than a source of atmospheric CO_2 (Brown et al., 1980). However, an atypical single case was taken and projected onto the conditions of the entire tropics. Conducting a detailed study on land use changes and the resulting carbon exchanges in the four tropical countries of Panama, Costa Rica, Peru and Bolivia, Detwiler, Hall and Bogdonoff (1982) find a best estimate for the release of carbon from tropical vegetation to range from 0.5–1.9 Gt/yr. They believe that these lower rates are more accurate than previous higher estimates because most tropical life zones store less carbon than once thought.

Temperate latitude forests – A team of experts has recently analysed the forest history in the temperate latitudes from the time before settlement up to the present (Armentano and Hett, 1980). They come to the conclusion that, as a result of the reforestation programmes, especially in Europe, the USA, China and perhaps also in the USSR, the forests have been a net sink for atmospheric CO_2 on the order of 1 ± 0.5 Gt C/yr during the last few decades. Probably 40–50% of this sink is in North America (Clawson, 1979). On the other hand, if the demand for wood energy and fibre were to intensify, the temperate zone forests could be significantly reduced and perhaps even lost as a carbon sink within the next 20–30 years (Armentano and Loucks, 1982).

Ecosystems in polar latitudes – Because of the large deposits of dead organic matter, the ecosystems north of 60°N are very significant for the carbon cycle. The net primary production is estimated to be about 0.5 Gt C/yr in the Arctic and 2.3 Gt C/yr in the Taiga. The rate of carbon accumulation is currently about 0.2 Gt/yr (Miller, 1982a). According to the present state of knowledge, it is not clear whether the northern ecosystems represent a negative or positive feedback for the global carbon cycle in the case of a potential CO_2-induced warming (Miller, 1982b). There is, however, a growing concensus that with a temperature rise, the Taiga would become a carbon source as decomposition processes increase, and the Tundra would become a carbon sink as more organic matter is deposited in the permafrost layer (Brown and Andrews, 1982).

Organic soils – As a result of intensive cultivation, soils can lose 1/3 to 1/2 of their original organic carbon content. Through the conversion of natural land into cultivated land during the last 100 years, at an annual rate of 1–2 Gt, about 150 GtC have been released to the atmosphere as CO_2 (Bohn, 1978). These numbers refer to soils in general, thus mainly to mineral soils. Organic soils release presently about 0.03–0.37 Gt C/yr to the atmosphere as a result of human activities, corresponding to a rate of 0.6–8% of the annual CO_2 increase in the atmosphere (Armentano, 1980). Carbon loss from erosion of agricultural soils is not well known and may be greater than currently assumed. The magnitude of the sink of organic soils unaffected by human activities is estimated to be about 0.045 Gt C/yr. An inventory classifying over 2,000 soil profiles from most regions of the world results in a present global soil carbon pool of about 1,484 Gt C (Post et al., 1982).

Charcoal formation - Seiler and Crutzen (1980) have pointed out that during the burning of forests not all of the carbon is released into the atmosphere. A part (presently about 1 Gt C/yr) is converted to charcoal which does not immediately release the CO_2 to the atmosphere. In this case there is a delayed release and not a real sink, unless the charcoal is prevented from further decay by being deposited in the river sediments.

River sediments - According to Schlesinger and Melack (1981), the world rivers can only store 0.37–0.41 Gt C/yr as particulate organic carbon in river sediments (black water rivers in the Amazon area) and in the shallow estuaries, thereby removing carbon from the atmosphere. However, Lieth et al. (1980) suggest a storage of 0.2–3 Gt C/yr. If this carbon sedimentation were predominantly through human influences (e.g. through the production of charcoal or erosion of soil humus) then one would have a plausible explanation for the whereabouts of the carbon from non-fossil sources and thus no difficulties in balancing the carbon cycle with observed and computed data. On the other hand, if we are dealing with natural processes, then we should be able to find large amounts of organic carbon in the river sediments and estuaries. These questions are still wide open. The carbon transport by rivers is, however, possibly an important component of the global carbon cycle that warrants a more detailed investigation (US DOE, 1981). The first results of such a detailed research programme, carried out under the auspices of SCOPE and UNEP, have just been published (Degens, 1982). Furthermore, Mulholland and Elwood (1982) postulate an increasing rate of carbon accumulation in the sediments of lakes and coastal areas as a result of man's activities, amounting to about 0.02 Gt C/yr and 0.2 Gt C/yr, respectively.

Stimulation of photosynthesis - As a result of laboratory experiments and greenhouse cultures, it is known that certain plants respond to an increasing CO_2 content in the air by increasing growth. Therefore, in carbon models, various authors (e.g. Zimen et al., 1977; Siegenthaler and Oeschger, 1978) have introduced **biota growth factors** of 0–0.4 as a CO_2 sink. Their modelling efforts seem to suggest that a model fit to the Mauna Loa data can only be achieved if the stimulation of land biota growth through CO_2 is included. Other investigations have, however, shown that the laboratory experiments are not globally representative of nature, since other factors, such as the supply of nutrients, water and solar energy are decisive. Van Keulen et al. (1980) and others, therefore, doubt that there is a **CO_2 fertilization effect.** Given the wide range of likely responses from the different types of ecosystems found on earth, it can be appreciated that the concept of CO_2 fertilization is highly controversial. (See also the carbon model calculations in IV.1.7, as well as V.4.4 and VI.3.)

Reduction of CO_2 sequestering as a result of air pollution - It is known from air pollution studies that gases such as sulphur dioxide (SO_2), nitrogen oxides (NO_x), ozone (O_3), as well as acid rain, lead to stunted growth, dropping of leaves, vulnerability to disease, and finally, to death of plants. On the basis of detailed investigations in the USA, Loucks (1980) suspects that the harvest yields (primarily grains and soya beans) will decrease by 28-34%, and forest stocks by 16-22% by 2000. The forests in the industrialised countries in the northern temperate latitudes, postulated above to be a carbon sink, could thus take up 1-2 Gt C/yr less.

Conclusions - The compilation of the present state of knowledge has clearly shown that land ecosystems play a critical role in the global carbon cycle that is far from being understood. A rough estimate of the total balance of all land ecosystems shows that in 1978 the net gains in the atmosphere (**sources**) of 1-4 Gt C/yr were opposed by net losses from the atmosphere (**sinks**) of 0.5-3 Gt C/yr (Hampicke and Bach, 1979). Since the two ranges overlap, one could postulate a balance on purely numerical grounds. Fig. IV.7 summarises the temporal changes of the individual contributions to the annual carbon transfer. This clearly shows the critical role the tropical forests and soils play as carbon sources and the woods of the temperate latitudes as carbon sinks. If the assumptions about the turn of the century are correct, then the land ecosystems would be a carbon source of about 2 Gt/yr, provided the above-mentioned sinks were effective to the anticipated extent.

IV.1.4.4 Role of the ocean

As we have seen in Fig. IV.1 the ocean is an important sink for atmospheric CO_2 on the order of about 2 Gt C/yr. Since some areas of the ocean take up carbon and others, however, release it, it is important for the modelling of the carbon cycle to consider the regions, as well as the direction and magnitude of the net exchange. The rate at which CO_2 is transferred into the ocean depends on four processes: 1. the speed of the transfer of the CO_2 from the atmosphere into the ocean surface water; 2. the absorptive capacity of the ocean in chemical equilibrium; 3. the transport of the CO_2 from the surface water into the deep sea; and 4. the CO_2 transport through biological processes, among which important sinks for atmospheric CO_2 are suspected.

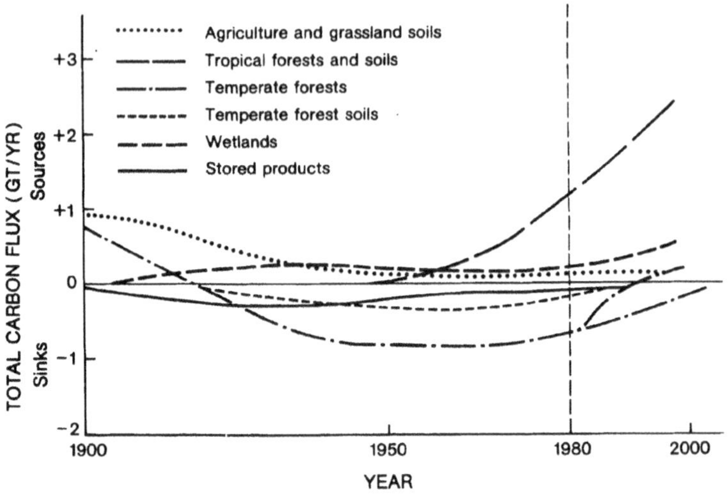

Fig. IV.7 : Total carbon fluxes from 1900, with projections beyond 1980.
Source: Loucks (1980).

CO_2 exchange between ocean and atmosphere - Diffusion through a thin liquid boundary layer controls the CO_2 exchange (Heimann et al., 1981). The CO_2 absorption rate of the ocean can be calculated with the aid of the transfer velocity, w. For a diffusion coefficient $D \sim 2.5 \times 10^{-5}$ cm^2/sec for trace gases, one finds a mean transfer velocity of $w = D/z$ of 2.2 m/day, which agrees well with the results of measurements of radioactive trace gases in the sea. In a series of gas exchange experiments Goldman and Dennett (1983) found that CO_2 chemical reactivity probably plays a minor role in the exchange of CO_2 between the atmosphere and the ocean, and that rather wind-induced turbulence appears to be the major factor controlling the ocean's uptake of atmospheric CO_2.

CO_2 absorptive capacity of seawater - Carbon is primarily found in seawater in inorganic form (so-called total CO_2 or ΣCO_2) and consists of 85-90% bicarbonate ions (HCO_3^-), 10-15% carbonate ions ($CO_3^=$) and 1% dissolved gaseous CO_2. Only the latter participates in the direct exchange with the atmosphere (Heimann et al., 1981). If the ocean water takes up more carbon, as a result of the CO_2 increase in the atmosphere, there is a shift of chemical equilibrium, generally expressed by the **Revelle** or **Buffer factor** $R \equiv (\Delta CO_2/CO_2)_{atm}/(\Delta\Sigma CO_2/\Sigma CO_2)_{sol}$ (Roether, 1980). The present value for average surface water is $R \sim 10$. This means that with an atmospheric CO_2 increase, e.g. of 10%, the sum ΣCO_2 (HCO_3^-, $CO_3^=$, CO_2) only increases by 1%, when complete equilibrium is achieved. Thus, if the ocean is to absorb more anthropogenic carbon, the deep sea must be involved.

An increasing atmospheric CO_2 concentration would cause a reduction of the pH of the ocean water by about 0.25 units for every CO_2 doubling (Baes, 1981). As a result the sediments could be dissolved, which could lead to a lower buffer factor and increased CO_2 uptake by the ocean. Since, however, the seawater is supersaturated with respect to calcite and aragonite, neither the dissolution of calcite and aragonite, nor the more soluble magnesium-rich calcite in the surface sediments is to be expected in the next 100 years (Broecker, 1974). It can be concluded that a significant reduction of the buffer factor is not to be expected and, therefore, a decisive increase of the CO_2 absorption capacity of the ocean surface water appears to be highly unlikely (Heimann et al., 1981).

CO_2 transfer in the ocean – The CO_2 absorbed in the ocean surface water is ultimately transferred to the deep sea layers and remains there. In this case, knowledge about the ability to absorb additional amounts of CO_2 and the rates of transfer into the deep layers of the ocean, on a time scale of several decades, are of decisive significance for the anthropogenic CO_2 problem.

The first relevant study, namely the Geochemical Ocean Sections Study (**GEOSECS**) was initiated in 1972, investigating seawater properties and the tracer tritium (H-3) from nuclear weapons testing. This was followed by the Transient Tracer in the Ocean (**TTO**) programme, which includes the North Pacific Experiment (**NORPAX**), the Mid-Ocean Dynamics Experiment (**MODE**) and its follow-up (**POLYMODE**). The major purpose of these projects is to measure ocean salinity, alkalinity, density, temperature, nutrient and oxygen content, and the tracers H-3, C-14, Kr-85, and Ar-39. From the penetration rate of the latter, information is derived about the potential CO_2 transfer rates in the deep sea. The results indicate that there may be significant differences in the transfer rates with latitude and depth between the Pacific, Atlantic and Indian Oceans (Millero et al., 1976; Stuiver, Östlund and McConnaughey, 1981; Takahashi, Broecker and Bainbridge, 1981). Specifically, the C-14 distribution indicates replacement times for Pacific, Atlantic and Indian Ocean deep waters (>1,500 m deep) of about 510, 275, and 250 years, respectively. The overall replacement time for the entire deep ocean averages about 500 years (Stuiver, Quay and Östlund, 1983). Additional observational studies under way include **TOPEX** (assessing sea surface elevation and hence wind stress and temperature patterns); **ENSO** (studying El Nino events and the role of the Southern Oscillation); **CAGE** (determining oceanic heat transport in the Atlantic Ocean); and **MIZEX** (studying energy interactions at the ice margins) (Riches, MacCracken and Luther, 1982).

In order to understand better the mechanisms for CO_2 transfers in the ocean, a short review of the **ocean circulation system**

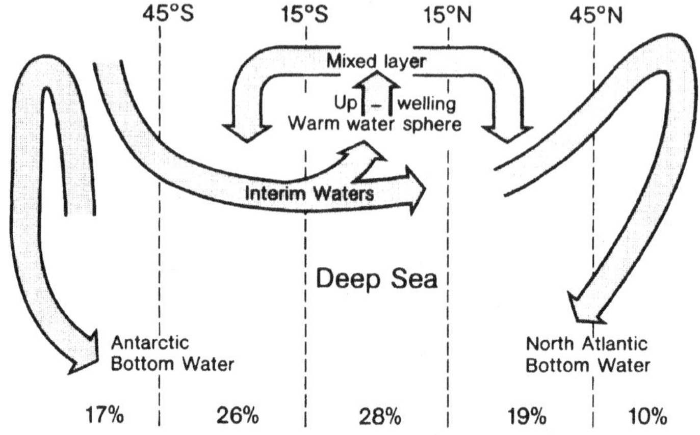

Fig. IV.8 : Diagram showing the patterns of the flow within the ocean. Upwelling occurs in the equatorial zone; downwelling occurs in the temperate zone and polar zone. The ocean area in each zone is shown as a percentage of the total ocean area. Sources: Dietrich et al. (1975); Broecker et al. (1980).

is appropriate. As the schematic illustration in Fig. IV.8 shows, the mixed layer, thermocline and deep sea exchange water through a surface and deep sea circulation system (Broecker et al., 1980). The surface circulation system is a result of the direct influence of the atmospheric wind field and extends to a depth of about 300-400 m, except in the area of the subpolar gyre, which extends into the deep sea. Within a week this surface layer is vertically well mixed. The ability to take up CO_2 is determined, as shown above, by gas exchange, carbonate chemistry, and the transfer into the deep sea (Heimann et al., 1981). In equatorial and middle latitudes there is a strong **thermocline** with temperatures of >8-10°C (Dietrich et al., 1975). Since it has a considerable absorption capacity over periods of 10-500 years, it can play a decisive role as a sink for anthropogenic CO_2. In the depths below 1,000 m there is a deep sea circulation largely driven by thermohaline processes that is largely independent of the surface circulation. It is mainly fed by the Antarctic Bottom Water originating in the Weddell Sea with a magnitude of 40 Sv* and by the North Atlantic Bottom Water coming from the Greenland and Norwegian Seas at a rate of 10 Sv (Fig. IV.8). The mass balance of the deep sea is maintained by **upwelling** water with typical velocities of 2-8 m/yr in central equatorial latitudes (Broecker, 1974). Finally, it must be noted that the deep

---

* 1 Sv = Sverdrup = 1 million m^3 water/sec.

sea plays a role in carbon storage only for very long time periods compared to the human time scale, but it is eventually able to take up the entire CO_2 from the combustion of all fossil fuel reserves.

CO_2 transport through biological activities in the sea - Organic carbon in the form of particulate and dissolved organic matter (known as **POC** = particulate organic carbon and **DOC** = dissolved organic carbon in the scientific literature) has been suggested as a carbon sink (see also Fig. IV.1). Particulate carbon includes the entire living biomass and entire dead organic matter, which sinks, together with the limestone shells as a continual particle precipitation, with the contained carbon into the deep sea (Smith, 1981).

The biological productivity of the ocean water is particularly high in the region of upwelling water because of the high nutrient content (N, P). The algal blooms that are particularly extensive there consume large amounts of atmospheric CO_2. This is the explanation of the high uptake of CO_2 by upwelling water, in spite of CO_2 supersaturation (Newell et al., 1978).

As a result of the increasing CO_2-induced warming, a reduction of the meridional temperature gradient and consequently a reduction of the ocean circulation are to be expected (Baes, 1981). A weakened circulation possibly reduces the biological productivity in the ocean because less nutrients are transported. On the other hand, the low nutrient content of the warmer water would be used more intensively, thus having the opposing effect. It is felt that the net result of these opposing effects is a reduced biological activity with an increase of the CO_2 partial pressure in the surface water, which could lead to a reduction of the absorption capacity for atmospheric CO_2 (Baes, 1982, 1983).

It is also suspected that, as a result of the introduction of dissolved phosphorous (from artificial fertilizers) in rivers into the sea, about 2% of the CO_2 production from fossil fuel combustion processes could be fixed (WCP, 1981b). In addition, Walsh et al. (1981) propose that, through the introduction of waste water in coastal zones, **eutrophication**, plant growth and CO_2 uptake could be stimulated, which would offer a solution to both the waste water and the CO_2 problems. Quantitative estimates are, however, not available. It would also be necessary to examine whether the oceanic ecosystem could possibly be damaged by waste water.

IV.1.4.5 Carbon budget

According to the present state of knowledge, we have the following approximate picture (Table IV.2 and Fig. IV.1). The data on the carbon amounts of about 5 Gt C/yr from fossil fuel combustion, the 2 Gt C/yr taken up by the ocean and the 3 Gt C/yr remaining in the atmosphere are relatively reliable. The most uncertain are the data about the net effect of deforestation and soil oxidation in the tropics. The estimates of the CO_2 emission from terrestrial biogenic sources vary between 2 and 5 Gt C/yr with presently a most probable value of about 2 Gt C/yr. Depending on which value one accepts, one must find a corresponding carbon sink in order to balance the budget. In this respect, in addition to the ocean, only the vegetation in temperate latitudes with a removal potential of about 1 Gt C/yr is considered by many as relatively certain at the present. Although the type of the sources and sinks, as well as their magnitudes, are still quite uncertain, it is possible to derive on a purely arithmetical basis a balanced carbon budget (Table IV.2).

Table IV.2 : Possible carbon balance, ca. 1978.

Source	*Carbon Gt/yr*
Combustion of fossil fuels (coal, oil, gas) gas flaring cement production, chemical industry (Rotty, 1980)	5.2
Deforestation and soil oxidation	
after Woodwell (1978)	5
Seiler and Crutzen (1980)	3
Hampicke and Bach (1979)	2
Bolin (1977)	2
Total	5 - 10
Sink	
Ocean (Broecker et al., 1979)	2
River sediments (Lieth et al., 1980)	2
Reforestation (Degens and Kempe, 1979)	1
Charcoal (Seiler and Crutzen, 1980)	1
Total	6
Remaining in the atmosphere	3
Total budget	0

On the other hand, we must be aware that this budgeting does not take account of the many subtle feedbacks, such as e.g. the CO_2-induced warming which accelerates oxidation of organic material at high latitudes. Moreover, there is, of course, the possibility that some source/sink relationships have been overlooked. For example, laboratory measurements, corroborated by some field measurements, show that termites may be a potentially large source of CO_2 of the order of 13 Gt C/yr ± 50%, more than twice the fossil fuel input (Zimmerman et al., 1982). Although other decomposition processes would eventually release the CO_2, termites help to accelerate the carbon cycling. The termite population per unit area is largest in tropical regions disturbed by man's activities. Shaver et al. (1982), who have developed a bookkeeping-type model that tabulates the storage of carbon on land and the fluxes to and from the atmosphere through vegetation changes, point out that current estimates of the carbon emissions from fossil fuels and the biota do not balance the observed CO_2 increase in the atmosphere and the computed CO_2 uptake by the oceans, because they think that the terrestrial ecosystem as a net source of carbon due to man's land use has been underestimated. If that were correct, then it would lead also to an underestimation of the airborne fraction, the oceans' long-term capacity to absorb the excess atmospheric CO_2, and, ultimately, to faulty projections of how much fossil fuel can be burned without exceeding critical atmospheric CO_2 levels (Dahlman, 1982). SCOPE 13 (Bolin et al., 1979) and SCOPE 16 (Bolin, 1981) should be consulted for further information on the state of knowledge of the global carbon cycle.

IV.1.5 Assessment of past CO_2 concentration

Before we risk taking a look into the future, it is appropriate to investigate briefly whether and how variations in the atmospheric CO_2 content have influenced the climate in the past. The consideration of CO_2 concentrations and climate indicators of past periods can provide us with important information for the interpretation of future CO_2/climate interactions. For the reconstruction of the history of the atmospheric CO_2 content we can apply indirect and direct methods of isotope analysis.

IV.1.5.1 Indirect methods: Analysis of tree rings

The natural global carbon cycle consists of about 99% of the isotope ^{12}C, about 1% of ^{13}C and of traces (10^{-10}%) of the radioactive isotope ^{14}C, which is formed in the atmosphere from nitrogen through the influence of cosmic rays (Freyer, 1979). The stable isotopes ^{12}C and ^{13}C are found in various reservoirs in small, differing amounts, that can be described by $\delta^{13}C$ (PDB). A standard substance (PDB) with a $^{13}C/^{12}C$ ratio of exactly 1123.72 x 10^{-5} serves as a reference case with a $\delta^{13}C$ of zero. The deviation of all other carbon-containing substances can then be computed with the aid of the formula $\delta^{13}C$ (PDB) = [$^{13}C/^{12}C$ (sample) − $^{13}C/^{12}C$ (PDB)]/$^{13}C/^{12}C$ (PDB) x 1000°/oo. The average $\delta^{13}C$ values (in ^{o}/oo) are −7 for present air, for coal −24, for oil −27 and for natural gas −41. The value for cement production of 0$^{o}/_{oo}$ is taken as PDB-standard (Tans, 1981). Since the $\delta^{13}C$ values from recent biomass or humus and coal are similar, it is not possible to distinguish between fossil and recent organic carbon using the $^{13}C/^{12}C$ ratio.

During the transfer from one reservoir into another, the carbon isotopes are fractionated so that different isotope compositions arise. For example, during the buildup of phytomass and humus ^{12}C atoms are preferentially removed from the atmosphere and the heavy isotope ^{13}C thus becomes enriched, i.e. the $\delta^{13}C$ value of the atmosphere increases. On the other hand, the decomposition of biomass (through deforestation) enriches the atmosphere with light ^{12}C isotopes, diluting the ^{13}C content and thus reducing the $\delta^{13}C$ value (the so-called **Suess Effect** for ^{13}C).

With the aid of isotope analysis of individual **tree rings**, one can look at the historical record of the ^{13}C content of the atmospheric CO_2 (Jacoby, 1980). Because of the small differences in isotope ratios, only isolated trees growing in very clean air can be analysed. The investigations by Freyer (1978), primarily for the Northern Hemisphere, show that from 1800 to 1960, i.e. a period of 160 years, the ^{13}C content of tree rings decreased by about 2°/oo. Since the reduction begins long before the consumption of fossil fuels had reached a significant level, one can conclude that the cause is the large increases in deforestation due to settlement in the temperate northern latitudes, especially at the beginning of the last century (IV.1.4.3). This systematic decrease is not confirmed in the Southern Hemisphere (Tasmania) (Francey, 1981). Reconstructing a 50-yr $\delta^{13}C$ record from juniper tree samples, collected at 10 different sites in Arizona in 1979, Leavitt and Long (1983) find indications that the biosphere may have acted as a source to about 1965 becoming a sink thereafter. Stuiver's (1982) $\delta^{13}C$ samples from 6 trees in different parts of the world indicate pre-industrial atmospheric CO_2 concentrations ranging from 230-260 ppmv − but some close to 290 ppmv, the value until recently assumed for the preindustrial CO_2 concentrations. Although there is still much noise in the sample reconstructions, tree rings appear to be a valuable repository of carbon information.

Atmospheric CO_2 contains, in addition to ^{12}C and ^{13}C, radioactive ^{14}C. With a decay time of about 5,700 years, fossil fuels do not contain any ^{14}C which leads to a ^{14}C dilution in the atmosphere (**Suess Effect**). A reduction of the ^{14}C content of the atmosphere therefore indicates exclusively the consumption of fossil fuels. In summary, we can see that, as a result of CO_2 emissions from fossil fuel combustion, both the $^{13}C/^{12}C$ and $^{14}C/^{12}C$ ratios are reduced but biogenic CO_2 emissions only reduce the $^{13}C/^{12}C$ ratio of the atmosphere. Thus it is possible to determine the relative proportions of the two most important anthropogenic CO_2 sources through simultaneous measurements of these isotope ratios.

In recent years, evidence is accumulating that carbon oxidation is a major cause for the decreasing oxygen content of the atmosphere (Dahlman, 1982). The systematic monitoring of changes in atmospheric oxygen would not only serve as an early detection of a potential **oxygen deficiency** but also supply an indirect measure of CO_2 emissions from fossil fuel and biogenic sources.

IV.1.5.2 Direct methods: Analysis of ice core samples

The ice deposits of Greenland and Antarctica are ideal indicators of the chemical composition and the physical conditions of the atmosphere at the time of ice formation (Dansgaard, 1981). The air bubbles and the CO_2 contained in them are locked in during the transition from firn to ice. Ice cores from various depths thus permit a reconstruction of the temporal course of the CO_2 content that is frequently presented, along with the record of the oxygen isotope $\delta^{18}O$, with which there is good agreement. As water passes from one phase to another, the proportions of the different stable isotopes (i.e. the $H_2{}^{18}O/H_2{}^{16}O$) are changed, which is strongly related to temperature. There is a depletion of ^{18}O in atmospheric moisture, and hence precipitation, when temperature is low, and vice versa, the ^{18}O concentration is high at high ambient temperature (Lawson et al., 1982).

With a new dry-extraction method to avoid contamination of the ice core samples, Delmas et al. (1980) have measured a CO_2 content of the Antarctic atmosphere that was about 160-200 ppmv 15-20,000 years ago, i.e. half that of today. Swiss scientists, in particular, have analysed 50-100,000 year old ice from deep cores in Greenland (Camp Century) and Antarctica (Byrd Station) (Oeschger et al., 1980; Berner et al., 1980; Neftel et al., 1982). The CO_2 concentrations from Greenland show a minimum of about 200 ppmv at the maximum of the last glacial period about 12,000 years ago and a maximum of about 500 ppmv in the warm period of the Holocene 6,000 years ago (Fig. IV.9). Global surface temperatures for this period, derived from climate models, show a range of variation of -2.2 to +1.8^{o}C, compared with the present global average. According to the curve, however, the CO_2 changes do not appear to cause the temperature changes, since they do not precede the relevant glacial or interglacial extremes. Thompson and Schneider (1981) conclude that climatic changes are probably triggered by some other mechanism and that the CO_2 effects only amplify them. Broecker (1982) has suggested that the deposition of phosphate-rich biological sediments results in a higher phosphate content of ocean surface water during glacial periods when the sea level is low. Thus, the enhanced biological productivity of surface water would be higher in glacial periods and hence fix more CO_2, thereby reducing the atmospheric CO_2 concentration. When the sea level is high, coral reef growth precipitates $CaCO_3$ from bicarbonate, thereby acidifying the sea surface water and releasing CO_2 to the atmosphere (Berger, 1982). Further mechanisms, as yet unknown, are conceivable.

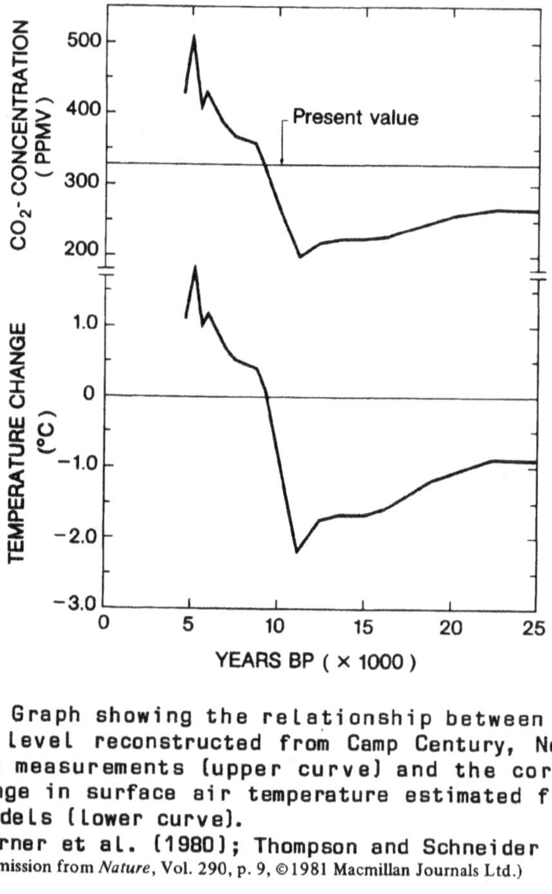

Fig. IV.9 : Graph showing the relationship between the atmospheric CO_2 level reconstructed from Camp Century, North Greenland, core measurements (upper curve) and the corresponding global change in surface air temperature estimated from current climatic models (lower curve).
Sources: Berner et al. (1980); Thompson and Schneider (1981).
(reprinted with permission from *Nature*, Vol. 290, p. 9, ©1981 Macmillan Journals Ltd.)

Ice core samples are also suitable for the reconstruction of more recent atmospheric CO_2 levels. The first preliminary results are rather interesting in that they suggest a pre-industrial CO_2 concentration of about 260 ppmv (Lorius and Raynaud, 1983), which is in good agreement with recent tree ring studies (see IV.1.5.1).

With present-day mass spectroscopy, even very small numbers of atoms of an isotope can be detected. Thus it is possible today to determine, in addition to ^{14}C, the ^{10}Be and ^{36}Cl atoms found in ice. These are also formed under the influence of cosmic rays, become impacted on aerosols, and are ultimately deposited by precipitation onto the ice (Oeschger, 1980; Dansgaard, 1981; Raisbeck et al., 1981).

IV.1.6 Assessment of future CO_2 concentration

One of the important goals of carbon cycle research is to develop a set of tools for the reliable assessment of the CO_2 introduced into the environment by mankind. To compute the CO_2 development over a period of decades to centuries, one requires a computer model of the fast carbon cycles showing the interactions between the sources and sinks, and the future anthropogenic CO_2 production rates as input.

IV.1.6.1 Carbon cycle models

The first carbon cycle models in the 1950s were **3 or 4 box models**, in which the atmosphere, biosphere, surface ocean and deep sea were treated as well-mixed boxes (Revelle and Suess, 1957; Oeschger and Heimann, 1981). Through lack of better information the fluxes between the individual reservoirs were made proportional to the reservoir sizes and these fluxes were described by a system of first order differential equations. Later, the 4 box models were extended to 6 box models, in which the atmosphere was subdivided into troposphere and stratosphere and the biosphere into a short-lived and a long-lived compartment (Bacastow and Keeling, 1973; Zimen et al., 1977).

Since it is the largest carbon reservoir and has, on the long term, the largest capacity for absorbing CO_2, the ocean is in many respects the core of every carbon cycle model. A significant improvement was brought about by the development of the **box diffusion model** (Siegenthaler and Oeschger, 1978). The transport of CO_2 takes place through turbulent mixing of ocean surface water into the deep sea and is described by a diffusion coefficient determined by tracer studies (see IV.1.4.4) to be 4,000 m^2/yr. According to the studies by Broecker et al. (1980), an even better agreement with observed isotope data is achieved using a vertical diffusivity of about 7,000 m^2/yr.

The most recent model calculations also take into account the upwelling water in the equatorial region and deep water formation in the polar latitudes (Hoffert et al., 1980; Michael et al., 1981), or upwelling processes at the equator and sinking processes in middle latitudes (Broecker et al., 1980; see Fig. IV.8; Hoffert et al., 1981). Flohn (1983) has pointed out that carbon dioxide and water vapour in the atmosphere can be significantly influenced by upwelling and downwelling processes which are important causes of local climate anomalies, especially in the tropics. Therefore, models which treat CO_2 in isolation, i.e. with sea surface temperature and hence water vapour held constant, fail to consider these potentially important feedback mechanisms. In order to more adequately model all processes involved, 3-D atmospheric circulation models will eventually have to be coupled to **3-D oceanic circulation models** (see IV.1.7.1 for further details). At present, however, all models, that is the simple box models and the box diffusion models as well as the recent **advection-diffusion box models**, in which the world oceans are divided into 12 reservoirs, interlinked with advective and diffusive transport, have difficulties in generating realistic values for any combination of realistic airborne fraction, depth of mixed ocean layer, oceanic C-14 content, or diffusivity in the ocean layer (Björkström, 1983). To appreciate the intricate reasons for this the reader is referred to the comprehensive treatise of SCOPE 16 (Bolin, 1981).

In an empirical study, Laurmann and Spreiter (1983) have shown that the carbon cycle models commonly in use predict almost identical increases in the CO_2 concentration trends in the next few decades. This is due to the fact that linear carbon cycle models are tuned to reproduce the CO_2 increase observed at Mauna Loa. The airborne fraction observed at Mauna Loa over the past 20 years therefore becomes the critical information for such predictions. Uncertainties in this quantity are due to errors in the past estimate of fossil fuel burning and related CO_2 emissions, potential natural fluctuations in the baseline CO_2 concentration, and uncertainties regarding biospheric and oceanic sources and sinks. Depending on the different assumptions, Oeschger and Heimann (1983) have found that the **effective airborne fraction**, defined as the ratio of CO_2 increase due to fossil fuel CO_2 alone to the integrated CO_2 production, might range from a low value of 0.38 to a high value of 0.72, compared to the apparent airborne fraction of 0.55 (see IV.1.3.1). Effective airborne fractions obtained from carbon cycle models considering only CO_2 uptake by the ocean, range from 0.60-0.70. Theoretical airborne fractions of 0.40 or less would be difficult to reconcile with the observed C-14 distribution in the ocean and would imply strong deficiencies of the models. A value of 0.55 or greater would not be in disagreement with present carbon cycle models. Under the assumption of a linear response of the system, the above model considerations are not incompatible with a high net biospheric CO_2 input before 1958 and the low pre-industrial atmospheric CO_2 level of 260-280 ppmv indicated in recent tree ring and ice core studies (see IV.1.5.1 and IV.1.5.2).

IV.1.6.2 Future atmospheric CO_2 content

For the assessment of the future course of the atmospheric CO_2 concentration we need a carbon cycle model and assumptions about the global growth rate of fossil fuel consumption, the temporal changes of the individual fuel shares, the variable capacity of the oceans for CO_2 uptake, and the temporal changes of the biogenic CO_2 sources and sinks. All of these are fraught with many uncertainties, so that, in the following, we restrict ourselves to a plausibility study showing, in the first part, the range of possibilities in the near future, and in the second part, the potential development, if all recoverable fossil fuel resources are used.

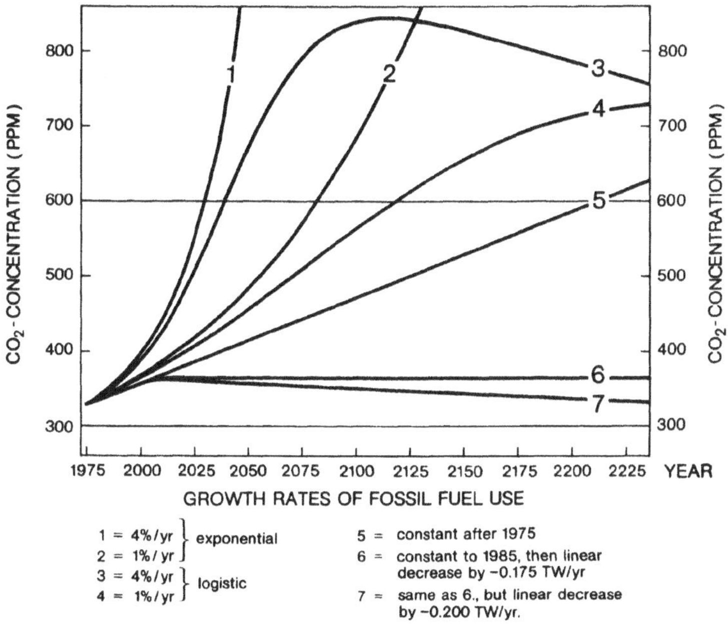

Fig. IV.10 : Projections of atmospheric CO_2 concentration for different growth concepts of fossil fuel use.

The somewhat schematic representation in Fig. IV.10 covers the approximate range of CO_2 projections that have so far been made. Prior to the energy crisis of 1973/74, one could only imagine a continuation of the previous exponential growth at a rate of about 4%/yr (curve 1, e.g. Machta, 1973). In the early 1980s the growth rate had dropped to about 2%/yr (Rotty, 1981), so that in future a 1%/yr growth rate (curve 1) would not seem implausible. Reviewing a good number of existing projections, Laurmann and Rotty (1983) conclude that over relatively short time periods, about the next 50 years, the increase of the CO_2 concentration of the atmosphere can be well represented by an exponential growth rate of fossil fuel consumption and a constant airborne fraction. Taking into account the fact that fossil fuels are in limited supply, and that they become progressively more expensive when some of them approach exhaustion, it is felt that future fossil fuel consumption can be better represented by a logistic function, roughly bounded by curves 3 and 4 (e.g. Zimen et al., 1977). In this case an exploitable fossil fuel reserve of about 3,000 TWyr was used as a base (see III.4). The many possibilities for a more efficient use of energy and the consumption of CO_2-free fuels mean that a stabilization of future fossil fuel use at the current level (straight line 5, e.g. Colombo and Bernardini, 1979) or even a decrease (curves 6 and 7, e.g. Lovins et al., 1981) no longer look implausible.

The starting point for all curves is 1975. For the curves 1-4 the shares of the individual fuels of 1975, i.e. oil 45%, coal 29%, gas 23%, gas flaring 2% and cement production 2%, were also used for the projections. For the calculation of curves 6 and 7 the probable change in the fuel shares is considered as follows: in 2000, 35% for both oil and coal, 30% for gas; in 2030, 40% for coal, 35% for gas and 25% for oil (JASON, 1980). For the calculation of the specific CO_2 emission per energy unit and energy carrier, the values in Table IV.1 were used.

With this information the future CO_2 emission can be calculated, which is then used as input for a box model of the carbon cycle to estimate the future CO_2 content of the atmosphere. The model used here (after Bacastow and Keeling, 1973; Zimen et al., 1977) consists of the 5 reservoirs atmosphere, short-lived and long-lived biosphere, ocean mixed layer and deep sea, that can act as both sources and sinks. The fluxes between the individual reservoirs are proportional to the initial carbon content. Given the large uncertainties in the ongoing discussion about the role of the biosphere (see IV.1.4.3 and IV.1.6.1), we have considered neither a biological growth factor nor a deforestation factor in the calculations, since it is possible that the two opposing influences are roughly in balance (see IV.1.4.5). The airborne fraction is variable, being calculated for each successive year by the carbon cycle model.

The results in Fig. IV.10 show that a high fossil fuel growth rate of the order of 4%/yr - whether exponential or logistic - will lead to a doubling of the atmospheric CO_2 content as early as 2025-2040. Only a strongly reduced growth (curves 2 and 4) gives considerable time gains, with a doubling not until the second half of the next century or the beginning of the 22nd century. If one could stabilize the fossil fuel consumption at the present rate of 8 TW/yr, that is a zero growth rate (straight line 5), then the CO_2 content would only double after 2200, and much additional time would be gained for countervailing measures. In addition, if it were possible through improved energy utilization and fast introduction of CO_2-free fuels, to reduce the fossil fuel consumption to about 1 TW/yr between 1985 and 2030 (curve 6), or to get along without fossil fuels after 2030 (curve 7), then one would have the guarantee that the CO_2 problem was practically eliminated. The estimates, for curves 1 and 3 can be considered as upper bounds and those for curves 6 and 7 as lower bounds so that future CO_2 concentrations will most likely fall within this area. In Chapter VI we shall address once more, with a somewhat less schematic assessment, the energy/CO_2/climate future.

In a second example we want to investigate how long the exploitable fossil fuel resources will last at different growth rates, what atmospheric CO_2 levels will be reached, and how long such high levels will be maintained. Keeling (1980) has made these calculations with a 6 box model (the atmosphere is divided further into troposphere and stratosphere), as described in

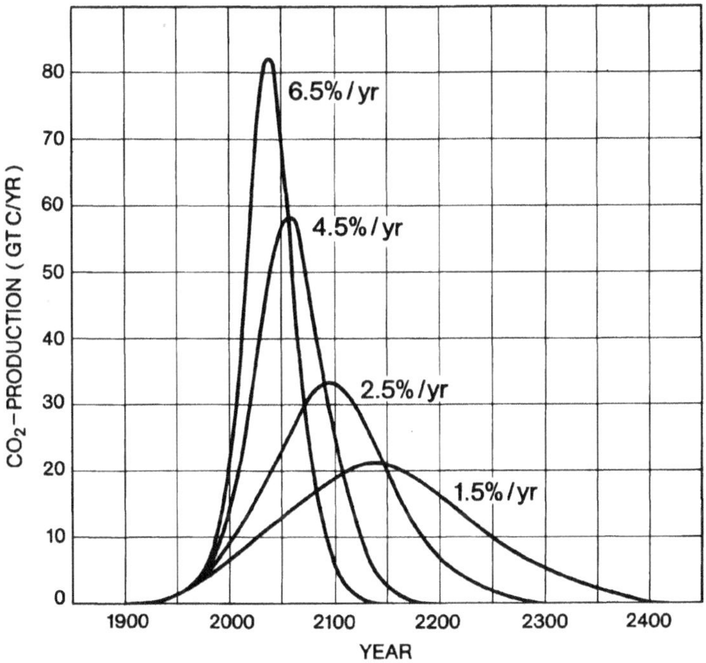

Fig. IV.11 : Industrial CO_2 production for various logistic growth rates. The ultimate production is fixed at 8.2 times the amount of CO_2 in the preindustrial atmosphere.
Source: Keeling (1980).

IV.1.6.1. The future fossil fuel consumption is assessed by a modified logistic function, assuming an exploitable amount of 5,600 TWyr (i.e. somewhat higher than in the previous example).

Fig. IV.11 shows the annual carbon emission into the atmosphere for various logistic growth rates of the commercial fossil fuel consumption. The present value (1980) is a little more than 5 Gt/yr. At the continuation of the present growth rate of about 2%/yr (see Fig. IV.4), it could increase by a factor of 6, and reach its maximum shortly after 2100. The areas under the curves are all equal and reflect the available fossil fuel amounts. The curves for the various growth rates show clearly how fast the fossil fuel reserves that appear to be so great could be exhausted.

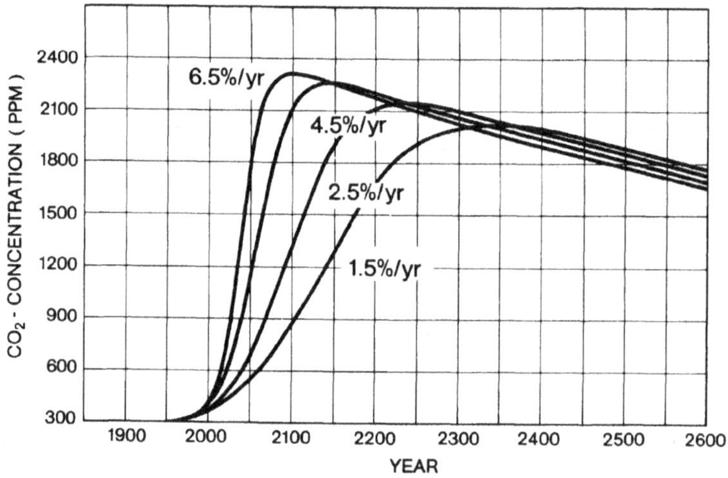

Fig. IV.12 : Predicted increase in the atmospheric CO_2. Curves are shown for the four simulated fossil fuel consumption patterns shown in Fig. IV.11.
Source: Keeling (1980).

Fig. IV.12 shows the expected CO_2 concentration in the atmosphere for the same growth rates. Some significant results emerge, namely that the differing growth rates lead to a clear difference in atmospheric CO_2 content only at the initial growth stages, that, with the consumption of all fossil resources, even at the smallest annual growth rate of 1.5%/yr, the CO_2 content increases about 7 times above the present value, and that independently of the consumption rate, the atmospheric CO_2 content remains about 6 times higher than today for centuries. This is the result of a shift in the carbonate chemistry of the oceans. As we have seen in IV.1.4.4, the ability of the sea surface water to respond to the rising CO_2 levels in the atmosphere will be substantially reduced. Only the atmospheric CO_2 slowly penetrating the deep ocean can increase the uptake capacity of the oceans again, after dissolution of the carbonate in the deep sediments. According to the calculations of Keeling (1980), this can only make a significant contribution after about 1,000 years.

IV.1.7 Climate response due to a CO_2 increase

An atmospheric CO_2 increase influences the energy balance of the climate system giving rise to a warming (see also Chapter II.2). For the investigation of the mechanisms that lead to a warmer climate, there are two complementary approaches: 1. **Palaeoclimatological analogue studies** and 2. **Climate model calculations** (Flohn, 1981a). Both methods have advantages and disadvantages. Analogue studies have the great advantage that they make use of observations of nature, i.e. without mathematical simplifications of the physical-chemical-biological climate system and with realistic time scales. Since, in this case, we are dealing with demonstrable events in the past, they could be useful in the evaluation of future climatic scenarios. It must be noted, of course, that the boundary conditions that influence a new climatic situation can be drastically different from those of the past (see Chapter V.1). Not only the gas and aerosol concentrations in the atmosphere but also the land-sea distribution and the altitude of the mountains have changed since the past epochs.

Therefore, we cannot simply take the conditions of the past and project them into the future. For the description of future climate, climate models have been developed that reflect the complex climate system as realistically as possible (see II.3 and IV.1.7.1). The great advantage of climate models is their flexibility and the fact that complicated and subtle feedback mechanisms can be looked at individually. Such diagnostic studies can increase our understanding of the climate system, which, in turn, can then be used to improve the modelling of climate. A disadvantage is the inherent difficulty to construct a model which is a perfect replica of the actual climate system. It is well to realise that present climate models are still plagued with many deficiencies. For example, cloudiness, heat transport in the oceans, the land surface and the cryosphere are still not sufficiently taken into account and the realistic inclusion of topography, land-sea distribution and seasonal changes is still at a preliminary stage. In spite of these deficiencies, which scientists at the many research institutes all over the world are continually trying to reduce through intensive study, we shall concentrate on climate models in the following, since they are the only tools available for the assessment of the different climatic effects resulting from a future CO_2 increase.

For the study of the influence of CO_2 on the model climate a **hierarchy of climate models** has been developed, ranging from zero-dimensional (0-D) energy balance models, through one-dimensional (1-D) radiative-convective equilibrium models, through 1-D and 2-D energy balance models, and 2-D statistical-dynamical models, to the 3-D general atmosphere-ocean coupled circulation models (3-D GCM) (see the historical review of CO_2/climate research in Appendix IV.1 and Chapter II.3). This hierarchy of climate models is applied in climatic research in two basically different but complementary ways: The perturbation of the equilibrium of the model climate, e.g. through a prescribed, **time-independent** CO_2 doubling, is assessed, thus allowing the **sensitivity** of the various mechanisms of the model climate to the perturbation to be studied. Since, however, in reality, the atmospheric CO_2 concentration is continually increasing, the response of the climate to a **time-dependent** CO_2 increase must also be studied. These two procedures are the basis for a selection of the recent model results assessing the climatic effects of a CO_2 increase.

IV.1.7.1 Equilibrium response experiments

So that the model results are more understandable, it is appropriate firstly to look briefly at the approach taken by the climate-model developers. Given the large number of differing models, we shall limit ourselves first to the work of the Geophysical Fluid Dynamics Laboratory (GFDL) of Princeton University which is representative of the most recent developments. We shall focus on the 3-D general circulation models (GCMs), because only they can simulate the geographical distribution of a CO_2-induced climatic change which is of greatest importance to mankind. We shall then review the other major modelling activities and supplement these with the tabulated overview given in Appendix IV.1.

Model structure of 3-D GCMs - These models are based on the **equation of motion** (this explicitly computes the acceleration of an air parcel caused by the Coriolis, pressure-gradient, and gravity forces); the **equation of continuity** (this computes the rate of change of mass of the volume of air); and the **energy equation** (this computes the rate of change of temperature). The complete system includes, in addition, the **equation of state** of gas and the **hydrostatic approximation** (see Appendix II.3 for the description of the entire system).

Knowledge of further specifications is important for the subsequent interpretation of the results. The calculation of the transfer of solar and terrestrial radiation also includes the effects of CO_2, water vapour (H_2O), ozone (O_3), and clouds. For CO_2 a uniform mixing in the atmosphere is assumed and for O_3 a distribution as a function of latitude, season and altitude is assumed. Cloudiness is predicted from condensation of the H_2O. For the computation of terrestrial radiation, the clouds are considered as black body radiators, and for the calculation of the solar radiation, scattering, reflection and absorption are prescribed as a function of altitude and cloud thickness. Snow depth and soil moisture on land are determined using budget equations. The differentiation of precipitation into rain or snow depends on the temperature at a height of about 350 m. Over the ocean, sea ice is predicted if the ocean surface temperature falls below $-2^{\circ}C$. Finally, albedo values are specified for land and ocean surfaces as a function of latitude and temperature.

Because of the high computer costs, the complete GCM cannot always be used for the many sensitivity experiments. Important simplifications must therefore be made, which we can follow through the previous model developments. The first 3-D GCM (Manabe and Wetherald, 1975) had an idealised topography (i.e. no relief differences and no vegetation), no seasonal changes (by removing the seasonal insolation), no heat transport by ocean currents (the ocean is treated as a swamp without heat capacity), constant clouds and a limited model sphere (only 1/3 of the Northern Hemisphere, in which an ocean-half to $66.5^{\circ}N$ and a land-half to $81.7^{\circ}N$ are considered). In the next sensitivity tests, about 5 years later, geographical differences (Manabe and Wetherald, 1980) and the influence of variable clouds on the model climate were investigated (Wetherald and Manabe, 1980), and the land/sea sector of 120° length extended to the North Pole and a simple scheme for cloud prediction was introduced. In a special study Meleshko and Wetherald (1980) examined the effect of the geographical distribution of high, middle and low clouds on global climate. In the most complete version up to now, the 3-D GCM of the atmosphere is coupled to a simple, homogeneous ocean-mixed-layer model of about 68 m depth (Manabe and Stouffer, 1980a, b). The extension to the entire globe and the introduction of realistic geography, with variable insolation but with fixed cloud distribution, allow the investigation of the seasonal and geographical changes in model climate, as a result of a CO_2 increase. In a special study, the seasonal changes, in particular, have been discussed, in which case idealised geography and, because of costs, a 120° sector (this time for both hemispheres) were used (Wetherald and Manabe, 1981).

Only a few studies (e.g. Chervin, 1980) have so far evaluated the temporal variability of a model atmosphere because of the large computer time required. The recent development of spectral GCMs using semi-implicit time integration schemes has considerably reduced the time required for integrating. Using such spectral GCMs, Manabe and Hahn (1981) show that in middle and high latitudes the observed geographical distribution of the variability of daily, monthly and yearly mean surface pressure and temperature is reproduced quite well, while in the tropics the variability of surface pressure tends to be underestimated.

Numerical procedure - The governing equations of the GCMs - the so-called primitive equations - are nonlinear, partial differential equations (see Appendix II.3) whose solutions require the fastest computers. For the numerical computations, the atmosphere is vertically subdivided into discrete layers. For each layer, the horizontal variations of the respective variables are determined, either at discrete grid points over the globe, as in the **grid point**, or **finite difference models**, or by a finite number of prescribed functions, as in the **spectral models.** Present GCMs have 2 to 9 vertical layers, a horizontal resolution of about 300-400 km, and a time step ranging from 10 to 40 minutes. The simulation of one day requires from one-half minute to a few minutes on the most advanced computers (e.g. the CRAY 1 and CYBER 205). Many of the subgrid-scale physical processes are not resolved by the GCMs due to their limited spatial resolution. They are therefore related to the variables on the scale resolved by the GCMs in a process called **parameterization** which is based on both theoretical and observational studies. The spatial resolution of GCMs is, of course, constrained by the speed, memory capacity and costs of the computer, so that a higher resolution desired for subgrid-scale processes and a greater regional differentiation must await the next computer generations (Schlesinger, 1983).

The time integrations are carried out using the model for the following three CO_2 concentrations: 300 ppmv (also referred to as 1 x CO_2 or control case), 600 ppmv (2 x CO_2 or doubling), and 1,200 ppmv (4 x CO_2 or quadrupling). A comparison of the differences shows the influence of the CO_2 increase in question on the climate. The runs start with an isothermal and dry atmosphere and the 1 x CO_2 experiment is integrated over a time period of 1,200 days. The state of the model-atmosphere at the end of the 1 x CO_2 integration forms the starting point for the integration over a further 1,200 days for the 2 x CO_2 and the 4 x CO_2 cases. A **quasi-equilibrium climate** is produced by averaging over the last 300 days of each integration.

The conventional 3-D GCM (without ocean heat transport), reaches quasi-equilibrium in about 300 days, in about 10 years after coupling with an ocean-mixed-layer model, and thousands of years if the entire ocean were considered. Since the coupled ocean/atmosphere models require a considerable amount of computer time, a **synchronous coupling** of a complete model is not possible. Therefore, an **asynchronous coupling technique** has been proposed, in which the atmosphere is computed over days or months and the ocean is considered in cyclical fashion over several years (Washington and Ramanathan, 1980). Presently it is, however, not clear whether this provides a realistic equilibrium, i.e. if the solutions of synchronous and asynchronous coupling are comparable.

Model verification – After a climate model has been developed, the first step is a comparison of computed with observed values. This appears to be easy, but in reality it is an extremely complicated matter, since in the case of the coupled ocean-atmosphere 3-D GCM, an attempt is made at getting as close as possible to the complex reality. A comparison of the data is therefore very time-consuming and expensive. In addition, it is very difficult to recognise and find the causes for the differences between model results and observations.

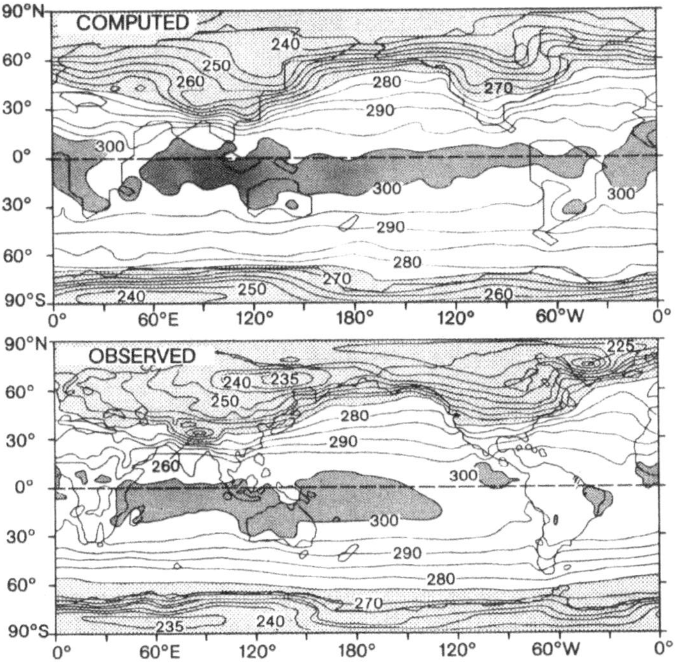

Fig. IV.13 : Geographical distribution of monthly mean surface air temperature (°K) in February. Top: the model simulation. Bottom: the observed distribution.
Source: Manabe and Stouffer (1980a,b).

In the past, model verification was only carried out for annually-averaged values. This is not, however, sufficient for testing the reaction of a climate model to changed conditions.

Our confidence in model results would increase if, for example, the course of the seasons simulated by the model, the geographical distribution of seasonal differences, the relative frequency and seasonal variations of various weather patterns, climatic variability from year to year, or climatic changes of the past, etc. could be verified. It is also very important that the results from sensitivity experiments, obtained by differing climate model types, are compared with one another.

Fig. IV.13 shows an example of model verification. The geographical distribution of the mean surface temperature for February, computed by a coupled ocean-atmosphere 3-D GCM, is compared with the observed distribution patterns (Manabe and Stouffer, 1980a,b). There are some unrealistic differences in the simulation, such as the extension of the temperature maximum at the equator into the eastern Pacific, which may be a result of the poor simulation of the cold upwelling water at the South American coast; or the temperature that is too high along the western periphery of the Antarctic which is as yet unexplained. In general, however, the model reproduces the seasonal temperature distribution quite realistically, so that it appears to be justified to use the model for assessing the response of the model climate to a CO_2 increase. From the many sensitivity studies, we shall look at a few typical examples that have been grouped in the following according to their thermal and hydrological responses.

Thermal responses - Fig. IV.14 shows the zonally-averaged latitude/altitude temperature distribution of the model atmosphere in response to a CO_2 doubling (2 x CO_2). This sensitivity test was made with a coupled atmosphere-ocean (swamp) GCM with variable cloudiness. If one averages the temperatures in the lower troposphere over the Northern Hemisphere, one finds the frequently cited mean temperature increase of about 3°C for 2 x CO_2. There is a marked temperature increase of up to 8°C in polar latitudes and a decrease of the same magnitude in the stratosphere above low latitudes. The cooling in the stratosphere is caused by the strong heat loss, as a result of both radiation to space and reradiation into the troposphere. The large temperature increase at the poles and resulting weakening of the **meridional temperature gradient** of significance to the atmospheric circulation is caused, among others, by the poleward shift of the strongly reflecting snow and ice surfaces, the increase of the transport of latent heat to the poles (see below the increase of the hydrological cycle), the inclusion of this additional heat energy in the stable stratification of the polar atmosphere, and the overlapping of the CO_2 and H_2O absorption bands, which amplify the CO_2 influence on the radiative flux (Bach, 1980a, b). It is interesting to note that the pronounced polar amplification of the warming is exhibited in most numerical modelling results except for those by Washington and Meehl (1983). This feature is

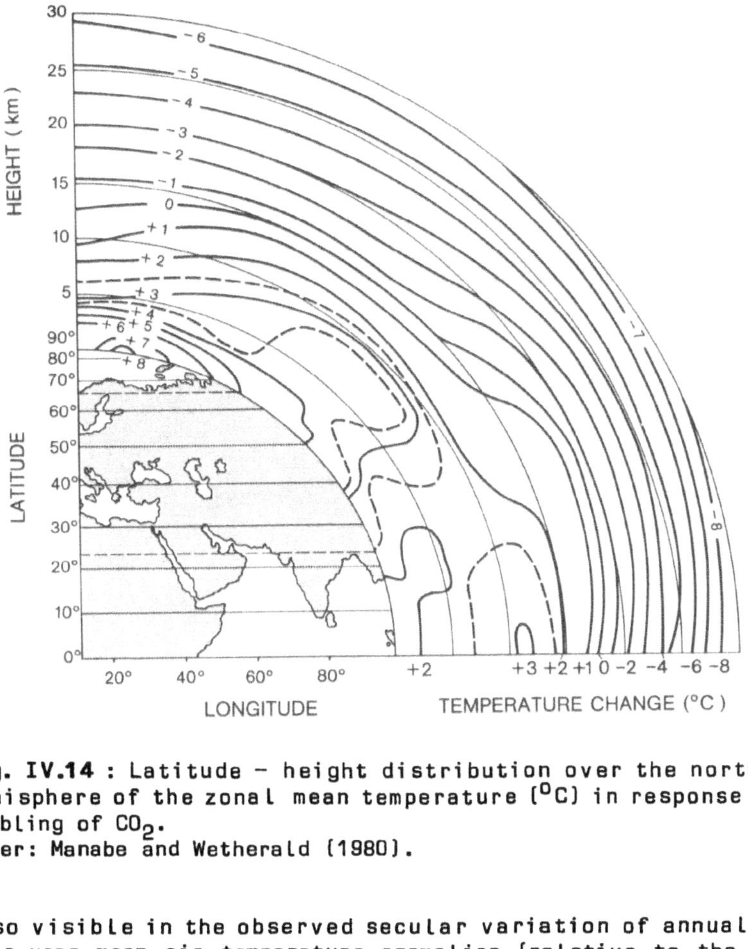

Fig. IV.14 : Latitude - height distribution over the northern hemisphere of the zonal mean temperature (°C) in response to a doubling of CO_2.
After: Manabe and Wetherald (1980).

also visible in the observed secular variation of annual and five-year mean air temperature anomalies (relative to the mean for 1881-1975) for different latitude bands in the Northern Hemisphere (Vinnikov et al., 1980).

As we have seen above, a CO_2 doubling leads to a marked magnification of temperature particularly in polar regions. Budyko (1974) studied the effect of a warming on the Arctic sea ice using a radiation balance model. The results showed that an increase of the summer temperature by 4°C would be enough to completely melt the sea ice. In contrast, Parkinson and Kellogg (1979), using a time-dependent sea ice model, which considered heat fluxes into and out of the ice, as well as the seasonal occurrence of snow and ice movement, found that even with a temperature increase of 6-9°C the sea ice forms again in winter. They point out that during the last million years there has not

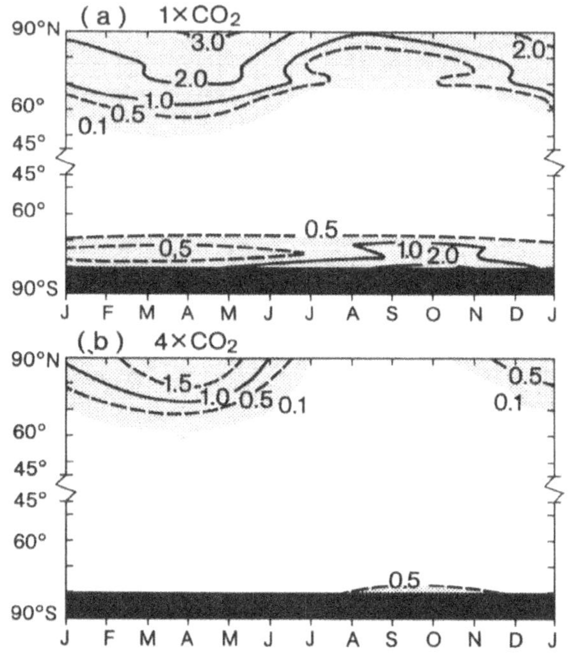

Fig. IV.15 : Latitude-time distributions of sea ice thickness (m). Top: 1 x CO_2 experiment. Bottom: 4 x CO_2 experiment. Shading indicates the regions where sea ice thickness exceeds 0.1 m. Source: Manabe and Stouffer (1980a).
(reproduced from *J. Geophys. Res.*, Vol. 85, published by American Geophysical Union)

been an ice-free Arctic Ocean for an entire year. Manabe and Stouffer (1980a) have investigated the influence of an increase of the CO_2 content to 1,200 ppmv (4 x CO_2) on the annual course of the thickness of the Arctic sea ice using the coupled atmosphere-ocean (mixed layer or slab model) global 3-D GCM. As the comparison with the standard CO_2 case (1 x CO_2) in Fig. IV.15 shows, not only is the ice thickness drastically reduced, but the sea ice disappears completely in the summer in the Northern Hemisphere and in winter in the Southern Hemisphere. During the summer months with little cloudiness the open water then heats up quickly, so that probably after a few years a new equilibrium temperature would be reached as a result of this positive feedback, so that in winter there would only be ice formation along the edges (Flohn, 1980). The decisive effect is that this would lead to a global redistribution of precipitation and therefore of the water balance (see also Chapter V.7). However, since heat transfer by ocean currents has been neglected in these model experiments, the simulation of the extent of the Arctic sea ice is probably not realistic, being overestimated in winter and underestimated in summer (Gilchrist, 1983).

Hydrological responses – All modelling results show an intensification of the **hydrologic cycle** of between 2 and 7% for 2 x CO_2 except for the Gates (1980) model, which gives a 1.5% decrease due to the neglect of the ocean response (Schlesinger, 1983a). The reason for this is the increase of reradiation to the surface (greenhouse effect), which increases the evaporation and thus the water vapour content of the air. Fig. IV.16 shows the response of the rate of change of the precipitation P minus evaporation E (P–E corresponds roughly to the run-off rate) to a CO_2 doubling. There is an increase of P–E along the east coast of the model continent, both in the tropics and subtropics, which suggests an intensification of the monsoon rains. The increase of P–E in subpolar and polar latitudes is caused in part by the poleward shift of the rain belt of the temperate latitudes (Manabe and Wetherald, 1980).

In contrast, between 40^0 and 50^0 latitude on the model con-

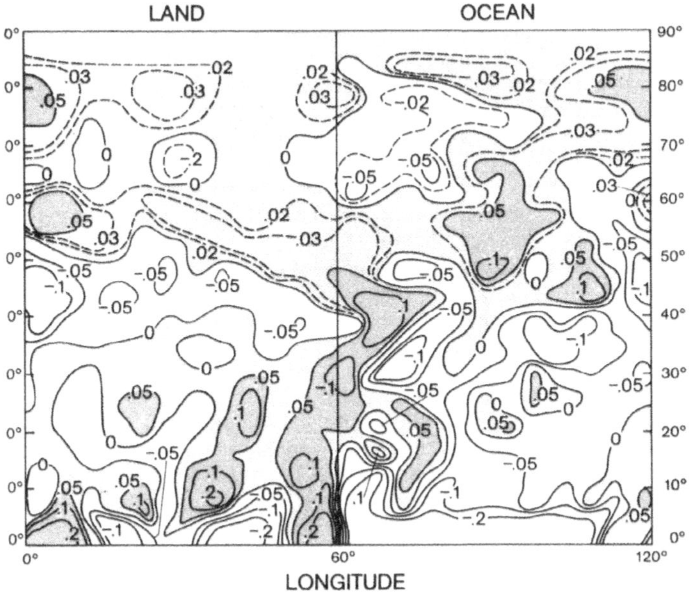

Fig.IV.16 : Horizontal distribution of the rates of change of precipitation minus evaporation (P – E) in response to a doubling of the CO_2 content. Units are cm/day.
Source: Manabe and Wetherald (1980).
(reprinted with permission from *J. Atmos. Sci.*, Vol. 37, p. 112, ©1980 American Meteorological Society)

tinent we find a decrease of P−E, that is, a reduction of the available soil moisture. The relative decrease of the precipitation in this region can be explained probably by the significant reduction of the eddy kinetic energy that the model determined in this region. In a more recent study using three different models, Manabe, Wetherald and Stouffer (1981) show that in the middle latitudes of the US continent in spring the soil moisture increases and in summer, in contrast, the dryness increases. Rosenberg (1982) cautions not to jump too quickly to conclusions because such important processes of the real world as the regional advection of sensible heat may not have been adequately considered. Other models, with a less sophisticated treatment of the ocean and the annual cycle, have not been able to duplicate the mid-latitude soil moisture deficit (Schlesinger, 1983). Should, however, the results from the more detailed simulations turn out to be correct, this might have grave implications for agriculture and hence **world food supply**, since this would affect the US corn belt (see also Chapter V.4). This example clearly shows again how important the **regional distribution of climate anomalies** due to a CO_2 increase is in the evaluation of potential socioeconomic effects and their political consequences.

In conclusion, the model results suggest the following responses (Bach, 1980b). An increase of the atmospheric CO_2 content leads to an

- increase of the air temperature, leading to an
- increase of the water vapour content, leading to an
- increase of the transport of latent heat to the poles; which leads to a
- reduction of the meridional temperature gradient and to a
- weakening of the intensity of the global atmospheric circulation; which further has an
- impact on the temperature and precipitation distribution and an
- impact on agricultural productivity and on water and energy supply and demand (see Chapter V).

Other modelling work and outlook − For the sake of showing the principle of CO_2/climate modelling as clearly as possible, I have relied, in the previous sections, almost exclusively on the work done at the GFDL in Princeton. There are, however, a number of other research teams involved in such modelling whose approaches differ in the way they treat the horizontal and vertical resolution, the geographic and topographic conditions, cloud prediction, the parameterization of sub-grid scale processes, and especially the ocean. Indeed, with the realization of the decisive role of the ocean in the maintenance of the global climate, modelling efforts have concentrated on the coupling of an atmospheric general circulation model (AGCM) with a realistic ocean model. (For a brief chronological overview, see Appendix IV.1.)

In summarizing the present and the anticipated modelling

development, the following categories can be distinguished, namely the coupling of an AGCM with
- a climatological ocean model
- a swamp model
- a slab model
- a variable depth mixed layer model
- a full oceanic general circulation model (OGCM).

The AGCM with a climatological ocean - Early experiments were carried out, e.g. with the 2-level Oregon State University atmospheric general circulation model (OSU AGCM) by Gates (1975) and with the 5-layer British Meteorological Office model (BMO GCM) by Corby et al. (1977). In these experiments sea surface temperatures and sea ice distributions were prescribed to be equal to their observed values and to follow observed seasonal cycles. The climatological ocean acts as if it were an infinite heat sink, i.e. as the atmosphere warms, there is no corresponding warming of the ocean and, hence, no return of energy to the atmosphere. These models are very useful when evaluating a model's performance in simulating climatic changes for which there are verifying observations (Schlesinger and Gates, 1980). In particular, the seasonal performance can be systematically evaluated. Gates, Cook and Schlesinger (1981) have also used this model to simulate the seasonal climatic changes in response to both a 2-fold and a 4-fold increase in CO_2 concentration. The 2 x CO_2 run resulted in an annual global mean surface air temperature increase of only 0.2°C which is about 1/10 of the warming obtained with a more realistic ocean model, such as the slab model (Manabe and Stouffer, 1980a, b). AGCMs, with a climatological ocean, cannot realistically simulate a climatic change due to a CO_2 increase (Schlesinger, 1983a).

The AGCM with a swamp model - This is the simplest thermodynamic ocean/sea ice model. It treats the ocean as a swamp of zero heat capacity without any heat transport but as an infinite source of water vapour. It can simulate neither annual nor diurnal cycles because heat cannot be stored in the oceans, and thus, only annual average values can be simulated. The prediction of a mean annual surface air temperature increase of about 3°C for 2 x CO_2 is considered to be in the right ballpark. First model experiments were carried out by Manabe and Wetherald (1975, 1980) and Hansen (cit. US NAS, 1979) followed, more recently, by Washington and Meehl (1982, 1983), and Schlesinger (1982).

The AGCM with a slab model - In this case a water column (slab) 60-70 m deep simulates the upper mixed layer. Because the sea surface temperature is treated as a prognostic variable (i.e. its change in magnitude with time is predicted) and heat storage in the ocean is considered, the slab model overcomes the heat storage problem of the swamp model and the order of magnitude problem of the climatological ocean model. One major unrealistic feature remains, however, in that the slab model has neither horizontal heat advection by means of ocean currents, nor vertical heat transfer between the various ocean layers, such as in upwelling and downwelling areas. Experiments with a coupled slab model have been conducted by Manabe and Stouffer (1979, 1980), Hansen (cit. US NAS, 1979), Pollard, Batteen and Han (1980), and Schlesinger and Gates (1981). The global mean surface air temperature increase for a 2 x CO_2 ranges from 2°C (Manabe and Stouffer) to 3.5°C (Hansen).

The AGCM with a variable depth mixed layer model - This is the latest in a series of OSU-experiments with progressively more complex upper ocean models coupled to AGCMs. Instead of one, this new model consists of two vertically homogeneous ocean layers of variable depth (Pollard, 1982). The upper well-mixed layer can entrain and detrain fluid within a 30-250 m thick layer. The lower layer, typically about 100-400 m deep and roughly representing the seasonal thermocline, does not mix with the motionless deep water. CO_2 perturbation experiments with this type of model are in progress.

The AGCM coupled with an OGCM - For more realistic modelling results, all of the above models will eventually have to be supplemented by a full OGCM. Work is under way (Bryan, Manabe and Pacanowski, 1975; Manabe, Bryan and Spelman, 1979; Holland, 1979; Gates, pers. comm., 1983). Much preliminary work remains to be done to remove errors due to unrealistic surface fluxes, given by the AGCM, those due to the underestimation of western boundary ocean currents, presumably caused by the coarse grid spacing, and those due to inadequate tuning of the entrainment parameterization (Pollard, 1982). Any realistic projection of a CO_2-induced climatic change, over a time span of the next 50 years or so, requires coupling of an AGCM to an OGCM that models the mixing of heat in the oceans between the surface and water down to a depth of at least 1 km (Dickinson, 1982).

Realism of the GCMs can be further improved by incorporating the remaining components of the climate system (see Fig. II.2). Changes in the geography of the land, ocean, and land ice are probably so slow that it would be unjustified to consider them in CO_2 doubling experiments which cover a time span of the next 50 to 100 years or so. This, however, would be quite different for the activities in the biosphere, as manifested by the ongoing changes in deforestation, reforestation, and in other land use modifications. The coupling of the above models with an interactive biosphere model, describing land use, vegetation, and soil moisture properties would, therefore, be the logical final step to complete the climate modelling efforts (Dickinson, 1982).

General circulation models are our main tools for the prediction of climatic impacts due to an increase in atmospheric CO_2

because only they can supply the information with the required high seasonal and spatial resolutions. Another major application of GCMs is to identify regions and times of maximum sensitivity to CO_2 increases. Comparison of model results is, however, severely hampered by the differences among the models' simulations because of differences in the geography/topography vs realistic land-sea geography and topography/ocean modelling (i.e. climatological ocean vs swamp vs slab vs variable mixed layer vs OGCM), and solar forcing (i.e. annually-averaged insolation vs annual solar cycle). Moreover, before the climatic changes simulated by different models can be compared, the statistical significance of the simulated climatic changes must first be established in order to avoid comparing simply the various models' noise levels (Schlesinger, 1983a). Statistical significance can be improved and noise levels reduced, if the models can be integrated over sufficiently long time periods. Much work still remains to be done so that the coupled ocean-atmosphere GCMs can be used with greater confidence in climatic impact assessment which is an essential part of the overall CO_2/climate research (see also Chapter V).

The salient points of the sensitivity studies for the 3-D GCMs, presented in this section, can be summarized as follows (US NAS, 1982; Manabe, 1983; Schlesinger, 1983a):

- The equilibrium global surface warming, due to a doubling of CO_2, is estimated to be $3 \pm 1.5^{o}C$ (for details and comparisons with other models than the GCMs see Appendix IV.1).
- In the stratosphere a cooling is indicated with relatively small latitudinal variation.
- Temperature increases near the Earth's surface could vary greatly with latitude and season, as follows:
  - In polar regions the warming could be 2-3 times greater than in the tropics; and over the Arctic it could be significantly warmer than over the Antarctic.
  - Over the Arctic, temperature increases could have large seasonal variations, with maximum warming in winter and minimum warming in summer; equatorward of 45° latitude the warming has a smaller seasonal amplitude.
- Also the global mean rates of precipitation and evaporation are estimated to increase. Latitudinally averaged model data allow the following qualitative inferences:
  - Annual mean runoff (precipitation minus evaporation) increases over polar regions.
  - Snowfall starts later and snowmelt begins earlier.
  - Summer soil moisture decreases in middle and higher latitudes of the Northern Hemisphere affecting the breadbaskets of major industrial nations.
  - Both coverage and thickness of sea ice decrease over the Arctic and circum-Antarctic oceans.

Finally, it is well to point out again that it is necessary

to identify the physical mechanisms of the CO_2-induced climatic changes and to assess their statistical significance, so that the results obtained from them can be accepted with greater confidence.

IV.1.7.2 Transient response studies

We now come to the second approach in climate model research, in which the reaction of the climate to a time-dependent CO_2 increase is investigated. First, however, we want to address the question of why in the past decades the temperature of the Northern Hemisphere has decreased in spite of the continual CO_2 increase. Then we shall discuss recent attempts to identify a CO_2-induced climatic change.

Natural climatic variations and the CO_2 influence - The climatic history of the earth is characterised by a large number of natural variations of many magnitudes (see Chapter II.1, Appendix II.2). In the Northern Hemisphere, data derived by Dansgaard et al., (1969) using the ^{18}O isotopic analysis in Greenland ice cores, showed a dominant 180-year temperature cycle (see small inset in Fig. IV.17), i.e. about every 180 years, the natural climate curve shows a warming or cooling. This 180-year cycle of the natural climate is indicated in the large diagram as a thin dashed line from 1800 to beyond 2000. The observed average temperatures (thick solid line) show a temperature increase of about 0.5°C from 1880 to the 1940s, followed by a temperature decrease of about 0.3°C and a slight increase in the last 5 years. Similar temperature variations are shown by more recent studies. Using monthly mean station data, gridded on a 5° latitude by 10° longitude grid and objective techniques, a similar surface air temperature trend has been obtained for the Northern Hemisphere (Jones, Wigley and Kelly, 1982) and for the Arctic Regions (Kelly et al., 1982a), although the latter show variations that are more rapid and of greater magnitude on shorter time scales. The warming trend until the 1940s is not only clearly visible in the Arctic Regions but also at remote mountain stations so that the argument that the warming was due to the heat island effect of growing cities does not appear to be justified (Schönwiese, 1983).

Since both the observed and the natural temperature trend almost overlap, it has been concluded that, at present, the natural climatic variations still dominate. If the temperature curve were controlled by the CO_2 effect alone, we would get the dashed-dotted curve, and combining both the CO_2 effect and the natural 180-year cycle might give the broadly dashed temperature curve. Since the natural temperature curve increases again towards the turn of the century, there is a possibility that at the continuation of the present CO_2 increase the resultant temperature might also increase again and reach magnitudes that have not occurred during the last 1,000 years (Bach et al., 1979). It should, however, be noted that in the above discussion other than CO_2 effects have not been taken into consideration so that any conclusions based upon this must be viewed as highly tentative.

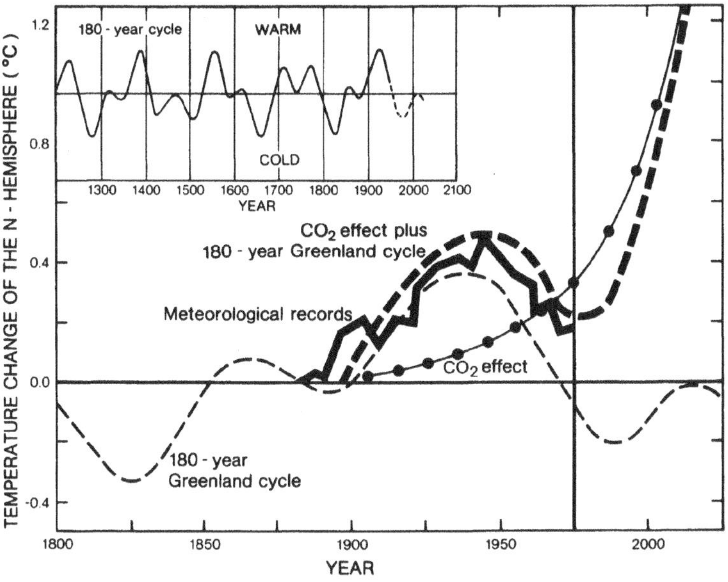

Fig. IV.17 : Recorded and estimated temperature changes (^oC) for the N-hemisphere plus the 180-year temperature cycle as inferred from Greenland ice cores.
After: Broecker (1975); Bernard (1980).

Detection of a CO_2-induced warming - Estimates made in IV.1.5 indicate that the CO_2 concentration in the atmosphere may have increased over the last one hundred years from about 260-290 ppmv to about 340 ppmv, suggesting that a warming of 0.7 ± 0.35^oC should have occurred (MacCracken, 1983). Temperature records and Fig. IV.17, discussed in the previous section, indicate that some warming has undoubtedly occurred over the past century. It is, however, difficult to apportion the warming over this period with any confidence among CO_2 and other gases, anthropogenic aerosols, volcanic activity or solar variability. In addition, the warming may have been masked by heat removal through melting of the arctic ice (Etkins and Epstein, 1982) and its detection may become delayed by several decades due to the **thermal inertia of the ocean** (MacCracken and Moses, 1982). If, as Schneider and

Thompson (1981) suggest, a surface temperature increase of about 1°C, averaged over a decade, is required for a **CO_2 signal** to be unambiguously detected against the **climatic noise** (i.e. the natural climatic variability), then the time of a signal detection should not be far away. When, in what regions, and in which season a CO_2-induced signal will first be detected are central questions which will now be addressed.

Madden and Ramanathan (1980) have computed the mean zonal surface temperature at 60°N for the period 1906-1977 from the data of 12 stations and compared these with the signal calculated by the 3-D GCM of Manabe and Wetherald (1975; see IV.1.7.1). They conclude that at 60°N a CO_2 signal will be detected first in the summer data.

The results for 60°N are not necessarily representative of other latitudes. Wigley and Jones (1981) have, therefore, calculated the monthly surface temperatures (noise) for 800-1,300 stations distributed over the Northern Hemisphere for the period 1941-1980 and compared them with the signal for the continents (there are too few stations in the ocean area for a representative description), computed with the 3-D atmosphere-ocean GCM of Manabe and Stouffer (1980a; see IV.1.7.1). The results show that, in contrast to previous notions, the CO_2 effect will not be detected first in winter in the Northern Hemisphere but in summer and in the middle latitudes, which is partly a result of the relatively low natural variability of the summer temperatures. Ramanathan et al. (1979) reached a similar conclusion using simpler model calculations. Kelly et al. (1982b) point out, however, that because of the uncertainties in the noise and the uncertainties in the signal due to model deficiencies, the relatively high signal-to-noise ratios, found in summer and in middle latitudes, may only reflect the inability of the models to simulate accurately the seasonal and spatial patterns of a CO_2-induced climatic change.

In addition to Madden and Ramanathan (1980), Hansen et al. (1981) have gone into the question of when a CO_2-induced warming should become visible in the climatic data. Fig. IV.18 shows the observed global temperature record from 1950-1976 (dotted line) and the range of natural temperature variability (noise) for a standard deviation of 1σ and 2σ, which includes 85% and 98% of all data, respectively. With the aid of a 1-D radiative-convective model (see Appendix IV.1; Manabe and Wetherald, 1967), the CO_2-increase (signal) is computed as a function of time for various model sensitivities (solid and dashed curves) for a so-called low energy growth scenario (2%/yr). The likely values for the climate model's equilibrium sensitivity for a CO_2 doubling are assumed to range from $\Delta T = 1.4°C$ to $\Delta T = 5.6°C$. On the basis of these assumptions the predicted CO_2-induced warming should rise above the noise level of natural climate variability between 1990 and 2000.

Calculations performed by Schlesinger (1983b) with a zero-

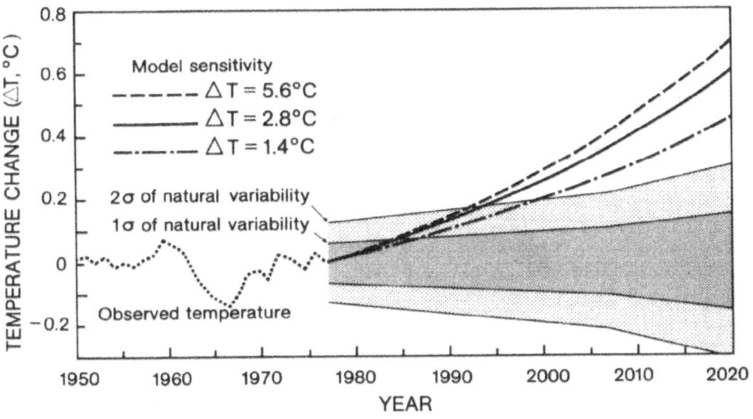

Fig. IV.18 : CO_2 warming versus noise level of natural climate variability. The effect of other trace gases is not included. Source: Hansen et al. (1981).
(reprinted with permission from *Science*, Vol. 213, p. 960, © 1981 AAAS)

dimensional climate model show that the detectability of a CO_2-induced warming depends critically on the interaction between the oceanic mixed layer and the deep ocean. The calculated change in surface air temperature from 1860 to 1981 is 0.37°C if only the sea surface is warmed, 0.32°C if also the mixed layer has warmed, and 0.17°C if heat has also been transferred to the deeper ocean. If a 1°C surface temperature increase averaged over a decade is taken as the CO_2 signal to be unambiguously observed against the natural climatic noise, then the time when the signal will be first detectable is estimated to fall between 2025 and 2053 depending on the assumed heat exchange rate in the ocean, the future CO_2 growth rate, and the relationship between the time-dependent CO_2 and temperature changes. In order to reduce the many uncertainties related to the rate of oceanic heat exchange, Schlesinger recommends to perform numerical simulations with coupled atmosphere-ocean general circulation models.

Using a maximum likelihood estimation technique on the global mean temperature data assembled by Angell and Korshover (1977) for the surface, the troposphere and the stratosphere from 1958-1976, Epstein (1982) finds that a statistically significant temperature change from an increase in atmospheric CO_2 is likely to be detected from global mean surface temperatures within ten years. According to him, a study of the joint behaviour of the tropospheric and stratospheric temperatures could make the CO_2 signal detectable as early as 1986.

Indeed, the simultaneous study of tropospheric and stratospheric behaviour appears to be very useful (Mitchell and MacCracken, 1980). For example, according to Fig. IV.14 a CO_2 increase should be accompanied by a strong warming at the poles in the troposphere and an equally strong cooling in the stratosphere. The surface temperature changes are affected by the thermal inertia of the oceans that tends to retard the warming of the troposphere, whereas the stratospheric cooling is primarily a radiative process and would not be retarded (Epstein, 1982). We can therefore scan the temperature changes in the troposphere and stratosphere for phase differences that might be useful as an early warning of a CO_2-induced climatic change (Dickinson, 1982). However, because the increase of chlorofluoromethanes (IV.2.1) causes an ozone decrease and thus a cooling of the stratosphere of the same order of magnitude, an unambiguous determination of the effects is not possible.

Climate models also predict an intensification of the hydrological cycle for a CO_2 increase (IV.1.7.1). The large amount of climate data should therefore be examined for changes in the specific humidity, and especially in the global mean precipitable water, as well as the occurrence of sea ice (Schlesinger, 1982). As a result of the shift of the precipitation zone from middle to polar latitudes, the runoff volume of Arctic rivers could increase and thus serve as a first indicator of a warming. Frequently, environmental indicators, such as changes in permafrost, in the tundra and in the vegetation, show responses in the climate system much sooner than measurements.

It may therefore be very rewarding to observe also the change of firn area and sea ice. Kukla and Gavin (1981) compared old maps and logbooks of whalers with new satellite pictures and found that the Antarctic sea ice must have been more extensive in the 1930s than the 1970s. Also the mean surface temperatures in the region of the North Pole snow melt (55°–80^oN) were higher between 1974–1978 than between 1934–1938. But, because of the high natural variability, the differing spacing between stations within these two periods, and also possibly other processes uninfluenced by CO_2, it is not possible at the moment to determine any clear cause-effect links between the observed CO_2 and temperature increases. Comparing temperature trends at 13 European stations over the last 25 years (1951–1975) with the preceding 25-year period (1926–1950) has not resulted in a statistically significant change, either in the seasonal mean temperatures or in the interannual variability of seasonal mean temperatures (Schuurmans, 1983). It is therefore important to continue to look for indicators with high signal to noise ratios to aid in early detection of a CO_2/climate effect (Thompson and Schneider, 1982).

At a DOE Workshop, a more systematic approach toward the first detection of CO_2 effects was devised (US DOE, 1982; Riches, MacCracken and Luther, 1982; MacCracken, 1983). A three-part strategy was proposed involving: 1. determination of whether a climatic or biospheric change has occurred, 2. identification of the role of other factors that might reinforce or mask the

CO_2-induced response, and 3. isolation of these elements of climatic change attributable to increasing CO_2 concentrations. One important new suggestion was to develop a unique "fingerprint" of CO_2-induced climatic changes, involving a set of perturbations distinctive from responses that would be caused by all other known influences, and to search for a correlated pattern of changes, and not just for a change in an isolated parameter. A WMO meeting in Moscow, on the detection of a possible climatic change, recommended as a research strategy to carry out diagnostic studies with improved climatic models and to improve the data base for detection studies (Kellogg and Bojkov, 1982). In the pursuit of CO_2 detection studies, the prime need is, however, not for additional measurements, as Kelly et al. (1982b) point out, but for the dissemination and analysis of the available climate data.

Comparison of time-dependent model calculations with the observations - As with equilibrium models (IV.1.7.1), our confidence in time-dependent climate models increases if they can realistically simulate the external perturbations to the climate system in the observations. Fig. IV.19 shows the observed global temperature trend compared with the results of the 1-D radiative-convective model described in the previous subsection with a model sensitivity of $\Delta T = 2.8^{0}C$ for the time period 1880-1980. It is clearly seen that, with a **perturbation** only by **CO_2**, the agreement is not good, but the addition of **aerosols from volcanic eruptions** and **solar variability** gives a notable improvement. The **damping effect of the ocean** is simulated using a mixed layer model. If only the heat capacity of the upper mixed layer (about 100 m) is considered, the amplitude of the computed temperature change is larger than that observed. In contast, if the heat diffusion into the thermocline (as far as 1,000 m depth) is considered with a diffusion coefficient of $k = 1\ cm^2/sec$, then all perturbing factors show a much improved agreement between computed and observed values. From this, Hansen et al. (1981) conclude that CO_2 and volcanic aerosols together have had a strong influence on the temperature change over the last 100 years. If this is true, then two important conclusions can be drawn. A large part of the global climatic variability over time periods of decades to centuries appears to be deterministic, which strengthens the hope that at some time climate models can be used to realistically predict the effects of a CO_2 increase over the coming decades. This conclusion can eventually play an important role in the consideration of the practicability and urgency of precautionary measures (see Chapter VI).

For the time being the results obtained by Hansen et al. (1981) must, however, be interpreted with the necessary caution, because Gilliland (1982) obtained a similarly excellent fit of the observed data with the modelled CO_2, volcanic and solar variations, although quite different formulations of the external forcings were assumed. Much work, therefore, remains to be done before it is justified to accept these results and those from other investigations as evidence that the effect of CO_2 and other forcing functions on climate is already understood.

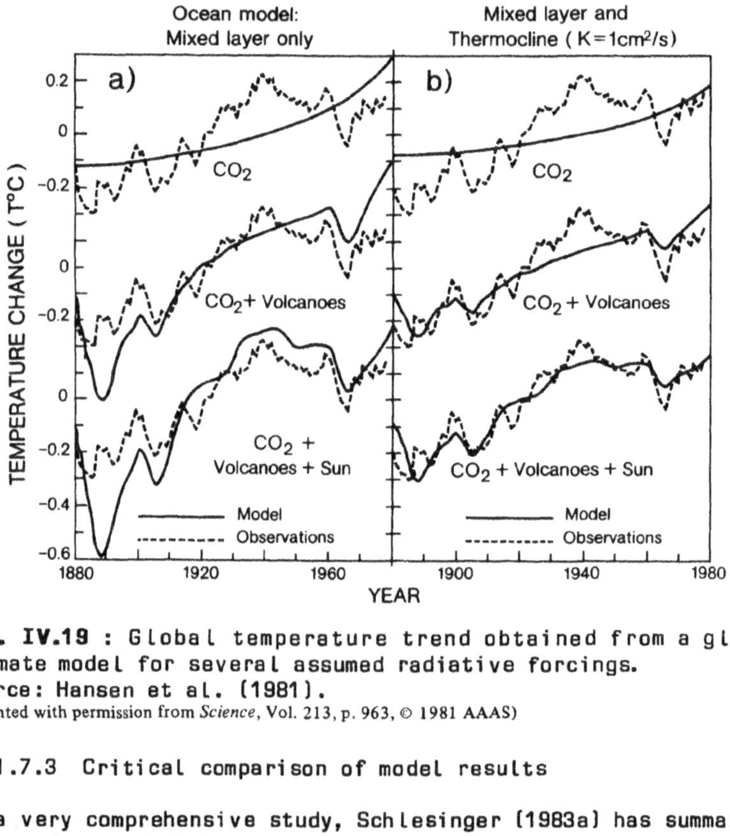

Fig. IV.19 : Global temperature trend obtained from a global climate model for several assumed radiative forcings.
Source: Hansen et al. (1981).
(reprinted with permission from *Science*, Vol. 213, p. 963, © 1981 AAAS)

IV.1.7.3 Critical comparison of model results

In a very comprehensive study, Schlesinger (1983a) has summarised the various modelling approaches for studying effects of a CO_2 doubling. He shows that, if the exceptions are discounted, the range of global mean temperature increase simulated by energy balance models is from 1.3 to 3.3°C, that by radiative-convective models is from 1.3 to 3.2°C, and that by general circulation models is from 1 to 3°C. In a second evaluation of the present evidence, a panel of the US National Academy of Sciences has concluded that a global warming of 3 ± 1.5°C due to a doubling of CO_2 in the atmosphere is presently the best estimate (US NAS, 1982). See also the listing in Appendix IV.1.

As explained by Watts (1982), the studies estimating a global mean temperature increase of only about 0.2°C must be considered as unrealistic because, in the case of Sellers (1974), the model was not run sufficiently long to reach equilibrium. In the case of Paltridge (1974), a strong negative cloud amount feedback was assumed but the important water vapour feedback excluded, and in Gates' (1980) 3-D GCM sea surface temperature and sea ice were prescribed. The reason for the small warming obtained by Newell and Dopplick (1979) using a surface energy balance model, and by Idso (1980) using a surface air temperature response function based on localized experiments, is the neglect of some important **feedback processes** as shown by Crane (1981) and demonstrated with the help of Fig. IV.20 by Ramanathan (1981) as follows:

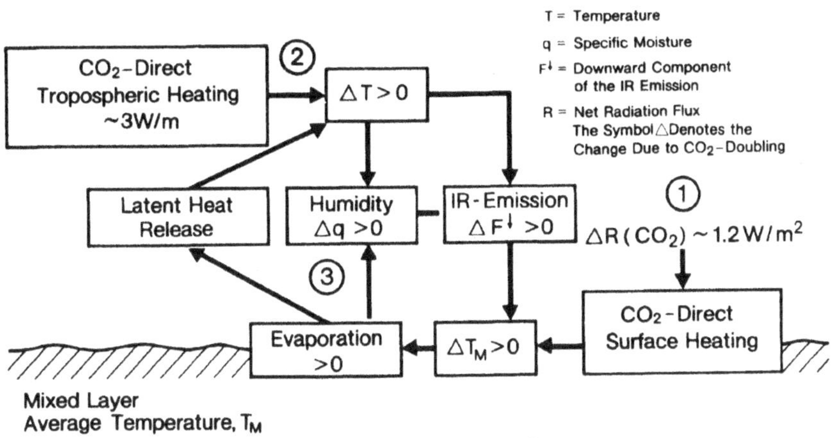

Feedback Processes

	Process (1)	Process (2)	Process (3)	Total
Flux (W/m^2)	1.2	2.3	12.0	15.5
Percent	8.0	15.0	77.0	
Change in Surface Air Temperature ΔT_S (Model dependent)	0.17	0.33	1.7	2.2

Fig. IV.20 : Schematic illustration of the ocean-atmospheric feedback processes by which a CO_2 increase warms the surface. All of the numbers correspond to hemispherically averaged conditions and apply to a doubling of CO_2.
Source: Ramanathan (1981).
(reprinted with permission from *J. Atmos. Sci.*, Vol. 38, p. 920, ©1981 American Meteorological Society)

A CO_2 increase warms the air layers near the ground through the processes labelled (1), (2) and (3). Processes (1) and (2) are pure radiative effects that heat both the earth's surface and the troposphere. In equatorial and middle latitudes the direct warming of the troposphere is about 2–3 times larger than the surface warming. This tropospheric warming, that was considered neither by Newell and Dopplick nor by Idso, increases the surface temperature through process (2). A warmer ground surface leads to increased evaporation and a warmer troposphere can take up more humidity. The net effect of the **positive feedback** between increased surface temperature, evaporation and humidity in the troposphere leads to an increased infrared (IR) counter-radiation from the troposphere, which, in turn, contributes to a further warming of the surface. Feedback process (3) also appears to have been omitted by Newell and Dopplick and insufficiently considered by Idso. However, as the tabulated information in Fig. IV.20 shows, precisely the largely ignored processes (2) and (3) are responsible for 92% of the temperature increase. Neglect of these important feedback mechanisms must lead necessarily to temperature changes that are too low and hence unrealistic. Moreover, it should be noted that the estimates of Newell and Dopplick are not global and therefore cannot be compared directly with the results of global models.

Taking the known feedback processes into consideration, 1–D to 3–D climate models calculate for a CO_2 doubling (600 ppmv), or

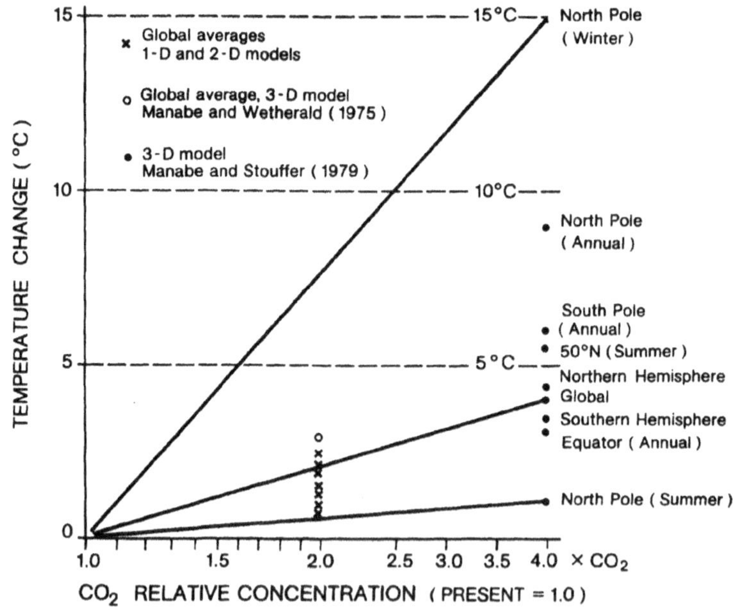

Fig. IV.21 : Global temperature change of model climates due to changes in CO_2 concentration.
Source: Boville and Döös (1981).
(reproduced with permission from *Nature and Resources*, Vol. XVII:1, ©1981 UNESCO)

a CO_2 quadrupling (1,200 ppmv), the temperature changes summarized in Fig. IV.21. According to this, the largest temperature changes are to be expected in winter at the North Pole and the smallest in summer at the North Pole. The North Pole and Northern Hemisphere would experience larger temperature changes than the South Pole and Southern Hemisphere.

These results must be interpreted with the necessary caution. Schneider and Thompson (1981) question the extent to which the results of equilibrium models for the identification of CO_2-induced climatic changes can be applied to the real world. In reality the CO_2 concentration does not increase as a step function (as in the time-independent equilibrium models from 300 to 600 ppmv) but increases continually (as in the time-dependent transient models). Their studies show that it does not appear to be justified, for example, for the CO_2 increase of 20% to be expected by 2000 simply to take 1/5 of the value calculated by an equilibrium model for a CO_2 doubling and to consider this to be the expected climate signal at this time. Depending on how the real climate system interacts with the ocean, there will be delays of up to several decades. In the investigation of the zonal temporal response to an external perturbation they found, in addition, significant regional deviations from the global average values in the equilibrium models. In contrast, Bryan et al. (1982) conclude from their findings that sensitivity studies of climatic equilibrium are suitable for giving rough indications for the prediction of the zonal distribution of sea surface temperature trends. On the basis of additional studies, Thompson and Schneider (1982a,b) conclude, however, that it is better to look for regional climatic variations with a clearly distinguishable signal to noise ratio in the results of time-dependent simulations based on coupled atmosphere/ocean models with realistic geography and plausible CO_2 growth scenarios.

A number of pressing questions remain (Riches, MacCracken and Luther, 1982). These include: What will the seasonal and regional distribution of changes be at various latitudes for both the equilibrium and transient type perturbations? How will the results differ between models calculating the seasonal cycle or annual averages? To what extent will improvement in the representation of the oceans and the cryosphere, a more realistic geography and topography, as well as a coupling to biospheric processes lead to different than the present results? Is the effect of water vapour, and especially clouds, more important than commonly thought? (Rossow, Henderson-Sellers and Weinreich, 1982). To what extent are the causes of differences between climate models understood, how sensitive are the model results to the input parameters and initial conditions, and over what time period must the models be integrated to give statistically significant regional climatic changes? And, as discussed in the following sections, what is, beside CO_2, the contribution of the other forcing functions?

IV.2 Climatic effects of other anthropogenic trace gases

In addition to CO_2 there is a considerable number of other trace gases produced by humans whose effects must be added to the greenhouse effect, because they have absorption bands in the atmospheric window region (7–14 μm)(Wang et al., 1976). The trace gases can influence the climate as follows: a) directly by influencing the radiation balance of the atmosphere, b) indirectly through photochemical transformation in the trace gas budget, both in the troposphere and stratosphere, and c) as aerosols that modify the radiation balance through scattering and absorption, cloud formation and precipitation. The latter will be discussed in IV.3.

Our understanding of the influence of complex photochemical processes on the radiation budget is presently still poor but is continually improved and modified (Ramanathan, 1980). New reaction schemes are continually added to the many that already exist. It is therefore extremely difficult to summarize the present state of the art of the influence of trace gases on the climate. The compilation given in Table IV.3 of the influencing factors should provide the reader with an aid in recognizing the complicated interrelations. It should be noted that the emphasis is on trace gases from anthropogenic sources, i.e. gases that are produced by human activity and that can be controlled by human action. The so-called background concentration (representative of an atmosphere not affected by direct sources) reflects the trace gas emissions from both anthropogenic and natural sources. Trace gases whose atmospheric concentrations are less than about 10 ppmv are usually referred to as "minor trace gases" (WMO, 1982). They are of concern because of their direct radiative effects and because of their potential perturbations to stratospheric ozone.

In the following sections the information in Table IV.3 is supplemented and the chemical-climatic interactions are described to the extent deemed necessary for an understanding of the assessment of possible climatic effects.

IV.2.1 Production and concentration

Natural processes like evaporation, condensation and sublimation during cloud and precipitation formation are the most important sources and sinks of **water vapour (H_2O)** in the troposphere. In some places anthropogenic sources, such as wet cooling towers can make a large contribution. The main sources of H_2O in the stratosphere are local thunderstorms (8 x 10^8t/yr), the Hadley circulation (2.2 x 10^8t/yr), methane oxidation (1.6 x 10^8t/yr) and highflying aircraft (7.8 x 10^7t/yr)(Bach, 1976). According to Ellsaesser (1979), the main sink is the freezing out of H_2O in the Antarctic winter, when 3–4.5 x 10^8t/yr are removed from the stratosphere. Because of the short residence time (about 10 days) of a water vapour molecule in the troposphere, its influence is only local or regional. In the dry stratosphere the residence time is several years. The temperature of the tropical tropopause seems to control the water vapour mixing ratio in the lower stratosphere (WMO, 1982). Its increase with altitude in the stratosphere appears to be due to production by oxidation of methane and molecular hydrogen.

Nitrous oxide (N_2O), also called laughing gas, forms naturally mainly from bacterial (microbial) denitrification in the soil as well as surface waters, and anthropogenically from the use of fossil fuels and artificial nitrogen fertilizers. The anthropogenic share of the annual N_2O-emissions is 4-8% (BMI, 1980). The present global nitrogen fertilizer consumption amounts to about 50 million t/yr (in the F.R. Germany it is about 1.3 million t/yr, corresponding to a 3% share). Compared to untreated forested mineral soil it was found that fertilizer caused N_2O emissions to increase up to fourfold from mineral soil sites in New York State and by ca. twofold from organic soil sites in the Florida Everglades Agricultural area (Duxbury et al., 1982).

The measured background concentration was about 310 ppbv in 1980 having increased at a rate of about 0.2%/yr since the first samples were taken in 1963 (Table IV.3). In the troposphere it is assumed that there is only a small sink due to absorption by desert sand. The main sink is transformation of N_2O in the stratosphere. The strong increase of population that is to be expected (see Chapter III.1) and the related need for increased food, fertilizer, and fuel use could lead, according to the estimates of Weiss (1981), to N_2O concentrations that would be 5-7% higher than today in 2000 and, according to Hahn (1979) even 100% higher than today between 2025 and 2040.

Chlorofluoromethanes (CFMs) are purely anthropogenic compounds that are used as aerosol propellents in spray-cans, as coolants in refrigerators, as solvents, and in the production of plastic foam. The best known are the Freons $CFCl_3$ (F-11) and CF_2Cl_2 (F-12). The global annual production of F-11 has increased from about 200 to 302,000 t in the past 40 years (1940-1980). Up to now a total of 4.9×10^6 t F-11 and 7.1×10^6 t F-12 were produced, about 86% of which have found their way into the atmosphere. For F-12, but not for F-11, the production rate has slightly decreased since 1974 (FAA, 1981). The Federal Republic of Germany contributed about 12% of the global production of CFM in 1977 (BMI, 1980).

The measured background concentration at the South Pole in 1975 was 0.09 ppbv for F-11 and increased with a growth rate of 10%/yr to 0.17 ppbv in 1980 (Rasmussen et al., 1981). The corresponding values for F-12 were 0.19 ppbv in 1976 and 0.28 ppbv in

Table IV.3 : Overview of the most important factors influencing climate.

Factor	Principal anthropogenic sources (1,2)	Atmospheric residence time (1-8,18,20)	Principal sinks and removal processes (1-8,17)	Background concentration in atmosphere (current) (2,8-10,17,18)	Future development (2,8,11-13,17,18, 19,21,22,23)	Potential impact on climate and surface air temp. change (8,10,11,14-16)	Magnitude and importance of impact
Carbon Dioxide (CO_2)	Fossil fuels, deforestation, soil destruction	4 yr (Trop.) 2 yr (Strat.)	Oceans Biosphere	340 ppmv	250-270 ppmv (ca. 1850) 340 ppmv (1982) Increase: 0.4%/yr	Warming in Trop. Cooling in Strat. 2 x CO_2 → 2-3°C global, 4 to 5 times greater at poles	Global, possibly one of the most important impacts
Water vapour (H_2O)	Combustion processes of high-flying aircraft	10 dy (Trop.) 2 yr (Strat.)	Precipitation	$10-3x10^4$ ppmv (Trop.) 3 ppmv (Strat.)	Increase due to growth in industry and air traffic	Greenhouse Effect and cloud formation/precipitation in Trop., influence on trace substances in Strat.	Local and regional in Trop., global in Strat.
Nitrogen Oxides (N_2O) (NO_x)	Fertilizer, fossil fuels, aircraft	150-175 yr 2-5 dy	Little absorption in Trop., photodissipation in Strat., dry deposition, oxidation to nitrate	310 ppbv 0.001-0.1 ppbv	N_2O: 310 ppbv (1980) Increase: 0.2%/yr	Greenhouse Effect in Trop., Impact on O_3-budget in Strat. 2 x N_2O → 0.6°C	Global, possibly very important
Ozone (O_3)	Indirectly produced through photochemical reactions with other substances	30-90 dy (Trop.) 2 yr (Strat.)	Catalytic reactions with other trace substances, such as NO_x, Cl_x, HO_x	0.02-0.3 ppmv (Trop.) 5-10 ppmv (Strat. at 30 km)	CO_2-induced cooling in Strat. reduces O_3-destruction	O_3-destruction in Strat. leads to cooling; 0.75 x O_3 → -0.4°C; O_3-production in Trop. leads to warming; 2 x O_3 → 0.9°C	Warming possibly exceeded by cooling
Chlorfluorocarbons ($CFCl_3$=F-11) (CF_2Cl_2=F-12)	Propellants, coolants and insulating materials	65-81 yr 135-182 yr	No known sink in Trop., sink in Strat. through photolysis	0.17 ppbv 0.28 ppbv	$CFCl_3$: 0.168 ppbv (1980) Increase: 10%/yr 9%/yr	Cooling through O_3-destruction; warming through CFM-increase, net effect a warming of 10xCFM → 0.15°C	Global, impact on temperature probably increasing
Methane (CH_4)	Fossil fuels, cattle raising, irrigation	4-7 yr	Oxidation with OH-radicals	1.65 ppmv	Increase through growth in fossil fuel use and agricultural activities Increase: 1-2%/yr	Impact directly on O_3 of Trop. and indirectly on Greenhouse effect; 2 x CH_4 → 0.25°C	Global, increasing in importance
Carbon monoxide (CO)	Fossil fuels, hydrocarbon reactions	0.1-0.5 yr	Oxidation with OH-radicals, soil organisms	0.05-0.2 ppmv	Increase through fossil fuel use and decrease through deforestation	Impact similar to CH_4, but smaller by order of magnitude	Regional, impact probably remaining small

Table IV.3 (Continued).

Ammonia (NH_3)	Cattle, fossil fuels, waste treatment	7-14 dy	Wash-out through precipitation, dry deposition, oxidation to NO_x	6 ppbv	Similar to CH_4	at 2 x NH_3 → 0.1°C	Similar to CO
Sulfur dioxide (SO_2)	Fossil fuels	4 dy	Dry and wet deposition, oxidation to sulfate/aerosols	0.05-1 ppbv	Further increase expected	Acid rain, aerosols influence cloud formation and radiation budget 2 x SO_2 → 0.02°C	Regional, similar to CO
Krypton-85 (^{85}Kr)	Reprocessing plants, nuclear power plants, atomic bomb tests	half-life 10.7 yr	Radioactive decay	20 pCi/m^3	Major increase (doubling in 10 years)	Decrease through ionization of electric resistance btw. earth and ionosphere causing increase or decrease of precipitation in certain areas	Global, of increasing importance
Aerosols	Industrial processes, grass and forest fires, expansion of desert, agriculture	10 dy (Trop.) 2 yr (Strat.)	Wash-out, dry deposition	10-50 μg/m^3	Similar to SO_2	Influence on cloud formation and precipitation, albedo increase on land, net effect: probably a slight warming	Regional, likely to increase in importance
Surface albedo	Overgrazing, deforestation, urbanization, energy plantations, irrigation projects				Strong increase	Change in surface albedo through desertification and deforestation, net effect: -0.2°C global average	Regional, may also become important globally
Heat release	All combustion processes		Outer space		Strong increase	Influence on evaporation and atmospheric circulation	Locally and regionally substantial; globally unlikely

Abbreviations:
ppmv = parts per million = 10^{-6}
ppbv = parts per billion = 10^{-9}
μg/m^3 = microgram (10^{-6}g) per cubicmeter
dy = days
yr = years
Trop. = Troposphere
Strat. = Stratosphere
2x = Doubling of concentration

Sources: *1. Almquist (1974a); 2. Bach (1976a); 3. Ellsaesser (1979); 4. Crutzen and Ehhalt (1977); 5. Ehhalt (1976/77); 6. Freyer (1979); 7. Ehhalt and Schmidt (1978); 8. Grassl (1980); 9. Almquist (1974b); 10. Ehhalt (1980); 11. Ramanathan (1980); 12. Isaksen and Stordal (1981); 13. Rasmussen et al. (1981); 14. Wang et al. (1976); 15. Groves and Tuck (1979); 16. Potter et al. (1980); 17. Ehhalt (pers. comm., 1981); 18. Crutzen (pers. comm., 1981); 19. Weiss (1981); 20. Ko and Sze (1982); 21. Zimmerman et al. (1982); 22. Ehhalt et al. (1982); 23. Fraser et al. (1983).*

1980 (Table IV.3). An increase over present concentrations by a factor of as much as 20 is seen to be possible in the next century (WMO, 1982). No sinks are known for Freons in the troposphere; it is assumed that they are only broken down in the stratosphere with a simultaneous endangering of the ozone layer. In contrast to earlier investigations (CEQ, 1975; US NAS, 1976) a more recent report of the US NAS (1979) assumes that the ozone destruction by CFM is twice as high with a value of 16.5%. Such a high ozone destruction would strongly increase the rate of skin cancer, hinder the pollinating activity of insects and thus limit food production, destroy fish larvae, and lead to a slight warming of the lower atmosphere.

It is important to note that new substances are continually added, such as F-22 (CF_2HCl) and methyl chloroform (CH_3CCl_3) used as coolants and solvents, or the various bromide compounds released by pesticide use. According to Yung et al. (1980) these could destroy the stratospheric ozone even more effectively than chlorine. Another radiatively effective halocarbon is carbon tetrafluoride (CF_4) which is mainly released during aluminium production. Because of its very long atmospheric residence time (>10,000 years), this relatively small source can lead to a significant impact in the long term (WMO, 1982).

The most important source of **methane** (**CH_4**) is the bacterial destruction of organic matter, especially in rice paddy fields, swamps and marshes, tundra and tropical rain forests, as well as termite activity, and the enteric fermentation in mammals. Methane emission from brackish marsh amounts to a loss of carbon equivalent to 3–10% of the fixed carbon. It is inversely related to salinity and sulphate concentration (DeLaune et al., 1983). The CH_4 production of the ecosystem amounts to a total of between 3.5 and 12.1 x 10^8 t/yr (Ehhalt and Schmidt, 1978; Zimmerman et al., 1982). As ^{14}C analyses show, the production from fossil fuel sources was only 20% of the natural emissions. As laboratory measurements indicate, termites may be a significant source of CH_4 production contributing as much as 1.5 x 10^8 t/yr (Zimmerman et al., 1982). 90% of the CH_4 sinks are in the troposphere acting mainly by reaction with hydroxyl radicals (OH). The CH_4 concentration was 1.5 ppmv in 1980 at the South Pole and 1.65 ppmv in the US Pacific Ocean at 47°N, reflecting the higher emissions in the northern hemisphere (Rasmussen et al., 1981). Observations made during 1978–1980 indicate a 1–2%/yr increase.

In terms of the emitted amount **carbon monoxide** (**CO**) is only second to CO_2. By far the largest anthropogenic source is automobile traffic; its release of 450 x 10^6 t/yr amounts to 70% of the total anthropogenic CO production (Freyer, 1979). The largest single natural source of CO is from oxidation of CH_4. The primary atmospheric sink for CO is its reaction with OH (WMO, 1982). The CO concentration in the Antarctic was 0.05 ppmv in 1980 and 0.15 ppmv in the US Pacific at 47°N, which again reflects the differing distribution of the anthropogenic CO sources in both hemispheres (Rasmussen et al., 1981). Since the CO production has been increasing strongly for many years but the CO concentration has remained constant for a long time, it is suspected that there are one or more effective sinks in the troposphere (possibly soils, plants).

The remaining trace gases such as **ammonia** (**NH_3**), **nitrogen oxides** (**NO_x**), and **sulphur dioxide** (**SO_2**) have directly probably only a regional influence because of their relatively short residence times (1-2 weeks), but could indirectly have global climatic effects through nitrate and sulphate aerosol formation (Husar et al., 1978). For the discussion of this aspect see section IV.3.

Finally, in Table IV.3 we have listed Krypton that has a concentration of 1.14 ppmv in the atmosphere (Grassl, 1980). A small fraction consists of radioactive ^{85}Krypton that is primarily released during the reprocessing of nuclear fuel elements, the operation of nuclear power plants and in nuclear weapons tests. With a continuation of the present energy policies an increase from the present 20 picocurie/m^3 to about 3 nanocurie/m^3 (i.e. a 150-fold increase) in the first half of the 21st century is anticipated (Umwelt, 1980). As a result of this the ionisation of the air could increase by 57%, which would have consequences for thunderstorm activity (see section IV.2.3).

IV.2.2 Climate-chemistry interactions

Many of the trace gases emitted into the atmosphere are chemically active so that they do not only influence the trace gas concentrations but also the radiation balance of the earth-atmosphere system. Fig. IV.22 shows the complex chemical-climatic interactions that occur between the troposphere and stratosphere. It is known that, as a result of infrared absorption by CO_2, N_2O, CH_4, and CFM, the greenhouse effect is increased and the near-ground air layer is directly warmed. Recently it has been found that the **ozone (O_3) content** of the troposphere is considerably increased by photochemical oxidation of CO, CH_4 and other hydrocarbons in the presence of NO_x and thus contributes indirectly to the temperature increase of the near-ground air layer (Ramanathan, 1980; Fricke, 1980). The importance of an O_3 increase in the troposphere is further underlined by the fact that under higher pressure the 9.6 µm ozone band broadens, so that tropospheric O_3 with only 10% of the entire ozone share reaches the same absorption effect as stratospheric ozone (Hameed et al., 1980).

The large number of the chemical-climatic interactions can be demonstrated by further examples. For example, an increase of NO_x could reduce the ozone destruction by chlorine (from chlorofluoromethanes)(Isaksen, 1980). The N_2O increase, decisive for the ozone destruction, depends again on the stratospheric chlo-

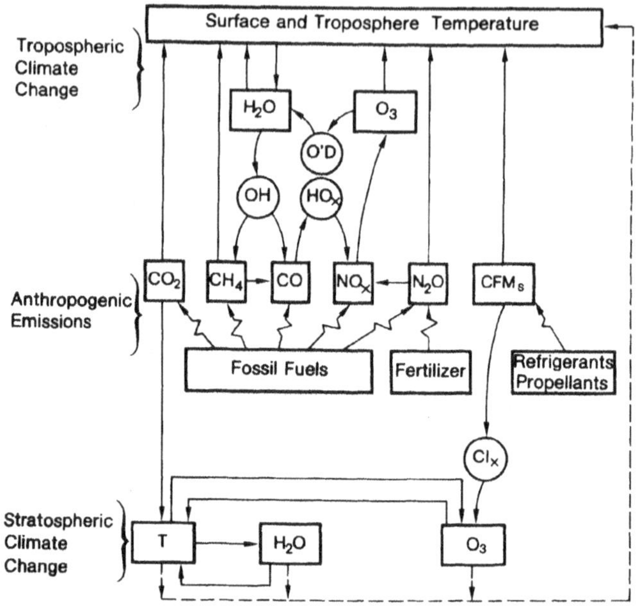

Fig. IV.22 : Climate-chemical interactions due to trace gases. Source: Ramanathan (1980).

rine concentration. An increase of the N_2O content would have drastic effects on the vertical distribution of O_3 which could have far-reaching climatic effects, since the influence of ozone on the climate is strongly dependent on changes in the vertical ozone distribution.

As we have seen in Chapter IV.1, the CO_2 increase in the troposphere leads to a pronounced temperature decrease in the stratosphere. This cooling, however, can lead to a marked O_3 increase in the upper stratosphere by affecting the temperature-dependent chemical rate coefficients related to O_3 production and loss (Groves et al., 1978; Groves and Tuck, 1979). According to the calculations of Haigh and Pyle (1979) a CO_2 doubling would lead to a total O_3 increase of about 7%. Of course, this effect again depends on the chlorine content of the stratosphere, since higher chlorine concentrations weaken the temperature effect on the stratospheric O_3 (Isaksen and Stordal, 1981). In the evaluation of the O_3 effect on climate, all chemical reactions, both in the stratosphere and the troposphere must be taken into consideration and it is this which makes a complete estimate of the climatic effects extremely difficult. At present tropospheric photochemistry is poorly understood so that simulated tropospheric O_3 perturbations are more speculative than for stratospheric O_3 perturbations (WMO, 1982).

IV.2.3 Assessment of climatic effects

Table IV.3 gives the potential climatic effects in terms of temperature changes in 2025, in most cases for a doubling of the individual trace gases, or in the case of CFM, for an increase by a factor of 10. The following estimates made by Ramanathan (1980) take into consideration the complex interactions in Fig. IV.22.

It is assumed that the fossil fuel consumption in 2025 will be four times larger than today, that the CO_2 content of the atmosphere has increased to 500 ppmv by then and reaches a new equilibrium level of about 600 ppmv after 2050. Assuming further that CFMs continue to be emitted at the same rate as in 1977, that the N_2O content continues to grow at the rate of 0.4%/yr and, finally, that CH_4 and O_3 increase by a factor of 4 in the troposphere in the period 1975–2025, Ramanathan finds the following results: the global warming of the troposphere caused by the greenhouse effect amounts to 1.9°C for the period 1975–2025 and 3.4°C for the period 1940–2050. One main conclusion of this study is that CO_2 with 60% has the main share of the warming of the troposphere but the share of the other trace gases of 40% is not inconsiderable and must be taken into account in a complete evaluation of the climatic effects (see IV.6).

The above results must be interpreted with the necessary caution because of our inadequate understanding of the various radiative/photochemical interactive processes and the neglect of feedback processes in 1-D modelling. Moreover, the overlapping effect of atmospheric CO_2, H_2O and O_3 on the radiative effect of CO_2 must be taken into consideration. Wang and Ryan (1983) find that the net effect of the overlapping gases is to increase the tropospheric warming but to decrease the surface warming caused by a CO_2 increase. Furthermore, CO_2-induced warming is likely to increase the CH_4 content of the atmosphere because bacterial processes in anaerobic environments become more productive at higher temperatures. This may lead to further warming of the atmosphere, since both CH_4 and O_3 (the latter produced in the troposphere by reactions of CH_4) have strong greenhouse effects resulting in a positive feedback to climatic change (Kellogg, 1983). But an increase in tropospheric temperature accompanied by a higher H_2O content is likely to decrease through chemical reactions the CH_4 and O_3 concentrations providing, in turn, a negative feedback. Model calculations, carried out by Hameed and Cess (1983) for a doubling of CO_2, show that the positive feedback due to an increase of natural sources of CH_4 is virtually compensated by the negative climate-chemical feedback. These results must, however, be considered as highly preliminary because of considerable uncertainties in our knowledge of the cycles of tropospheric gases.

Despite the above uncertainties, present studies suggest that the combined climatic effects (i.e. surface warming and stratospheric cooling) of the minor trace gases can be as large as those expected from a CO_2 increase (WMO, 1982). There is, however, a major difference between the effects of CO_2 and the minor trace gases in that, with the exception of O_3, the direct radiative effect of the latter on the stratosphere is negligible. In fact, it is the O_3 decrease (due to reaction with CFMs) which makes the cooling of the upper stratosphere due to minor trace gases comparable in magnitude (about 10 K) to that caused by a CO_2 increase.

Because of our limited knowledge, only a few statements can be made about the potential climatic effects of ^{85}Krypton (Boeck, 1976; Grassl, 1980). If the estimate in IV.2.1 of 3 nCi/m^3 were reached, this would correspond to the order of magnitude of ion pair formation produced by cosmic radiation. An increase of ^{85}Krypton would increase the ionisation rate and lead to a decrease of the electrical resistance between the earth's surface and the ionosphere. This could influence the distribution of electrical loading in clouds and thus affect cloud physics and precipitation formation with far-reaching effects on precipitation frequency and distribution in various regions, which again would have serious consequences for food security. These potential consequences could be avoided if the ^{85}Kr were contained at the various sources by appropriate measures.

IV.3 Climatic effects of aerosols

Aerosols can influence the radiation balance and thus the climate in many ways: through scattering (forward- and backward-scattering) and absorption of solar radiation (change of the planetary and surface albedos), and through emission of long-wave radiation. The indirect effects result from the influences on cloud formation and thus on precipitation and from changes of the optical characteristics of clouds (Table IV.3). Aerosol injections into the stratosphere, for example through frequent volcanic eruptions, have been variously related to the occurrence of glacial periods as a possible causal factor.

The damaging effects are not only limited to climate. As a result of the washing out of sulphate and nitrate aerosols, the notorious **acid rain** is formed leading to acidification of soils and water bodies. In principle one could counteract this by adding limestone. In addition to the damage to ecosystems food production in particular is badly affected. It is also important to note that, as a result of a pH decrease the solubility of dangerous **heavy metals** increases, such that they can no longer be so effectively filtered out of the soil and then accumulate in foodstuffs.

IV.3.1 Sources, emissions, concentrations

Aerosols can be subdivided according to their origins as **primary** and **secondary**, as well as into **natural** and **anthropogenic aerosols** (Bach, 1976; Dittberner, 1978). Primary natural aerosols are, e.g., sea salt (spray), blown dust or volcanic ash. The primary anthropogenic aerosols include, e.g. dust from industrial combustion processes or fires connected with agriculture. The secondary aerosols arise from transformation of gases like SO_2, H_2S, NO_x and hydrocarbons. Naturally formed aerosols include, e.g. hydrocarbons from plant exhalation (terpenes) or sulphates from volcanic eruptions. Sulphate and nitrate aerosols in **acid rain** and the notorious **photochemical smog** (first analysed in Los Angeles) are typical examples for secondary anthropogenic aerosols.

Table IV.4 summarises the sources, causes and emissions of aerosols with a particle diameter <5 μm. Particles of this size are particularly climatically effective because of their optical characteristics. The global emissions are estimated to be about 2 bill. t/yr, of which about 35% are from anthropogenic aerosol production (the total production of all particle sizes is about 3-4 bill. t/yr). Almost 99% of all aerosols are found in the troposphere. Their residence times depend on their size (they range over four orders of magnitude, from 1 μm to 0.1 mm) and their location. The average residence time in the troposphere amounts to about 10 days, but in the stratosphere, depending on the altitude, several years. In the troposphere they are therefore more of a regional phenomenon and in the stratosphere a global phenomenon. Except for volcanic eruptions, only a very small amount of tropospheric aerosols in the vicinity of jet streams over the intertropical convergence and the antarctic polar front reaches the stratosphere (Hogan and Mohnen, 1979). From the stratosphere they diffuse only slowly back into the troposphere, where they reach the earth's surface again through a series of processes, such as, for example, washout or dry deposition. The background concentration of aerosols is between 10-50 $\mu g/m^3$. In the vicinity of urban and industrial agglomerations, or forest and grass fires, this can increase by several orders of magnitude (Bach and Daniels, 1975).

IV.3.2 Climatic effects

A semi-quantitative documentation of the magnitude of explosive eruptions, termed the 'Volcanic Explosivity Index', is now available with location and date for over 8,000 eruptions from 1500 to the present (Newhall and Self, 1982). This is a useful supplement to Lamb's (1977) 'Dust Veil Index' which used atmospheric opacity, temperature and volcanological information to assess the amount of dust injected into the stratosphere by ca. 250 erup-

Table IV.4 : Estimates of global particle production (diameter <5 μm) from nature and man-made sources.

Source	Cause	Emissions Mill. t/yr	%
Natural (excluding volcanoes)			
Sea salt	Ocean spray	500	46
Sulfates	Decay of biomass	335	31
Wind-blown dust	Wind	120	11
Hydrocarbons	Exhalations	75	7
Nitrates	Decay of biomass	60	5
Aerosols and dust	Forest fires (through lightning)	3	0
Total		1093	100
Volcanoes			
Sulfates	Eruption	42-255	63
Aerosols and dust	Eruption	25-150	37
Total		67-405	100
Anthropogenic			
Gases (subsequently converted to particles in the atmosphere)	Fossil fuel use	311	45
Aerosols and dust		54	8
Wind-blown dust	Human impact	180	26
Gases (subsequently converted to particles in the atmosphere)	Agricultural burning	79	11
Aerosols and dust		62	9
Aerosols and dust	Fuel wood	4	1
Aerosols and dust	Forest fires (intentional)	2	0
Total		692	100
Grand Total		1852-2190	

Adapted from: Bach (1976) and Dittberner (1978).

tions. Large volcanic eruptions, whose emissions go far into the stratosphere can have a significant influence on the climate. As Fig. IV.23 shows, after the eruption of Mt. Agung on Bali in 1963, the temperature in the stratosphere increased by about 7°C within one year, while the surface temperature in the tropics decreased by 0.5°C (Toon and Pollack, 1980). In the stratosphere the temperature increased through the enhanced absorption of solar energy by the aerosols. In the surface air layer the

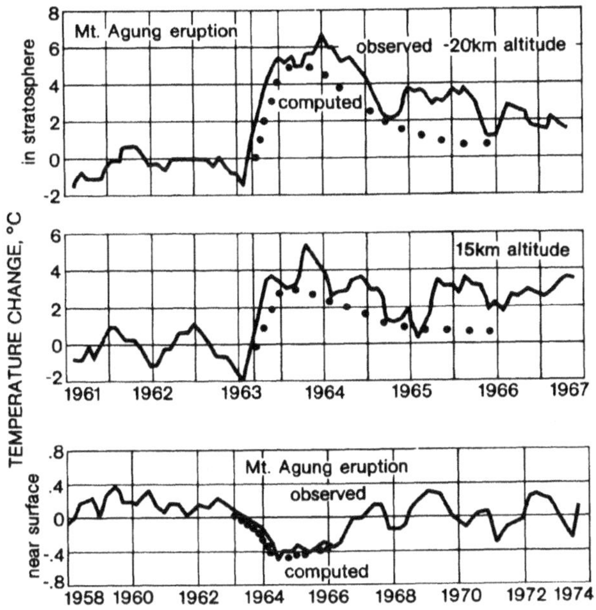

Fig. IV.23 : Stratospheric and surface air temperature change in the tropics after the Mt. Agung eruption on Bali in 1963. Dots are based on calculations with a one-dimensional radiative model. Source: Toon and Pollack (1980).
(reprinted with permission from *Science*, Vol. 199, © 1980 AAAS)

temperature decreased because part of the solar radiation was reflected into space by the aerosols or was absorbed in the stratosphere and thus did not reach the surface. It is interesting that in the stratosphere the temperatures still had not reached the values experienced prior to the eruption after four years and in the tropics the influence at the surface disappeared only after about five years.

The eruption of Mt. St. Helens supplied a good opportunity to examine, for the first time, the local, short-term effects on surface temperature. Robock and Mass (1982) found that immediately after the eruption the daytime temperature was reduced by ca. 8°C, while the nighttime temperature was increased by about the same amount. The net effect was a warming for the first few days, while on climatological scales a cooling might be expected. Quite a different effect is expected from the eruption of Mexican volcano El Chichon of April 1982, which probably lofted a 20 times more massive cloud of debris into the stratosphere than did Mt. St. Helens (Kerr, 1981, 1982). Preliminary sampling results indicate that the cloud has been circling the globe about every

10 days, that it has spread unexpectedly from the initial layer of 22-26 km to a height of about 32 km, and that it appears to have put at least as much SO_2 into the stratosphere as Mt. Agung. It is the injection of sulphurous gases which convert to sulphuric acid droplets, rather than of dust that will largely determine the radiative effects. At 26 km over Guam the stratospheric temperature increased by 3°C making it the warmest since recording began in 1958 (Kerr, 1983).

Model calculations made by Arking et al. (1983) with a multi-layer energy balance model indicate a Northern Hemispheric mean surface air temperature decrease of 0.2 to 0.5°C between $1^1/2$ and 2 years after the aerosol layer has reached its peak, that is, toward the end of 1983 and early 1984. Preliminary results obtained by MacCracken and Luther (1983) with a seasonally varying zonal climate model coupled with a 56 m oceanic mixed layer also indicate slightly lower temperatures in the summer of 1983 and the winter of 1984 in the Northern Hemisphere. Beside these delayed effects some more immediate ones can perhaps not be ruled out. The coincidence of the tropical eruption of El Chichon and the unanticipated warming of equatorial Pacific surface waters nourished this suspicion. Such an El Niño situation (see also V.6) began developing in May rather than, as is usual, during October to November. The last out-of-season El Niño occurred in 1963, shortly after the Mt. Agung eruption (Kerr, 1983).

In general, climate model calculations about the effects of aerosols are somewhat controversial. Charlock and Sellers (1980) calculated for a doubling of the optical thickness (a measure of turbidity), depending on the model used, a surface air temperature decrease of 1.5-1.8°C, and, by considering the ice-albedo feedback effect, a decrease of 3.2°C. A similar temperature decrease (-1.8°C) was also calculated by Rasool and Schneider. In contrast, Reck found a temperature increase (+0.1°C at 85°N and +1.1°C at 85°S)(cited in Bach, 1979).

It is very difficult to estimate the total effect of aerosols on climate because it does not only depend on the surface albedo and the distribution of the various aerosol types over continents and oceans but also on the various vertical profiles of the aerosols and the mean solar zenith angle. The complexity of the problem and the state of knowledge are made clear by the example of the following chain of arguments (Grassl, 1979). In the cloud-free atmosphere an aerosol increase can either increase or decrease the albedo, depending on the surface albedo and the optical characteristics of the aerosols. In a cloudy atmosphere, three different effects of aerosols that can change the albedo must be distinguished. The total effect is that the albedo increases for thin cloud layers but is reduced for thick cloud layers. Since one cannot determine the variable extinction characteristics of the aerosols with sufficient accuracy, the changes of albedo that they cause cannot be estimated with sufficient accuracy at present and the consequent climatic changes cannot be satisfactorily assessed.

Based on the present state-of-the-art, the complicated effective mechanisms can be summarised as follows (Eiden, 1979; Kellogg, 1980): a. scattering and non-absorbing aerosols cool the entire planet; b. scattering and absorbing aerosols warm that part of the atmosphere in which they are found; c. aerosols in general lower the net albedo of the surface air layer over land but increase it over ocean; d. since most aerosols are found over land, the total effect probably is a slight regional warming; and, e. in some regions aerosols appear to increase atmospheric stability and therefore affect the formation of convective precipitation (Bryson and Baerreis, 1967). This would have consequences for the water budget and agriculture.

IV.4 Climatic effects of land use changes

Land use changes through overgrazing, deforestation and desertification, through urbanisation and industrialisation, as well as through energy production projects and irrigation and drainage systems, all affect the climate system. The related changes of albedo, aerodynamic and hydrologic characteristics of the earth's surface can influence both the energy and water budgets. Local and, to some extent, regional climatic effects are already detectable. Given the projected societal developments (see Chapter III) climate effects on a global scale can be expected in the future.

Changes of the **surface albedo** have a strong influence on the energy budget of the entire earth-atmosphere system. Typical mean albedo values are: 14% for tropical rainforest, 20% for agricultural land (meadows, crops); 31% for desert; 20% for urban areas and 6-30% (depending on the sun's elevation) for water surfaces (Baumgartner et al., 1978). The most important processes that can cause albedo changes are summarised in Table IV.5.

Changes of the aerodynamical characteristics of the earth's surface have a strong influence on the atmospheric circulation. Using the **roughness parameter** the aerodynamic characteristics of the various surfaces can be specified. Typical values (in cm) are: 100-400 for forests and cities; 1-25 for grassland; <1 for deserts and ~0.01 for sea (Baumgartner et al., 1978). Evaporation and precipitation, and therefore the **hydrological cycle**, are strongly influenced by land use changes, as shown in the following examples.

Table IV.5 : Causes of large-scale changes in land surface albedo.

Processes that cause increases in albedo	
	Desertification
	Deforestation
	Overgrazing in semi-arid regions
	Burning of grassland in semi-arid regions
	Ploughing of fields
	Phytoplankton growth on sea surface
	Soil salinization through irrigation
	Land reclamation from the sea
	Production of sea salt
	Drainage of swamps
Processes that cause decreases in albedo	
	Overgrazing in regions with moderate to heavy rainfall
	Man-made lakes and irrigation
	Urban and industrial agglomerations
	Snow removal
	Deposition of particles on ice and snow
Albedo changes unclear	
	Pollution of sea surface (e.g. oil slicks)
	Energy farms (e.g. solar farms, biomass plantations)
	Removal of glaciers (e.g. for drinking water)

Adapted from: Munn and Machta (1979) and supplements.

IV.4.1 Overgrazing – desertification – deforestation

Excessive overgrazing favours desertification. It is true that the large deserts of the earth are not of anthropogenic origin but their expansion is the result of human activities. In the Sahara alone, the desert expands by 1–2 km/yr, corresponding to an increase of more than 20,000 km^2/yr* (Flohn, 1978). It is generally acknowledged today that the extension of the Rajasthan desert between India and Pakistan and the recent desert formation in the Sahel are predominantly a result of overgrazing (Bryson and Baerreis, 1967). In at least three places it has been proved that in fenced areas protected from grazing animals the original vegetation, with a completely different albedo than the surrounding desert, is soon established (e.g. near Jodhpur in the Rajasthan desert with about 270 mm rain per year; near Khartoum in the

* This estimate is probably too small because the UN Conference on Desertification in Nairobi in 1977 gave a value of 60,000 km^2/yr.

Sudan with 160 mm; and near Nefta in Tunisia with 70 mm)(Flohn, 1975).

To explain the **desertification** phenomenon, Charney (1975) postulated the following **bio-geophysical feedback mechanisms:** excessive overgrazing increases the albedo, this reduces the thermal convection and thus suppresses cloud and precipitation formation, which again favours desertification. With a general circulation model (see Appendix II.3) he computed the effects of an albedo increase from 14 to 35% in the Sahara north of 18° latitude. Fig. IV.24 shows that over a period of 6 weeks the precipitation decreased by about 40%. Simultaneously, the convective cloud formation decreased by 40% and the mean position of the intertropical convergence zone shifted by 4 degrees latitude to the south. Elsaesser et al. (1976) obtained similar results with a zonally-averaged two-dimensional climate model.

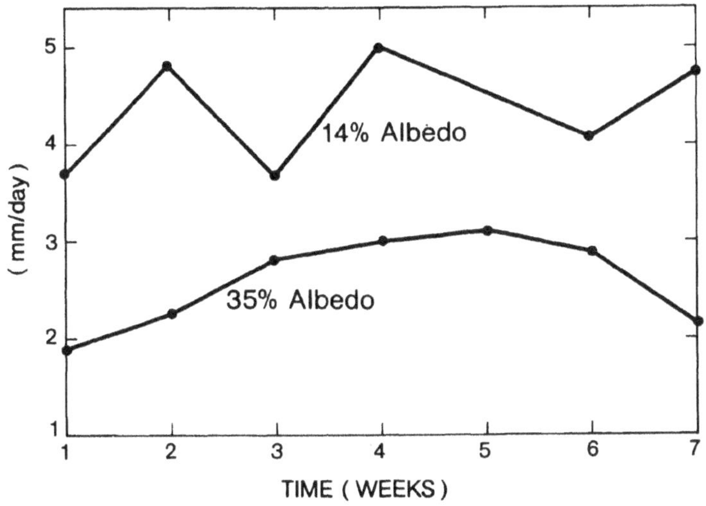

Fig. IV.24 : Precipitation calculations for various values of surface albedo in the Sahara at 18°N.
Source: Charney (1975).
(reprinted with permission from *Quart. J.*, Vol. 101, p. 201, ©1975 Royal Meteorological Society)

Overgrazing and trampling of the vegetation in the surroundings of the few water holes can also influence the production of ice nuclei necessary for precipitation formation. Vali et al. (1976) found that certain bacteria are involved in the decay of plant remains and these are at the same time very effective ice nuclei. The disappearance of the vegetation removes the source of nutrition for the bacterial ice nuclei, which leads to a reduction of convective precipitation.

In the past 8,000 years 11% of the total land surface was converted into agricultural land (Flohn, 1973). During the last 4,000 years the proportion of forest land, measured by the total area, was reduced from about 90% to about 24% (Flohn, 1975). The **conversion** from **forest** into agricultural or grassland increases the albedo from 0.12–0.15 to 0.18–0.22 and thus reduces the net amount of available solar energy from about 63 W/m^2 to about 46 W/m^2. As a result, the evaporation decreases by about 3%, while the surface runoff of precipitation increases by about 5%.

The **destruction** of the **tropical rainforest** continues undiminished in Asia, the South Pacific (Ranjitsinh, 1979), as well as in Africa and in Central and South America (Myers, 1981). With a continuation of the present trends only small remnants will be left in 50 years from the 9 million km^2 available today (Myers, 1979). According to Myers the forest destruction is carried out by three groups: legal and illegal lumberers, who annually cut 55,000–90,000 km^2; share croppers and forest farmers, who annually destroy 200,000 km^2; and, finally, the cattle farmers, who mostly take over the land cleared by the forest farmers, but who also cut an average of 20,000 km^2/yr. The estimates of the Global 2000 (1980) report are 160–180,000 km^2/yr.

The first consequences of the large forest cutting in the Amazon area can already be detected. As a result of the strong deforestation the runoff of the Amazon has increased by 8% between 1962–69 and 1970–78 (Gentry and Lopez-Parodi, 1980). Increased soil erosion and the loss of valuable topsoil are related consequences.

Present plans are for deforestation and subsequent agricultural use of 50% of the Brazilian Amazon area. Lettau et al. (1979) have therefore estimated the resulting potential changes of the water budget with their climatonomy model. The results show that the precipitation with a magnitude of about 2,000 mm/yr increases by 140 to 190 mm/yr and the evaporation increases more strongly by 360 to 410 mm/yr, so that the runoff rate decreases from 1,075 mm/yr to 1,018 mm/yr. The soil moisture would be reduced over the entire Amazon basin, which would very quickly have negative effects on agricultural yields.

Baumgartner and Kirchner (1980) have investigated the effect of a conversion of the total forest area into grassland. Their results show that the average global surface albedo increases by 16.7 to 17.4% and that the atmospheric CO_2 content would increase strongly because of the reduced assimilation and the increased carbon emission from biomass decay.

Estimates made with a simple carbon cycle model and a one-dimensional vertical radiation model (see also Appendix II.3) show that a 50% conversion from forest into agricultural land would increase the atmospheric CO_2 concentration by 95 ppmv, which could lead to a mean global temperature increase of 0.6°C (Niehaus, 1977). Simultaneously, however, this conversion would cause an increase of the mean global albedo by 1%, which could lead to a global cooling of 0.13°C. Potter et al. (1980) have used a detailed two-dimensional zonally-averaged climate model to consider the complicated effect of desertification in the Sahara and deforestation of the tropical rainforest. Assuming that at 20°N the desert albedo increases from 16 to 35% over an area of 9 x 10^6 km^2 and that between the equator and 10°S over an area of 7 x 10^6 km^2 the albedo increases from 7 to 16%, the average surface temperature was reduced by 0.6°C in the Northern Hemisphere and by 0.2°C for the global average (see Table IV.3, surface albedo).

IV.4.2 Urbanisation and industrialisation

The transformation of a natural landscape into an urban and industrial landscape is accompanied by an increase of the gas, aerosol, water vapour and heat emissions from the various combustion processes. If fields and forests are replaced by buildings and streets, the surface albedo and heat storage capacity are changed, the roughness increases, the water runoff is accelerated and snow melts away faster. The term **"heat island effect"** (see IV.5) appropriately characterizes the total effect.

At present urban and industrial areas cover about 1 million km^2 (about 0.2% of the total earth area). Annually, about 20,000 km^2 are added, changing the global albedo by about 2.5 x 10^{-5} which is negligible on a global scale (Sagan et al., 1979). Although the effects of land use changes are presently limited to the local and regional climate, it is not impossible that future conurbations that grow to **megalopoli** and **oecumenopoli** (see Chapter III) could also influence the global climate (Oke, 1980).

IV.4.3 Energy conversion and use

Here we are interested in energy conversion plants that need large areas. These include **solar farms** and **solar towers** for thermal electricity conversion and **bio-plantations** for the production of liquid fuel (alcohol) or electricity production. Changes in the energy and water budgets are to be expected, if, e.g., a previously overgrown surface is built upon, or if a previously unvegetated surface is irrigated and used as bioplantation. The roughness parameter would also be changed by the installation of **heliostats** (mirrors on stilts) that are several metres high.

Calculations for a hypothetical solar power plant in Spain, with a capacity of 30 GWe, wet cooling towers and a built-up area of about 1,000 km^2, show that cloudiness and precipitation would increase in the region (Williams and Bach, 1979). The study "Solar Sweden" envisages that in 2015 half of the Swedish energy demand would be supplied by energy plantations covering 3 million ha (Johansson and Steen, 1978). Since, at present the Swedish forestry already intensively uses 23 million ha, no problem is foreseen.

IV.4.4 Water engineering projects

The heat and water budgets and thus the climate can also be influenced by the construction of dams, by river regulation and irrigation systems, as well as drainage of swamps and salt production through evaporation of seawater. According to the calculations of Hummel and Reck (1977) the dams and artificial lakes that cover 300,000 km^2 could already have caused a temperature increase at the earth's surface of about 0.4°C. It is also suspected that large water stores and irrigation systems in the tropics increase the local precipitation (Sellers, 1977). The proposed diversion of the large Siberian and Canadian rivers towards the south would lead to an increase of the salt content of the Arctic Ocean. The consequence would be that a large part of the Arctic Ocean would remain ice-free also in the winter, which could lead to unforeseeable regional and global climatic changes (Bach, 1979/80).

IV.5 Climatic effects of waste heat

During all energy production and conversion processes, waste heat is released. In contrast to CO_2 emissions, **all** energy carriers release waste heat - fossil, nuclear and solar energy sources. It has been argued that with the expected population and energy growth (Chapter III.1,2,5) and the continuing concentration of urban and industrial areas into enormous energy centres with ever-increasing energy consumption, perturbations of the climate system could occur not only on the local and regional scale but also on the global scale (Williams and Krömer, 1979; Bach, 1980a). To clarify the extent of potential effects, we compare the magnitudes of natural and anthropogenic heat emissions, report of recent model studies assessing the potential climatic effects of waste heat and discuss some possibilities for reducing it.

IV.5.1 Natural and anthropogenic energy fluxes

The radiation flux from the sun reaching the upper boundary of the atmosphere, the so-called **solar constant**, amounts to 1.36 kW/m^2. If this were distributed over the entire planet it would correspond to a radiative power of $(1.36\ kW/m^2 \times 5.1 \times 10^{14} m^2)/4 = 173 \times 10^{12}$ kW (Table IV.6). Using a planetary albedo, i.e. the fraction of solar radiation reflected by the earth and its atmospheric shell, of 28% (see Fig. II.3), the solar input is reduced

Table IV.6: Comparison of natural and artificial heat production.

	Area m^2	Total power kW	Power density W/m^2
Solar radiation from the sun			
at the top of the atmosphere	} 5.1×10^{14}	173×10^{12}	340
global mean at the earth's surface		81×10^{12}	160
Waste heat from urban areas			
Cincinnati	2.0×10^8	5,000,000	26.2
Berlin (West)	2.3×10^8	5,000,000	21.3
Los Angeles	3.5×10^9	74,000,000	21.0
Ruhr District	6.5×10^9	111,000,000	17.0
Waste heat from selected objects			
Wet cooling tower of a 1000 MWe nuclear power plant (closed cycle)	20,000	1,680,000	84,000
Oil refinery with a capacity of 5-6 mill. t of oil/yr	100,000	400,000	4,000
Cement manufacture with 800,000 t/yr	80,000	48,000	600
A 1 km section of the Upper Rhine River with a 1°C warming above equilibrium temperature	200,000	4,600	23
A 1 km freeway section with a daily traffic density of 30,000 cars and 7000 trucks	50,000	2,300	46
A 1 km main railway track with 200 trains/day and 95% electric traction	10,000	120	12
A one-family residence (4 persons)	100	5	56

Adapted from: Bartholomäi and Kinzelbach (1979); Bach (1980).

to 123 x 10^{12} kW, from which about 47%, i.e. 81 x 10^{12} kW or 160 W/m^2 on the global average, is absorbed at the earth's surface and is available energy (Table IV.6).

A frequent starting point is to consider waste heat to be significant for climate from a global point of view, if it reaches about 1% of the solar constant S_o. The global energy consumption increased between 1860 and 1980 with an annual growth rate of about 2%/yr (Marchetti, 1977) from about 0.8 TW to about 9 TW. Considering the assumptions for population, economic and energy growth etc. (see Chapter III), the global energy consumption could be about 36 TW in 2030, with a global population of 8 billion (see the high IIASA scenario in Table III.8). Compared with the solar constant this would correspond to 0.0002 S_o. Using the Stefan/Boltzmann law, the **sensitivity** of the **equilibrium temperature** T_e to a change in the solar constant can be approximated as follows: $\delta Te = 1/4\ [(1-\alpha)So/4\sigma]^{1/4}\ [\delta So/So]$ (Manabe and Wetherald (1975)). For a planetary albedo of α = 28% and Stefan/Boltzmann constant $\sigma = 0.579 \times 10^{-7}$ $W/m^2/K^4$ (K is the absolute temperature), one obtains a T_e of about 0.01°C. Since, however, a change of the solar constant of 1% gives a change of temperature of 0.64°C, it is usually concluded that the energy amounts produced by humans are negligible. It should be noted that in this rough estimate without an atmosphere, no feedback processes were considered and that the energy amounts are evenly distributed over the entire globe which is unrealistic.

IV.5.2 Climatic effects of heat emissions

It is, therefore, useful to compare the heat emissions of various power densities, since the climatic effects depend on both the size of the emitting surface and the radiated power density. Useful reference values are the density of net terrestrial radiation of about 100 W/m^2 and the mean global potential energy of the atmosphere of about 2.4 W/m^2 (Fig. IV.25).

Observations have shown that power densities of about 50 W/m^2 extending over an area of several 10^{12} m^2 increase the mean sea surface temperature by 1°C and thus have a notable influence on the atmospheric circulation (Sawyer, 1974). For such an analysis it is useful to know that 10^{12} m^2 is about four times the area of the F.R. Germany and the present mean power densities for such areas are only about 1 W/m^2. However, as Fig. IV.25 shows, with the expected growth of large conurbations and future irrigation and dam projects, between 2000 and 2050 power densities between 40 and 80 W/m^2 (today 10–50 W/m^2) over areas of 10^{12} m^2 magnitude are not impossible.

With the aid of **three-dimensional general circulation models** (see also Appendix II.3) the potential regional and global climatic effects of heat emissions from large urban areas and energy centres have been investigated. The first experiments were made with unrealistically high input data, i.e. 150 or 300 TW, corresponding to 17 or 33 times the present global energy production (Washington, 1972; Williams et al., 1977). In later studies the present power density of Manhattan, New York of 90 W/m^2 extending over the region from Boston to Florida and to the Great Plains was used to study climatic effects (Llewellyn and Washington,

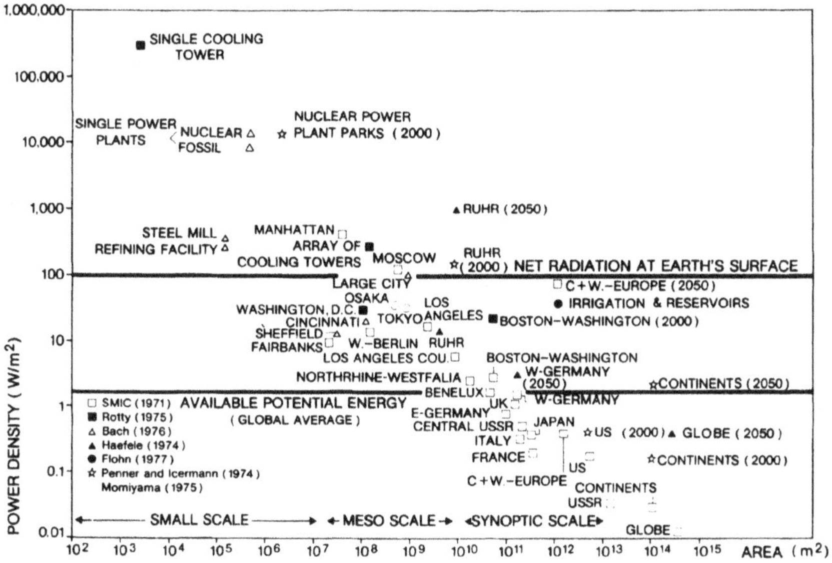

Fig. IV.25 : Power densities related to man's activities ca. 1970 and projected.
Source: Bach (1980b).

1977). In these **sensitivity tests** there were statistically significant temperature increases of 12°C in winter and 3°C in summer in the vicinity of the heat emission areas. More detailed investigations have shown that not only the temperature but also the vertical wind velocity, the precipitation and the soil moisture were changed, but **teleconnections**, i.e. effects over long distances, could not be found (Chervin, 1980).

In summary, for the expected future energy development a global climatic effect, as a result of heat emission is unlikely, but the local and regional climatic effects which are already produced by very high waste heat power densities, will be even more noticeable in the future (see also direct heat emissions in Table IV.3). A reduction of waste heat emissions is therefore advisable.

IV.5.3 Possibilities for waste heat reduction

Waste heat is "wasted" from two points of view. On the one hand, because of the energy required to get rid of it and, on the other hand, because of the missed opportunity of using this important energy potential (Kunkel, 1981). The continually increasing prices for primary energy have already brought about a change of philosophy.

In industry and conventional power plants the degree of efficiency with which energy is utilized has already been optimized to some extent (Schikarski, 1980). However, there are a number of energy policy developments that would allow a drastic reduction of the waste heat amount. For example, increased installation of **combined heat and power plants** with an efficiency of more than 80% could provide the heat and electricity needed in a region with a smaller primary energy amount than the conventional power plants with an efficiency of 40% at most. Also, the use of **heat pumps** for heating and warm water preparation could greatly reduce heat losses. While the ratio of useful heat to primary energy input is 0.32 for electric storage heating, it is 0.63 for a coke central heating, 0.98 for an electric heat pump and 1.33 for a natural gas heat pump (Herrmann, 1978). Economically acceptable measures, such as the improvement of efficiency, district heating and cogeneration, the introduction of heat pumps, the improvement of insulation and many others, are aspects of **energy saving** that are synonymous with a reduction of waste heat (for further details, see Chapter VII.4).

IV.6 Assessment of overall effects – critical threshold values – prospects

IV.6.1 Individual climate factors and the greenhouse effect

It is certain that mankind can influence local and regional climate through his activities. An influence on the global climate is presently not detectable in the observed data. The climatic effects are essentially a result of three influencing factors: changes of the composition of the atmosphere, modification of the surface characteristics of the earth, and direct heat addition. The previously identified factors that change climate include carbon dioxide and other carbon, nitrogen and sulphur compounds, chlorofluoromethanes, ozone and aerosols, as well as land use changes and waste heat. Table IV.3 in section IV.2 summarises these factors and the potential climatic effects.

All of these factors affect the radiation budget and can cause a varying warming or cooling, depending on the size of the perturbation. Fig. IV.26 shows the computed temperature changes for the perturbations indicated and permits a ranking of the most important influencing factors (Hansen et al., 1981). In order to have a uniform basis for comparison, all perturbations were computed using a 1-D radiative-convective model with constant relative humidity and constant cloud temperature. For a **CO_2 doubling** from 300 to 600 ppmv, for example, this climate model gives a temperature increase of 2.8^{o}C.

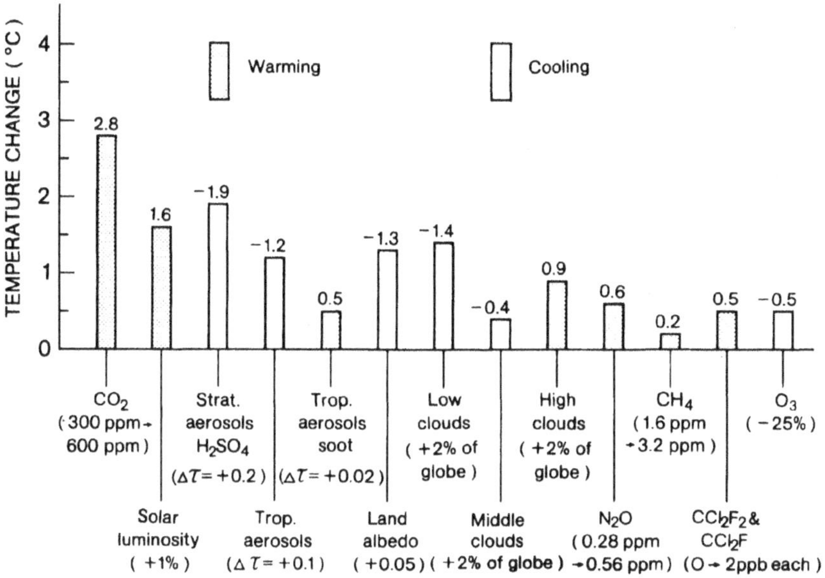

Fig. IV.26 : Potential radiative perturbations of climate.
Source: Hansen et al. (1981).
(reprinted with permission from *Science*, Vol. 213, p. 965, © 1981 AAAS)

An increase of **solar radiation** by 1% would increase the global temperature by 1.6°C. Since the effects of small radiation variations are linear, a change of only 0.3% would be enough to bring about a mean global temperature change of 0.5°C, which would roughly correspond to the warming effect of the entire CO_2 introduced into the atmosphere from 1880–1980. The newest measurement techniques are now fine enough to measure changes of the "solar constant" of only a few tenths of a per cent. These must definitely be considered in the search for a cause of climatic variability.

Aerosols that remain for many years in the stratosphere after a volcanic eruption have been shown to cause a noticeable cooling in the near surface atmospheric layer. Depending on their type and composition in the troposphere, aerosols can lead to a warming, if they consist mainly of highly absorbing soot particles, or to a cooling, if they are composed mainly of sulphate aerosols with a high albedo. As we have shown in Chapter IV.3 the climatic effects of aerosols are known with little certainty. Probably they are globally but not regionally negligible. Only a global measurement programme with detailed consideration of aerosol characteristics can provide further insights.

Settlement in the past and deforestation, overgrazing and desertification at present are reflected by changes in the **land**

surface albedo. Since the effects on climate could be considerable, strengthened efforts at a more detailed description of the rates of change, for example, through a satellite observation programme, are urgently required. Depending on their optical characteristics, extent and vertical distribution, **clouds** can cause a warming or a cooling effect. It is possible that the various feedback mechanisms compensate each other. This must be clarified.

The other **trace gases** that have been listed all lead, with the exception of ozone destruction, to a warming and thus strengthen the greenhouse effect of carbon dioxide. This analysis has shown that for the assessment of global climatic changes on a time scale of decades to centuries, all of the above-mentioned factors, especially the traces gases, must also be considered (see also Smith, 1982).

During the decade 1970–1980, the measured incremental increases have been 12 ppmv for CO_2, 150 ppbv for CH_4, 6 ppbv for N_2O, 190 pptv for CCl_2F_2, and 135 pptv for CCl_3F, respectively. Applying the same 1-D model used for the calculations in Fig. IV.26, Lacis et al. (1981) estimate a decadal greenhouse warming of 0.14°C for CO_2 and 0.10°C for the other considered trace gases. Hansen, Lacis and Rind (1983) conclude that the contribution of the sum total of the trace gases to the greenhouse effect is now comparable to that of CO_2. There is a consensus among climatologists that a substantial climatic change due to the greenhouse effect will become apparent during the next decade or two. At present we are left in the somewhat uneasy situation that we have clear evidence that significant climatic effects are imminent but are lacking the knowledge, or the tools, to accurately describe them. Major areas of uncertainty include the equilibrium climate sensitivity, the contribution of clouds, transport and storage of heat in the oceans, and the intricate feedback mechanisms within the climate system. According to Hansen and co-workers, the main research needs include both global monitoring and measurements of local processes guided by theoretical studies and climate modelling.

IV.6.2 Combined greenhouse effect and critical threshold values for the CO_2 and temperature increase

To calculate the combined greenhouse effect, Flohn (1978,1979) compares the **"real" CO_2 content** (without trace gases) with a **"virtual" CO_2 content,** in which the CO_2 effect is increased by 50% or 100%, due to the other trace gases. The equivalent level of "real" CO_2 is obtained by subtracting the increase due to the other trace gases. This means that a certain temperature increase is reached at a correspondingly lower CO_2 concentration and hence at an earlier point in time.

Flohn (1981a,b) recently combined this concept with some

model-dependent parameters from previous climate model results and derived critical threshold values for a certain atmospheric CO_2 content and the corresponding temperature change. The procedure is described in Appendix IV.3 and the results are illustrated in Fig. IV.27. The comparison with the conditions in the various warm phases of the past allows a meaningful assessment of critical threshold values. According to Flohn (1980a,b), if we want to keep our climate relatively stable, the equivalent CO_2-content (i.e. the other trace gases are considered) must not increase above 400–450 ppmv. This first **critical threshold range** corresponds to a warming of 1–1.5°C that has not occurred since the Middle Ages about 1000 A.D. **Catastrophic climatic changes** would be expected when the equivalent atmospheric CO_2 content reaches 600–700 ppmv (Flohn, 1981c). This would lead to a mean global temperature increase of 4–5°C, as occurred in the early Tertiary (5–3 MYBP). On the basis of palaeoclimatic indicators, Flohn postulates not only an ice-free Arctic Ocean but also a shift of the climatic zones and possibly irreversible climatic changes (see also section V.7.2).

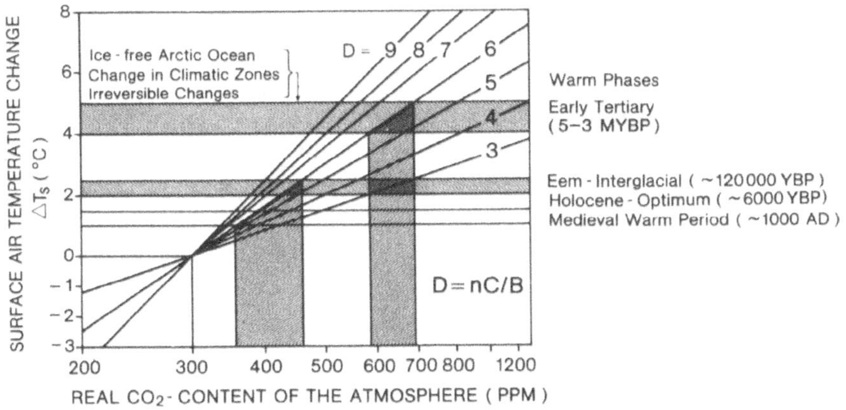

Fig. IV.27 : Critical threshold values for a CO_2 and temperature increase related to model-dependent sensitivity parameters. Source: Flohn (1981b).

IV.6.3 Outlook

In this extensive chapter I have attempted to illustrate the present state of scientific knowledge about the CO_2/climate problem. The use of this knowledge in the political decision-making process is necessary for a rational solution of the CO_2 problem. This is made clear by the frequently asked questions that are presented below, which serve as a starting point for the following Chapters V and VI which look at the effects of CO_2 on society and the measures to be taken to avoid these effects.

Is there presently a climatic effect of CO_2? This is a question with scientific content and policy implications. As a result of simulations of climate and comparison with analogue cases, we know that a CO_2/climate effect exists. However, at present it is smaller than the natural climatic variability and consequently not detectable in the observed data. This fact does not exactly favour the public interest in this important problem.

When and where will the CO_2 effect be detectable? The relatively large heat capacity of the ocean surface has the effect that a detectable CO_2-induced warming of the lower atmosphere is delayed by one to two decades. Model calculations show that for energy scenarios that could be realistic, the **CO_2 effect** should be detectable in the climate data by **2000** at the latest. In contrast to previous beliefs, recent information suggests that the **CO_2 signal** will be distinguishable from the noise of the observed data first in **summer** in the **middle latitudes.** The question of whether we can wait with our countermeasures until we have unambiguous climatic evidence of the CO_2 effects at the turn of the century depends on the answer to the following question.

How urgent is the CO_2 problem ? The average exponential growth rate of fossil fuel consumption was about 3.4%/yr in the past 100 years and, since the energy crisis of 1973/74 it has fallen to less than 2%/yr. The development of a large-scale synthetic fuels programme and the increasing return to coal, as a result of exhaustion of gas and oil reserves, could again lead to a greater rate of CO_2 increase. In addition, with the increasing population pressure and larger demand for agricultural land, even to the extent of a possible loss of large tropical rainforest areas, we must expect a considerable increase of the biogenic CO_2 emissions. Even if the growth rate remained at the low level of about 2%/yr over the coming decades, which many feel to be unlikely, the present atmospheric CO_2 content of 340 ppmv would still be considerably increased during the course of the next century. The first critical threshold value is, as we have seen above, at a mean global temperature increase of 1-1.5°C with a corresponding equivalent CO_2 concentration in the atmosphere of 400-450 ppmv (for explanation, see IV.6.2). Major adverse climatic events are expected to increase significantly with a global temperature increase of 4-5°C and an equivalent CO_2 level of 600-700 ppmv (which includes the expected rate of change of the other trace gases). Without corrective measures these threshold values will be reached sooner or later during the course of the next century. There is the danger that the measures to be taken must be more drastic the longer their introduction is delayed. Thus the **CO_2 question** has a **high level of urgency.**

When should countermeasures be introduced? If the consequences of the CO_2 increase expected in the next century are to be avoided, it is appropriate not only to cut the fossil fuel consumption but also to keep deforestation and reforestation in balance. One of the most effective measures is the more efficient use of energy, since it can immediately contribute to reducing the fossil fuel consumption (see Chapter VI.4). In addition, it is most sensible from an economic point of view. In contrast, the transition to energy carriers that produce little or no CO_2 requires several decades. A smooth transition to new energy carriers, that the global economy can cope with, requires time. Thus it can be concluded that the latter measures should also be introduced now.

The willingness to think about taking countermeasures depends strongly on how we judge today the **future potential threat.** Therefore, in the following Chapter we look at the potential effects of a CO_2-induced climatic change on environment and society.

Decisions will have to be made, and they should be made in the light of the best perceptions of the likely consequences.

Working Group report of the International Workshop on the Interactions of Energy and Climate, Münster, Germany

V. IMPACTS OF CLIMATE CHANGE ON SOCIETY

If we agree that a global warming is likely to result from the CO_2 increase and other factors, as shown in the previous Chapter, then it would be advisable to estimate already the possibly extensive consequences of such a climatic change for society. This is all the more urgent because, as we have seen in Chapter III, the long time required for the adjustment of the social systems means that decisions made today determine events in the distant future. The long-term prediction of climate and climatic variability (see Chapter II) and the long-term prediction of human behaviour (see Chapter III) will not be possible in the near future (Roberts et al., 1980). Nevertheless, far-reaching decisions are made daily. These should, however, be based on the best knowledge of the possible consequences. **Climate impact studies** can provide the necessary background for the objective evaluation of possible future risks and the need for and appropriateness of precautionary measures.

V.1 Climate impact programmes

Methods to assess impacts have been applied for some time in various environmental areas (Bach, 1980a). In recent years, both international (US NAS, 1978; WMO, 1980) and national bodies (e.g. the Climate Research Programme of the FRG, 1980) have emphasized the central importance of climate impact studies within climate programmes.

V.1.1 Nature and methods of impact studies

Climate impact studies investigate the interactions between climate and society. The purpose of such analyses is to estimate the consequences for society and environment of climate changes and climate variability, advertently or inadvertently caused by

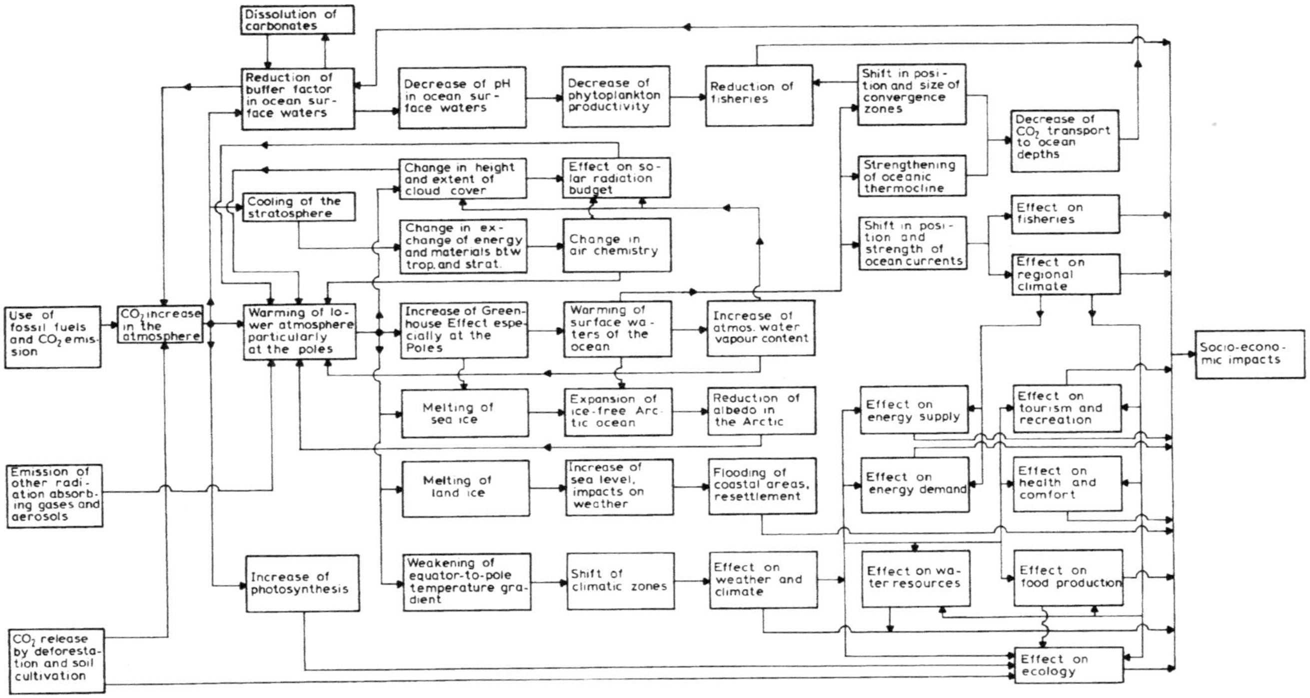

Fig. V.1 : System showing the CO_2/climate-interactions. With additions from: Markley and Carlson (1980).

human activities. They are **systems analyses**, which provide a synthesis of all relevant factors and thus make it easier for decision-makers to reach a rational decision.

Fig. V.1 shows the flow diagram of such a system with the various cause-effect linkages. By assuming various scenarios for the demand for fossil fuels, deforestation and soil cultivation, the atmospheric CO_2 increase can be derived. This CO_2 increase, together with the changes caused by other gases and aerosols that absorb in the infrared spectral region, affect all sub-compartments of the climate system and, as a result of complex feedback mechanisms, lead to a warming of the lower atmosphere and a cooling of the stratosphere. The resulting alteration of weather and climate leads to regional and seasonal changes of temperature, precipitation and soil moisture distributions etc. These then give rise to impacts on the ecologic and socioeconomic sectors, such as ecosystems, energy, food and water supplies, fisheries, land use, health and recreation. These impacts will be discussed separately in the following paragraphs. Finally, the socioeconomic impacts feed back on the total societal system. This requires political decisions so that the climate impacts can be reduced, if not altogether avoided, by precautionary measures. This important aspect of climate impact analysis is considered in Chapter VI.

Up to now there are only a few methods and these are unsatisfactory because of the complex problem and the difficulty of describing developments in the climatic and socioeconomic sectors with any certainty. In cases where it is difficult to describe future developments, **the scenario analysis** is particularly suitable as a research tool.

V.1.2 Climate scenarios

To avoid misinterpretation, it should be made clear that scenarios are not intended to be predictions. Their strength is rather that they present and show the limits of plausible and self-consistent future possibilities. Basically there are two procedures. One can use **historical climate research** (see also Chapters II.1 and IV.1.5) and analyse observed data or indirect climate indicators (see Appendix II.1) in order to derive from the climatic past analogues of the future (CO_2-induced) climatic changes. In the past natural climatic variations were the cause of climate anomalies. One cannot conclude from this that a future CO_2-induced warming would lead to similar anomalies, although this is not entirely impossible. The second procedure involves **climate modelling** (see Chapters II.3 and IV.1.7) which provides estimates of climatic changes as a direct result of a CO_2 increase or other influences. The comparison with past climatic events can provide additional verification.

The following research topics and methods could provide the

basis for climate scenarios (Roberts et al., 1980; Flohn et al., 1980):

- the early and middle Pliocene (about 5-3 million YBP*) when the mean global temperature was about 4°C higher than today (Fig. II.1; Appendix II.2);
- the Eem-Sangamon Interglacial (about 125,000 YBP) directly before the last major glacial with a mean global temperature about 2-2.5°C higher than today;
- the Altithermal or Hypsithermal warm period in the Holocene (about 6,000 YBP), the warmest period since the last glacial with a mean global temperature about 1.5°C higher than today (Kellogg and Schware, 1981);
- the medieval warm period (about 900-1050 AD), that was about 1°C warmer than today;
- the warmest years and seasons at various latitudes on the basis of reliable observations (Williams, 1980; Wigley et al., 1980; Rocznik, 1981);
- scenarios on the basis of dynamical (Gates, 1980; Manabe and Stouffer, 1980) and stochastic (Hasselmann, 1979, 1981) climate models (see also Chapter IV.1.7 and Appendix IV.1);
- a combination of these methods including numerical modelling, extreme warm and cold year periods, dynamical/empirical reasoning and palaeoclimatic reconstructions of the Hypsithermal (Pittock and Salinger, 1982);
- the translation of palaeoclimatic information into units that record changes in the biosphere, such as the transformation of tree ring widths into estimates of tree growth and productivity, pollen numbers into estimates of forest composition and biomass, and marine plankton counts into the availability of nutrients for fish and other marine life (Webb III, 1982); and
- the assessment of spatial similarities between modern July temperature variability over Canada north of 50°N for the period 1943-1972 and Holocene (about 6,000 YBP) midsummer temperature derived from palynological data covering a few centuries (Diaz and Andrews, 1982).

The studies of Williams, Wigley et al. and others, reached conclusions that have an important bearing on future impact analyses. They showed that temperature and precipitation changes as a consequence of a warming have very large regional differences, that they are unevenly distributed seasonally, and that not all regions have the same magnitude and direction of change.

---

* YBP = years before present

V.1.3 Socioeconomic scenarios

We do not yet have the necessary tools for describing the chain of effects: CO_2-increase – change of climate – socioeconomic effects – reaction of society. There are, however, already a number of studies that have used conventional economic methods, such as **cost-benefit analysis** and various **econometric methods** (Ft. Lauderdale Conference, 1980).

For the more quantitatively oriented branches of the national economy, the time period significant to the CO_2 problem is unusually long. Classical cost-benefit analysis attempts to partially eliminate the increasing uncertainties over such a long time period by choosing an appropriate discount rate. This method appears, however, to be unsuited for the purpose, since it means that future costs and benefits are always considered to be smaller than present costs and benefits and because the choice of a discount rate assumes a constant future behaviour of society and thereby certainly wrongly estimates the expected changes (Smith, 1982).

Some have felt that these deficiencies can be avoided with the **"willingness-to-pay" method** (d'Arge et al., 1980). Here one asks how much the present generation is prepared to pay to protect following generations from CO_2-induced damages, or to provide compensation for these damages. Since this method developed from the original cost-benefit analysis, it is very questionable whether it reduces the uncertainties about the projected effects and countermeasures. The method can certainly be used to answer questions of the present day but not questions of the future such as the CO_2 problem. It is necessary to improve further the above methods, or develop new procedures before they can be usefully applied in impact analyses or in the decision-making process.

V.2 Ecosystems

Here we want to consider the impact on the less-managed ecosystems or biomes (Odum, 1971). The main types include forest, savanna, grassland, tundra, alpine semi-deserts and deserts. The biomes are strongly determined by the existing natural factors, such as soil type or the availability of particular nutrients in the soil, and they respond rather sensitively to the influence of climate and human intervention.

V.2.1 Influence of mankind on the biomes

Globally, Man's activities have reached an order of magnitude that can significantly influence the regional and global processes within the biosphere (Bolin, 1979). Compared with natural rates of change, these interventions are so abrupt that they disturb the natural equilibrium. The response of the natural system to reachieve balance is, however, too slow, so that once the equilibrium has been disturbed, the perturbation can persist for a long time.

There are hardly any natural ecosystems that have not been influenced by Man's activities. Some of the most threatened biomes are tropical rain forests, because of land requirements and the opening of commercial pastures (Myers, 1979), and the increasing demand for timber (Baumgartner, 1979) and fuelwood, especially for food preparation (Revelle, 1980). Likewise, steppes and semi-deserts are threatened where overgrazing and poor soil cultivation methods destroy the sparse vegetation and the upper soil layer (Otterman, 1974; Mensching, 1978).

V.2.2 Response of the biomes to climate changes

The individual biomes are strongly dependent upon climate and can be classified according to the mean annual temperature and mean annual precipitation, as shown in Fig. V.2. Our knowledge about the ecosystem's response to a CO_2-induced climate change is mostly based on analogues drawn from historical observations and climate modelling.

In a very carefully made study, Nicholson and Flohn (1980) describe the climatic and environmental changes at the end of the Pleistocene and in the Holocene. They show that the warm phases in the Holocene coincided with two moist periods in the Sahara, leading to a reduction of the desert area. In the first moist period, about 9,500 yBP, there was a relatively dense game-hunting population in the Sahara, and in the second moist period, about 6,000 yBP, cattle-rearing nomads roamed the desert.

The question then arises whether the atmospheric CO_2 increase caused by human activities would lead to a similar warm and moist period. Can climate history be repeated? That depends on whether the boundary conditions from the past could be exactly reproduced today. It is, however, clear that, since the two Holocene warm periods, three significant boundary conditions have changed considerably, namely the distribution of ice and snow, the changes and destruction of the natural vegetation cover and the composition of the atmosphere.

With these changed conditions, the following could be expected with a warmer and moister climate in the Sahara-Sahel belt and the Mediterranean area: a possible extension of the tropical summer rain into the southern Sahel region; no persistent precipitation increase, however, in the central and northern Sahara; and in the Mediterranean countries, a precipitation decrease. Nicholson and Flohn (1980) emphasize that the present dry climate in the southern Sahara-Sahel zone can only be improved in the long term by significant changes in the natural climate system with a strong convergence of warm water currents at the equator and by a drastic reduction of the overgrazing.

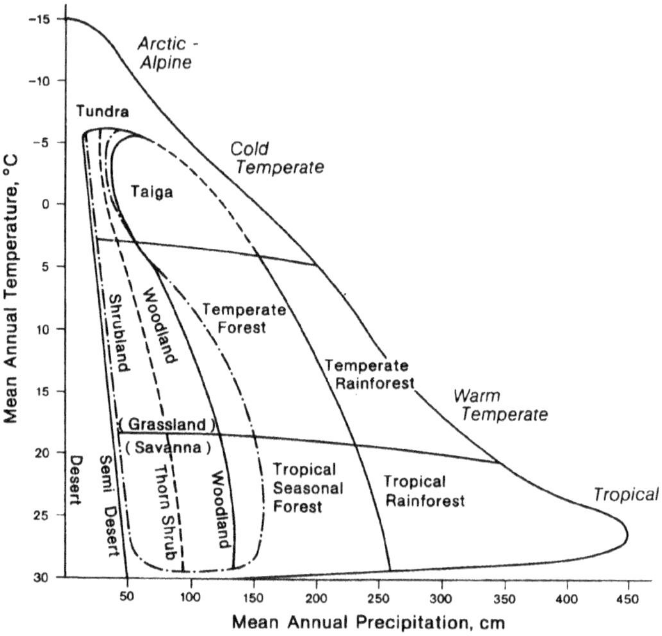

Fig. V.2 : World biome types in relation to mean annual precipitation and temperature. The dot-and-dash line encloses a wide range of environments in which either grassland or one of the types dominated by woody plants may form the prevailing vegetation.
Source: Whittaker (1975).
(reprinted with permission from *Communities and Ecosystems*, 2nd. edn., ©1972, 1975 Macmillan Publ. Co.)

Further analogue studies include those reported by Budyko (1982) for Europe and Asia of the Pliocene (ca. 5-3 million yBP), which would, according to him, apply to the potential warming expected in the 2020s. They indicate an increase in air temperature over present conditions by 12-15K in July and 15-20K in January in the Arctic region, and by 2-5K in July and 15-20K in January in mid-latitudes, and a shift of the zero isotherm in January northward by 10-15 degrees of latitude. This could produce temperature conditions in the northwest of the European part of the USSR that would be similar to those in present-day central France; those of the northern part of Siberia would be similar to southern Poland, and those of the central parts of western Siberia would be similar to those in the middle Danube lowlands. One should take note, however, that these values are considerably higher than those obtained by other research workers.

Climate modelling provides a further important method for estimating the effects of a climate change. Unfortunately, even the most sophisticated present climate models (e.g. those presented in Chapter IV.1.7.1) are not yet realistic enough to provide reliable predictions, with the temporal and spatial detail required for the assessment of most CO_2-induced climatic impacts (US NAS, 1982). But, by suggesting the scales and ranges of seasonal and regional variations, they can still be valuable in scenario analysis of possible climatic changes. This type of important work has just begun. At this point it may suffice to present some preliminary results from a different approach by Sergin (1980). He introduced into the hydrodynamic and thermodynamic system of equations for the atmosphere and the ocean some dimensionless similarity parameters which are determined by the climatic elements of the system and the characteristic space and time scales of the processes involved. The results of Fig. V.3 show how the annual average precipitation in Europe, parts of Africa and Asia, could change with a temperature increase of 2^oC. According to the results of this study, a warming would be accompanied by a reduction of precipitation in the dry areas of the Mediterranean area and Near East. In contrast, the precipitation could increase in the southern Sahel area, and in central, north and north-east Europe as far as Siberia, where precipitation is already sufficient.

What does this mean for ecosystems? Forests need more rainfall than many other vegetation types. Especially in semi-arid areas, the trees are very sensitive even to small precipitation decreases (Walter, 1973; Kellogg and Schware, 1981). The individual tree sorts are optimally adapted to the prevailing local climate. Adaptation to climatic changes that are too fast (caused by a possible CO_2 increase within a few decades) is unlikely (Phares, 1980). As a consequence, a significant change of the global climate in any direction would reduce the yields of the forests for a long time.

A CO_2-induced warming would have a large influence on the tundra (Kellogg and Schware, 1981). As a result of a warming, the permafrost and tundra would shift further north allowing trees to grow further poleward. However, the upper swamp and peat layers would simultaneously dry out and the oxidation would release the stored carbon as CO_2 into the atmosphere, thus amplifying the warming trend. This positive feedback loop through: warming - stronger drying out - increased CO_2 release - further warming, must be taken seriously especially in northern latitudes because there the influence of a warming on the global climate system is particularly large.

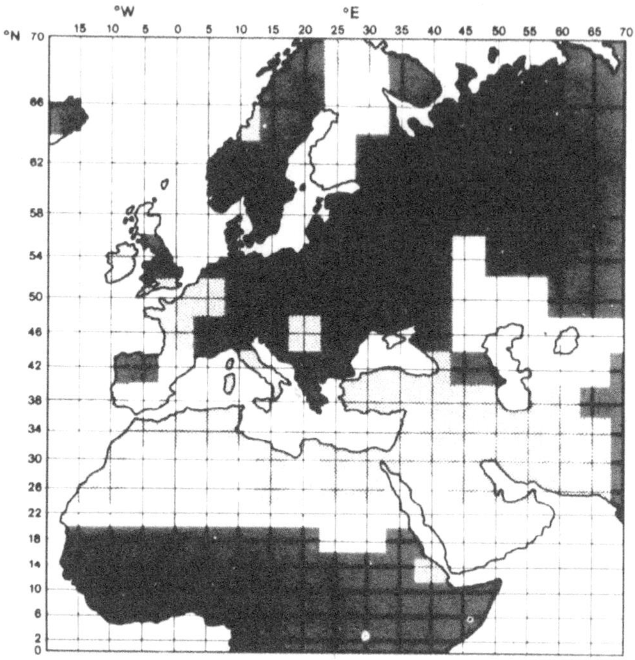

Fig. V.3 : Estimate of the average annual precipitation distribution for a warm climatic period in comparison with recent times. Hatched areas have an increase, dotted areas have a decrease, and blank areas have no change in precipitation. Source: Sergin (1980). (reprinted with permission from *Climatic Constraints and Human Activities*, J. Ausubel and A. K. Biswas (eds.), ©1980 Pergamon Press Ltd.)

V.3 Energy use

Climate is an important factor in energy supply and demand. This has become clear since energy became expensive. We need more energy for heating in particularly cold winters; we need more energy for air conditioning during summer heat waves; and we need a lot of energy to protect us from storms, snow, ice, heavy rainfall and other elements of the weather and climate.

Weather and climate also influence energy extraction. The exploration and mining for oil and natural gas in the offshore areas, the mining of coal, the transport of oil and liquefied gas over the oceans, and the distribution of fuels and electrical energy to consumers, can all be influenced by climate. This has often enough led to serious supply difficulties. Even when the most advanced technologies are used, the influences of weather and climate cannot be excluded, and the vulnerability appears rather to increase with the sophistication of the energy supply technology.

Climate is, however, not only a **hazard** for the energy supply, it is also a **source** of energy. Individual climatic elements, such as solar radiation, wind and precipitation etc. determine the amount and availability of renewable energy sources, such as solar, wind, bio-, and ocean energy, as well as hydropower. The availability of renewable energy sources depends on regional climate. The related unreliability is compensated by the large variety of supply options.

This short review has shown that weather and climate can influence our energy supply. Appendix V.1 shows a compilation of the climatic impacts on the various aspects of energy supply and demand. The differentiation between short-term and long-term effects was found to be practical. As shown in the CO_2/climate systems analysis in Fig. V.1, the expected warming due to the CO_2 increase should result in long-term (over decades) weather and climate anomalies, which could influence energy supply on the short-term and variously on a regional scale.

The following paragraphs consider how the climate affects energy demand and supply and what additional costs could arise through climate variations.

V.3.1 Climate and energy demand

The proportion of the total energy used to compensate for climate influences through heating or cooling amounts to about 30% in North America and about 50% in Europe (McKay and Allsopp, 1980). As shown in Fig. V.4, the main energy demand for heating lies between the 2°C and 18°C **annual isotherms.** In both the northern and southern hemispheres, the main industrial areas lie in this zone with 85% of the global energy demand. When the annual mean value is greater than 18°C cooling is necessary instead of heating, especially in the regions between the 18° isotherms of both hemispheres. Polewards of the 2° isotherm, the heating requirements increase more strongly but the population is small here so the total demand is negligible. A CO_2-induced warming would shift the isotherms polewards, which would lead to a reduction of heating requirements there but also to a simultaneous increase of the cooling requirements in southern latitudes (Kellogg and Schware, 1981).

The heating and cooling requirements of a region can be described using the **heating-degree-day** method. The mean daily temperature is derived and the number of degrees above or below a base temperature are cumulated. For example, the base temperature for heating (i.e. the temperature threshold at which heating is required) is between 15°C and 19°C in most countries. It depends on a number of factors, such as, for example, the house construction (especially the insulation), the condition of the building, and varying lifestyles. Heating-degree-days have proved to be good indices for the climatically-required heating and cooling. It should be noted that for determining the cooling demand, the relative humidity must be considered in addition to the temperature and that the heating requirement is primarily met

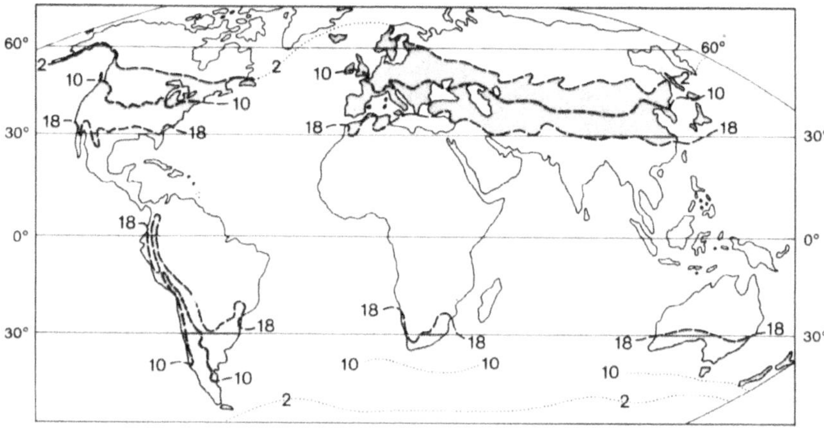

Fig. V.4 : Heating demand and mean annual temperature.
Dotted area lies between 2°C and 18°C.
Source: McKay and Allsopp (1980).

by oil, gas or coal, while the cooling requirement is met exclusively by electricity.

A few examples serve to show how strongly the heating energy demand is influenced by climate. In the USA, the winter of November–March 1976/77 had 22% more heating-degree-days than the mild winter of 1975/76, corresponding to a temperature difference of about 1.8°C (Quirk and Moriarty, 1980). That caused an additional heating oil requirement of about 56 million tons, which corresponded to about twice the amount of oil that was imported into the USA in 1978 from Iran. The impact of climate on fuel demand for heating and hence on cost, can, however, be somewhat reduced through efficiency improvements, as demonstrated by Thornes (1982), in a case study for Birmingham University. He found a reduction in heating fuel requirements of 22% during the heating season 1981/82, as compared to 1974/75, despite the very cold weather in December and January of 1981/82. These examples show that climate anomalies can affect considerably the energy planning.

V.3.2 Climate and energy supply

The continually increasing costs of conventional energy sources and the serious climatic and environmental problems that are associated with their use, make the increased use of **renewable energy resources** more and more attractive (Lovins, 1978; Hayes, 1979; Sorensen, 1979). As shown in Fig. V.5 the supplies of most of the renewables are associated with particular climate zones.

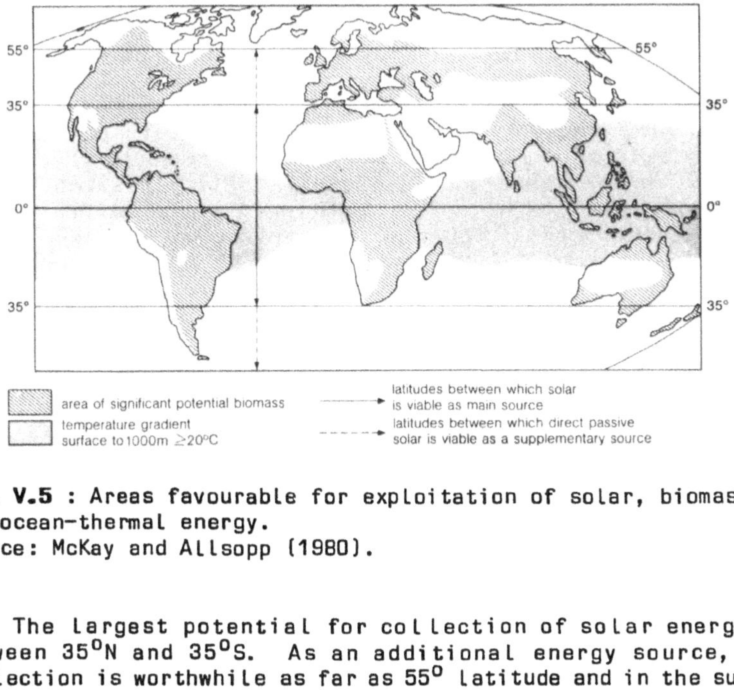

Fig. V.5 : Areas favourable for exploitation of solar, biomass and ocean-thermal energy.
Source: McKay and Allsopp (1980).

The largest potential for collection of solar energy is between 35°N and 35°S. As an additional energy source, the collection is worthwhile as far as 55° latitude and in the summer months even further northwards. Because of the reduction of energy supply on cloudy days, storage and a wide variety of options are needed. The use of ocean temperature gradients for energy supply is worthwhile when the surface water temperature is about 25°C (see the dotted areas in Fig. V.5). The use of wave energy is restricted to ice-free water; the use of wind energy has best results in coastal zones and mountainous inland areas that are affected by the prevailing wind systems such as the westerlies, the trade winds or the monsoons. The supply of hydropower, but not its use, requires an area with heavy precipitation; the growth of biomass for the supply of bioenergy is dependent upon certain temperature and precipitation limits (see the hatched areas in Fig. V.5). If there are sufficient soil nutrients and soil moisture, the biomass production could double with a temperature increase of 10°C within a mean annual temperature range of -10°C to +20°C (Lieth, 1973).

In the absence of detailed model studies, it is not possible to estimate the extent to which a CO_2-induced climate change would influence, either positively or negatively, the supply of renewable energy sources, since, as we have seen in Chapter IV.1.7, the resulting temperature, precipitation and soil moisture distributions, among others, could differ greatly, both seasonally and regionally. If renewable energy sources are included in a total energy supply system, one must allow for the fact that weather processes and climate anomalies are connected over very large distances (teleconnections) and could simultaneously influence the energy supply options. For example, during the 1976/77 winter, already referred to above, a stationary blocking anticyclone caused an unusually cold period in the eastern half of the USA and relatively warm and extremely dry weather in the western half (Quirk, 1981). The empty water reservoirs in the west could not produce any more electricity and from the eastern half, which suffered a long cold period, no more oil and gas reserves could be diverted to the west. Such frequently occurring situations can be prevented by developing a great variety of energy supply options.

V.3.3 Climate and energy costs

The cold winter of 1976/77 caused expenditures of about \$6 billion for the additional energy requirements in the eastern half of the USA. The simultaneous drought gave rise to costs of about \$400 million in California alone for the additional energy requirements (Quirk, 1981).

The question remains, how a CO_2-induced climate change could affect energy costs. D'Arge (1979) has looked at this question and has calculated the changes in domestic energy costs for various regions of the USA and a range of temperature changes (Table V.1). The results show, for example, that for a temperature increase of 2°C (the assumed value for a CO_2 doubling), only the northeast coast at Portland and the west coast at San Francisco would experience a cost reduction, while all other towns and regions would be faced, contrary to expectations, with a considerable price increase (see, for example, Tucson with \$45 per year and household). The reason for this is the assumed rapid increase, in southern regions, of air conditioning units which run on more expensive electricity.

In summary, we can see that with a warming, the temperature zones and corresponding energy demand regions do not shift uniformly and simultaneously polewards. Instead, quite different patterns of regional distribution can develop (see Chapter IV.1.7). Varying seasonal temperature changes can likewise lead to quite different cost levels. In addition, the effects of other climatic elements, such as snow amount and duration, precipitation and wind become increasingly significant. If the estimate of costs for the US is realistic, then a cost increase and a shift in energy demand from the primary energy carriers oil and gas to the secondary energy carrier electricity (because of air conditioning), is to be expected. The US cost estimates cannot, however, be automatically applied to the Federal Republic of Germany because of the much cooler summers in the latter.

Table V.1 : Changes in expenditure for energy in selected U.S. cities for given changes in temperature (dollars per year).

City	Latitude N	Mean Annual Temperature in °C (1977)	Change in spending per customer given temperature change -2°C	-1°C	+1°C	+2°C
East Coast						
Portland, ME	43	6.9	+13.7	+ 6.2	- 4.9	- 8.6
New York, NY	40	11.3	-12.7	- 7.3	+ 9.2	+20.3
Baltimore, MD	39	13.5	- 7.7	- 4.5	+ 5.6	+12.4
West Coast						
San Francisco, CA	37	13.7	+18.0	+ 8.5	- 7.5	-13.9
Los Angeles, CA	34	17.5	+ 3.7	+ 1.4	+ 0.5	+ 0.02
South and Southwest						
Shreveport, LA	32	18.3	-23.2	-12.2	+13.3	+27.9
Tucson, AZ	32	20.8	-47.4	-20.7	+22.0	+45.3
Bakersfield, CA	35	19.9	-27.8	-14.4	+15.4	+31.8
Great Lakes Area						
Madison, WIS	43	7.6	+ 2.5	+ 0.8	+ 0.1	+ 1.1
Rocky Mountains						
Cheyenne, WYO	41	7.9	+ 2.9	+ 0.9	+ 0.3	+ 1.6

Plus denotes an increase, minus denotes a decrease of costs.

Adapted from: d'Arge (1979).

V.4 Food security

One of the most important human problems is the provision of sufficient food for the growing world population (Bach et al., 1981). Considering the fact that at present about 400 million people are undernourished and that the world population will increase by a further 2 billion in the next 20 years (see Chapter

III.1), the guaranteeing of food supply for the human race appears to be an almost impossible task.

Production, storage, distribution and consumption of foodstuffs depend on many interrelated factors. The most important influencing factors include arable land, work force, capital, energy, technical development, population size, social structure, economic development and, not least, the climate (see Fig. V.1), that represents one of the most uncertain variables in the whole food production system (Schneider and Bach, 1981).

The need to increase agricultural yields by introducing particularly high-yielding, but at the same time climatically-specialized types, has increased the vulnerability of food production to climatic variations. In addition, climate anomalies can simultaneously occur in the various producing lands. This can lead to a global food shortage and in the poor developing countries to famine.

V.4.1 Climate and the world food situation

Since some 70% of the world nutrition comes from grains (Coakley and Schneider, 1976), the changes in the world food situation are demonstrated by the altered world grain trade. As shown in Table V.2, a significant change has taken place in the past 40 years, in which the most populous regions have turned from being net exporters to strong net importers of grain. In the USSR alone (i.e. without the other East European countries), total grain imports have dramatically increased from a low of 755,000 tons in 1971 to the present (1982) all time high of 46 mill. tons, which amounts to nearly 30% of the indigenous grain production (Brown, 1982). The imports for Africa, where in recent times hunger was often a great problem, especially in the Sahel region, are probably so low only because the grains offered on the world market are too expensive for the poor countries of this region. During at least the last 10 years, only two regions managed to produce a grain surplus, namely USA/Canada and Australia/New Zealand. Climate anomalies and climate variations in these "bread baskets" would have catastrophic consequences for a large part of the world population.

The safest way to insure against a climate-induced hunger-catastrophe is the establishment of grain stocks in the consuming countries (Schneider and Mesirow, 1976). A compilation of the world grain reserves shows, however, that they have been dramatically reduced in the last 20 years (Brown, 1978, 1981). In 1960 the world population could have been fed for 102 days with the available grain stocks, whereas the stocks would have lasted only 40 days in 1980.

The world food situation does not look at all encouraging and this will not change much in view of the rapid population increase. Therefore, the question whether, in addition to the natural climate influences that are already effective, a CO_2-induced climatic change can influence the food situation is of more than academic interest.

Table V.2 : The changing regional patterns of world grain trade (mill. tons), 1934-1980.

Region	1934-38	1948-52	1960	1970	1976	1978	1980
North America	+ 5	+23	+39	+56	+94	+104	+131
Latin America	+ 9	+ 1	0	+ 4	- 3	0	- 10
Western Europe	-24	-22	-25	-30	-17	- 21	- 16
Eastern Europe and USSR	+ 5	0	0	0	-25	- 27	- 46
Africa	+ 1	0	- 2	- 5	-10	- 12	- 15
Asia	+ 2	- 6	-17	-37	-47	- 53	- 63
Australia and New Zealand	+ 3	+ 3	+ 6	+12	+ 8	+ 14	+ 19

Plus sign indicates net exports, minus sign indicates net imports.

Adapted from: Coakley and Schneider (1976); Brown (1980, 1981).

V.4.2 Climate and harvest yields

Model calculations show (Chapter IV.1.7) that the CO_2-induced warming increases with latitude and the zones suited for agriculture are thereby shifted polewards (US NAS, 1977a). There is a rule of thumb that a change of the summer surface air temperature of 1°C lengthens or shortens the vegetation period by about 10 days (Kellogg, 1978). Moreover, although the relationship may not be linear, it is assumed that a 1°C temperature increase might result in a decrease of about 0.3% in organic matter and about 0.02% in the nitrogen content of soils (Tucker, 1982).

The acceleration of the hydrological cycle that is expected with a warming (see Chapter IV.1.7) has opposing effects. On the one hand, some regions could profit from increased precipitation, while, on the other hand, the increased evapotranspiration could also reduce the yields. It is generally valid that those cultivated plants, that give the highest yields, have become optimally adapted in the course of time to the climatic conditions and soil characteristics where they grow (Andreae, 1980). Highly developed agricultural systems have always taken advantage of this. A shift of the climate zones could, therefore, lead to considerable losses because it forces change of location and, therefore, the loss of the optimal cultivation conditions. The world's main area with grain surpluses would be particularly affected, i.e. the Midwest of the USA with its zonally arranged monocultures of corn (maize), wheat and peanuts. Newman (1982) has shown that the North American Corn Belt would shift in a SW-NE direction about 175 km/^{o}C change in growing season temperatures. Serious crop failures have been shown to result, for example, from moving a soybean variety as little as 160 km north because the crop would flower at the wrong time (Haynes, 1982). To think that crop species could simply be moved in response to temperature changes to less suitable soils in the north formed by glacial erosion would be overly simplistic. A reduction in yield is highly likely.

It is important to note that every type of cultivation responds differently to the individual climatic influences. Here the results for just the three most important grains are summarised. More detailed descriptions are given, for example, by Bach (1978, 1979a), Schneider and Bach (1980) and Bach et al. (1981).

Agricultural productivity depends on such things as technical innovations (e.g. new equipment, irrigation methods, cultivation of new and more resistent species, varieties of fertilizers and pesticides); environmental stress (e.g. pests); social factors (e.g. agricultural and economic structure, land ownership) and, not least, on climate variability. Using a semiempirical model for describing the **US corn production** with historical climate data since 1890 and by fixing the technological factors to the state in 1973, McQuigg et al. (1973) succeeded in isolating the climatic influences. The results in Fig. V.6 show

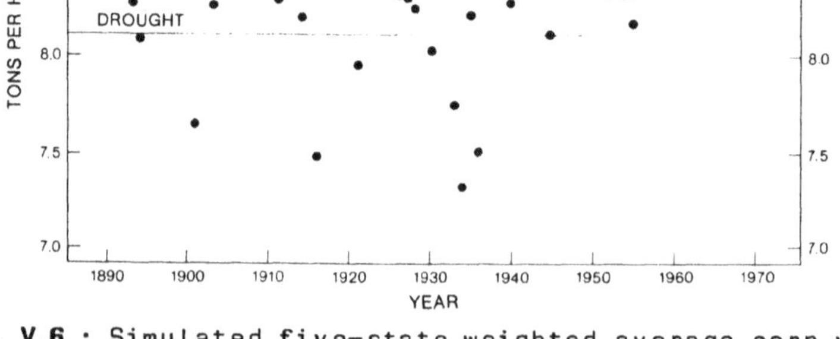

Fig. V.6 : Simulated five-state weighted average corn yields using 1973 technology and harvested average: Ohio, Indiana, Illinois, Iowa, Missouri.
Source: McQuigg et al. (1973).

clearly the very low corn yields (less than 8 t/ha) for the drought periods of 1930-40 with the notorious dust storms. In contrast are the higher yields (more than 9 t/ha) for the so-called high-yield period of 1957-1972. During the drought period precipitation was under and temperature was above the long-term mean value, while in the high-yield period this was exactly the opposite. There are considerable differences of opinion regarding the proportion of the yields from the high yield period that were due to the favourable weather and the share due to other factors. McQuigg et al. assert, however, that the probability of the occurrence of a further 15-year period with such a favourable climate and with similarly high yields is 1 to 10,000.

How would a CO_2-induced climate change affect the corn production? Table V.3 shows, for the US corn belt, that with a temperature increase of 2°C - that is expected during the course of the 21st century - regardless of whether precipitation increases or decreases, the corn production would decrease by 20-26%. Changes of this magnitude must have serious economic consequences. In general, in the American "bread basket", a cooler and moister climate would increase corn yields, and warmer and drier conditions would reduce yields (Benci et al., 1975).

Wheat is, like corn, one of the most important grains in the world. It differs from the other grains in that it needs a longer growth period and somewhat higher minimum growing temperature and the various types respond differently to temperatures that are too high or too low. During the growing period between flowering and grain development, wheat is particularly sensitive to heat (>35°C).

A similar change in temperature and precipitation to that above for corn would lead to about 10% yield reduction in the US wheat belt (Table V.3). Other important regions for wheat production, such as Kazakstan in the USSR, would suffer a yield reduction of even 20% for a temperature increase of 1°C and a precipitation decrease of 10% (Ramirez et al., 1975).

Rice is the principal food for the poorest and most populated regions of the world and guarantees survival for a third of the human race. According to the model calculations in Chapter IV.1.7, the temperature increase in the low latitudes would be small and the hydrological cycle would be accelerated. This could lead, based on the results shown in Table V.3, to an increase between 10 and 16% in the global rice yield (Stansel and Huke, 1975).

According to this review, it appears that a possible CO_2-induced climate change would benefit the world rice production and be disadvantageous to the corn and wheat production in the USA and USSR. The European grain harvest would probably be less strongly affected because of the smaller moisture variations. It should, however, be realized that this assessment is based on statistical models which predict crop yields from various climatic variables using empirical relationships derived from historical yield and climate data. The coefficients of these models are not universal constants but rather statistical estimates and are therefore conducive to several sources of error (Katz, 1977).

It is possible that the agricultural technology could adapt to climate variations by cultivation of new grain sorts. In this case, it is important that the climatic change does not occur **too fast**, so that the plants have enough time to adapt.

Table V.3 : Possible effects of climatic change on the average yields of corn, wheat, and rice in the major growing regions expressed in percent of average yield.

Change in Precipitation (%)	Change in corn yields (%) in U.S. corn belt[1] for a temperature change (°C) of				
	-2°	-1°	0°	+1°	+2°
-20	+19.8	+ 8.4	-2.9	-14.2	-25.6
-10	+21.2	+ 9.8	-1.5	-12.8	-24.2
0	+22.7	+11.3	0	-11.3	-22.7
+10	+24.2	+12.8	+1.5	- 9.8	-21.2
+20	+25.2	+14.2	+2.9	- 8.4	-19.8

	Change in wheat yields (%) in U.S. wheat belt[2] for a temperature change (°C) of					
	-2°	-1°	-0.5°	+0.5°	+1°	+2°
-20	0	+0.6	-0.4	-3.4	-5.5	-9.7
-10	+2.5	+1.8	+0.7	-2.2	-5.8	-8.6
+10	+3.2	+3.1	+1.4	-1.7	-3.6	-7.8
+20	+3.2	+2.0	+1.0	-2.0	-3.7	-8.7

	Change in world rice yields (%) for a temperature change (°C) of					
	-2°	-1°	-0.5°	+0.5°	+1°	+2°
-15	-19	-13	-8	- 4	0	+ 3
-10	-17	-11	-6	- 2	+ 2	+ 5
- 5	-13	- 7	-2	+ 2	+ 6	+ 9
+ 5	- 9	- 3	+2	+ 6	+10	+13
+10	- 5	+ 1	+6	+10	+14	+17
+20	- 3	+ 3	+8	+12	+16	+19

Plus sign indicates yield increase; minus sign indicates yield decrease.

[1] The U.S. corn belt includes: Indiana, Illinois, Iowa, Missouri, Nebraska, Kansas.

[2] The U.S. wheat belt includes: Indiana, Illinois, Oklahoma, Kansas, South and North Dakota.

Source for corn yield (Benci et al. 1975); for wheat yield (Ramirez et al. 1975); for rice yield (Stansel and Huke, 1975).

Agriculture is not so much damaged by a general climate variation but more by an increase in the weather extremes and annual variability that are expected to accompany the change.

V.4.3 Climate and pest management

The above statements did not consider the impacts of insects, weeds, plant diseases and nematodes, which are major plant pests (Haynes, 1982). The annual damages in agriculture and forestry are, however, considerable. Even in the USA, despite all chemical and non-chemical control methods, about 37% of the agricultural and about 25% of the forestry products are lost (Pimentel, 1980).

The increase of annual pests is strongly temperature dependent. For example, some female insects produce 500–2,000 offspring within 2–4 weeks. A warmer and longer growth period could lead to 1–3 additional generations and thus to an exponential increase of the pests. Pest mortality would be reduced by winters that are too warm. Reduced snowfall and stronger frost would have the opposite effect.

Higher temperatures could also unfavourably influence insect control, because the effectiveness of pesticides decreases. The same holds for weed removal, since the physiological activity of the weeds is much reduced when the conditions are too warm and dry and the toxic power of the herbicides cannot be fully effective. With warm and moist conditions plant illnesses increase very fast and lead to considerable harvest losses. It is likely that a milder climate (with milder winters and a longer growing season) makes the problem of plant protection more difficult. It has also been hypothesized that weeds may be able to compete more effectively for nutrients, light and space, with enhanced CO_2 concentrations (Dahlman, 1982). While one may be able to control weeds through agricultural pest management, this may not be feasible in non-agricultural ecosystems.

Climate factors can, however, also play a useful role in so-called **integrated pest management.** For example, a type of nematode that does not cause damage and that reproduces especially well with high temperature and humidity could be bred and then used for the destruction of damaging insects (Klingauf, 1981). Some of the possibilities for reducing the risk of climatically-induced yield reduction as a result of pests are biological/climatological control methods such as those described here, improved methods of cultivation and the maintenance of genetic diversity (Hekstra, 1981). Plant breeding, which helps to improve adaptation to temperature and water stress, can be supplemented by novel crop management which improves genotypic resistance to environmental stress (Howell, 1982).

V.4.4 CO_2 fertilisation and plant productivity

The scientific community is strangely divided over the beneficial aspects of a CO_2 increase on plant growth. On the one hand Goudriaan and Ajtay (1979) were not able to find any indication of a change in the net CO_2 assimilation rate due to an observed CO_2 increase. Similarly, Van Keulen et al. (1980) and Strain and Armentano (1982) suggest that too much optimism regarding the beneficial effects of a CO_2-enhanced atmosphere appears to be unwarranted, because under natural conditions nutrient supply (especially N), water, light and habitat space seem to be the main limiting factors for the rate of biomass production. Pimentel (1981) thinks that it is highly improbable that the reduced yield, due to a CO_2-induced warming, could be compensated by a CO_2-induced potential growth in biomass.

On the other hand, exposing plants such as corn and soybean, as well as trees, such as loblolly pine and sweetgum, in open top cylinders in the field over a 3-month period to CO_2 concentrations ranging from 340-910 ppm, Rogers et al. (1982) could show that growth, water use efficiency, and the total dry weight

transpiration in the case of cotton (Baker, Allen and Lambert, 1982). Moreover, conducting an intensive literature search, Strain (1982) finds 31 reported responses, of which the most important ones include changes in tolerance to temperature and

wheat, to a large change in photosynthesis and little response in transpiration in the case of cotton (Baker, Allen and Lambert, 1982). Moreover, conducting an intensive literature search, Strain (1982) finds 31 reported responses, of which the most important ones include changes in tolerance to temperature and carbon allocation to organs, increased tolerance to air pollutants, changes in germination and seedling development, and changes in shadow shape and habitat structure. Very importantly, since each type of organism responds differently, it is probable that ecosystems will change structurally and functionally in the future, thereby causing major shifts in the array of ecosystems on the landscape (Strain and Armentano, 1982).

Since it is difficult to maintain an artificially enriched CO_2 atmosphere around crops against the forces of turbulent transfer, open field studies remain, however, inconclusive (Rosenberg, 1982). In order to reduce some of the uncertainties surrounding the response of vegetation to increased CO_2, the US DOE has initiated an intensive research programme whose objectives are to assess the potential for increased crop productivity and to predict the change in yield of major crops as a function of enhanced CO_2, to estimate the net carbon storage in the biosphere due to CO_2 fertilisation, and to identify the potential effects of CO_2 on competition, composition and other relationships among agricultural crops and natural plant communities (Dahlman, 1982).

V.4.5 Learning from past experience

A comparison of the climatic history of central Europe with the **grain price index** (Flohn, 1978a, 1981a) and the **grape harvests** (Pfister, 1981) shows clearly that climatic extremes in the past 400 years have always led to price increases. History shows that climatic anomalies often resulted in famines and social unrest. For example, the droughts and poor harvests of 1788/89 in France were one causal element of the French Revolution; and the cold and wet years between 1845 and 1850 gave rise to the potato famine in Ireland and forced almost an entire nation to emigrate to the USA, and they were an influencing factor in the Revolution on the European continent in 1848. Even with today's widespread technology, climate continues to affect the agricultural and price sector. In the USA, which is so far the only country where, since 1980, all climate damages in the agricultural sector are systematically collected monthly, the hot and dry period of 1980 caused a total loss of about $19.3 billion (US DOC, 1980).

It is often claimed that a global warming would produce winners and losers and from this some draw the conclusion that the positive and negative effects are balanced. It is further claimed that the cultivation of new grain types and consequent higher yields could compensate for losses due to climatic effects. Experiences in the recent past give no reason for such optimism (Bach, 1979b). On the contrary, if grain surpluses are available, they go primarily to the countries that can pay for them. In the past, this meant the USSR and the People's Republic of China, and in the Sahel states of Africa the famines continue. It is in these marginal agricultural lands that people already at a subsistence level will suffer most under a potential climatic change so that efforts to improve both their nutritional and income levels will be severely hampered (Tucker, 1982). There is, unfortunately, very little hope for a betterment, since the population pressure is fast increasing precisely in the developing countries (see Chapter III) and the potential for yield increases from energy, capital and socio-political considerations is very limited. Thus, the losers were, up to now, predominantly the developing countries.

If the CO_2-induced climate variability in the Midwest of the USA should occur, as indicated in Chapter IV.1.7, then many industrialised countries would be affected, since grain exports on a large scale are only to be expected from the US breadbasket. The main consequences for Europe would be, above all, reduced feedstock imports with general price increases, especially for meat.

V.5 Water resources

Water is one of the vitally important resources of Man. All living processes depend upon a continual water exchange between living things and the surroundings. About 75% of the Earth's surface is covered by water or ice. In spite of this, water suitable for human use is scarce, since more than 99% of the water is too saline (sea water) or too far away (polar ice caps).

The water available for Man's use is subject not only to large temporal variations but is also very unevenly distributed between the individual regions. The quality of the water plays an additional role in its usefulness. At present, half of the world population has neither enough water nor water in a satisfactory hygienic state (Lindh, 1981). The UN have therefore proposed that 1981-1990 should be declared as the **International Decade for Drinking Water Supply and Hygiene**, with the aim of improving the precarious state of the water supply in many parts of the world.

With the increasing world population, a large increase of the demand for water is to be expected. A CO_2-induced global climate change could make the water supply doubtful as a result of a change in the hydrological cycle (see Chapter IV.1.7 and Fig. V.1). The hydrological cycle is an important part of the climate system (see Chapter II.2 and Fig. II.2) and determines the regional water availability. Through precipitation and moisture storage water is added to the ground, through evaporation, transpiration, and run-off water is removed. A change of the hydrological cycle would affect the drinking water supply, energy production and agriculture, in particular.

V.5.1 Water reserves and water demand

The water reserves are distributed over the hydrosphere and lithosphere (in liquid form), the atmosphere (in gaseous form) and the cryosphere (in solid form). As a result of the continual changes of the three states of water, a **hydrological cycle** is established and this operates within the climate system (see Fig. II.2) and quite significantly influences weather and climate processes. Land and sea receive precipitation from the atmosphere. Part of the precipitation that falls on the land returns directly by evaporation and indirectly by transpiration from life forms into the atmosphere. The other part is stored in lakes, groundwater, glaciers or ice caps, or flows in rivers into the oceans. The water entering the ocean via precipitation and run-off returns to the atmosphere via evaporation and is partly transported again to the land and thereby closes the water cycle.

Table V.4 shows the distribution of water reserves between the individual stores, as well as the global water consumption in 1965 and the estimated global water demand in 2000. Above all, the very high water consumption in agriculture for artificial irrigation is striking. If an increase in the water demand by a

Table V.4 : Estimated global water needs and water resources.

Water resources (km^3)			
	Soil moisture	21×10^3	Continent
	Rivers and lakes	116×10^3	
	Groundwater ($\leqq$ 750 m depth)	4×10^6	
	Ice	27×10^6	
	Ocean	$1,370 \times 10^6$	

Water use and water needs (km^3/yr)	1965			2000		
	Withdrawal	Return	Evaporation	Withdrawal	Return	Evaporation
Municipal water supply	98	56	42	950	760	190
Industry	200	160	40	3,000	2,400	600
Energy	250	235	15	4,500	4,230	230
Irrigation	2,300	600	1,700	4,250	400	3,850
Total	2,848	1,051	1,797	12,700	7,790	4,870

Adapted from: Flohn (1974) and Lvovich (1977).

factor of 15 is assumed, water consumption rates for industrial production and energy conversion could be comparable with those for agriculture in 2000. The ultimate size of the water demand depends to a large extent on whether a very water-use-intensive programme for processing fossil fuels (synthetic fuels production) is begun (Bach, 1979c), and whether predominantly wet or dry cooling is used in power stations (Bach, 1980b). It is also important to note that in agriculture most of the water is lost to the atmosphere after being used once as a result of evaporation, whereas in other sectors the water can be recycled and used more than once (Lindh, 1979). This raises the important question of water availability.

V.5.2 Water availability

Most grain plants and trees consume large amounts of water. For example, 1 hectare of corn in the US corn-belt requires about 4.7 million litres of water during the growing period (Pimentel, 1980). In dry areas 1 hectare of corn requires even 12 million litres of water to produce 5 tons of corn. If the necessary **irrigation** uses groundwater from a 100 m deep well, the additional energy costs amount to 25,000 kWh, increasing the price of corn by 400%.

According to Table V.5, all important industrial nations and heavily populated developing countries have a low water availability with the exception of North and South America, Australia and New Zealand. Even in these areas a very low water availability can regionally occur. According to this tabulation up to the year 2000 a population increase will be accompanied by a decrease of the water availability. The population increase, assumed here for the Federal Republic of Germany and the German Democratic Republic contradicts, however, all current prognoses. The developing countries are in a particularly precarious situation, since with the large population increase the urgently required water for irrigation of the fields is lacking, so that food security is made questionable. In Europe and particularly in the Federal Republic of Germany, the water management is not so much a problem of amount but more a **problem of quality.**

There is apparently at present no shortage of cultivable land. Altogether, about 32 million km^2 are available, although about 19 million km^2 are in arid and semiarid areas with 6-12 dry months (Flohn, 1973). If the land were, in addition, artificially irrigated, then this would require, assuming an additional water supply of 20 cm per dry month, about 36,500 km^3/yr or about 33% of the precipitation.

In the future the water needed for irrigation must come from **desalination of seawater.** As shown above for groundwater use, the additional energy costs for desalination make an agricultural use on a large scale impossible. If one wanted to desalinate half of the above water demand of 18,000 km^3 with the present usual energy requirement of 47 kWh/m^3, this would require about 90 TW, or more than ten times the present global energy demand and is therefore out of the question (Flohn, 1973). If only 10% of the water required for irrigation in 2000 should be produced by desalination, then an installation with a desalination capacity of 14 billion m^3 per day would be needed. With present plant costs of about 5,000 German marks per m^3, a global investment of about 70,000 billion marks would be required and this would be unfeasible (Hauser, 1977).

There are, in addition, several other technical possibilities to increase water availability, namely **cloud seeding**, the **transport of icebergs** from Antarctica, or the **diversion of rivers.** The statistical evaluation of a long series of attempts with cloud seeding has shown that, depending on the prevailing meteorological conditions, sometimes more and sometimes less rainfall fell and often no effect was found (US NAS, 1973). Moreover, the Florida Area Cumulus Experiment - phase 2 (FACE-2) of 1981 failed to confirm the statistically weak positive results of FACE-1 in 1978 which had provided some evidence that seeding cumulus clouds enhances rainfall (Woodley et al., 1982).

The transport of enormous icebergs by towing them behind several freighters has been considered by the rich Arabian oil countries. The transport of a large melting icemass could possibly influence the meridional (north-south) heat transport. Such a project must require an immense amount of energy.

Table V.5 : Per capita water availability (thousands of m^3/yr) for selected countries, 1971 and 2000.

Country	1971	2000	% change, 1971-2000 water	% change, 1971-2000 population	Country	1971	2000	% change, 1971-2000 water	% change, 1971-2000 population
Europe					*Africa*				
Netherlands	0.8	0.6	-25	+ 30	Egypt	0.1	0.05	-50	+111
Hungary	0.8	0.7	-13	+ 20	Tunisia	0.9	0.4	-56	+126
Belgium + Luxembourg	0.9	0.8	-11	+ 7	Burundi	1.0	0.6	-40	+ 81
East Germany	1.2	1.2	-	+ 1	Rwanda	1.8	0.8	-56	+135
West Germany	1.4	1.3	- 7	+ 9	West Sahara	1.8	0.9	-50	+ 94
Poland	1.7	1.3	-24	+ 35	Malawi	2.0	1.0	-50	+108
Rumania	1.8	1.3	-28	+ 40	Marocco	2.1	0.9	-57	+132
Czechoslovakia	1.9	1.5	-21	+ 27	Algeria	2.2	1.0	-55	+111
Bulgaria	2.1	1.7	-19	+ 23	South Africa	2.9	1.3	-55	+128
Great Britain	2.7	2.0	-26	+ 34	Kenya	3.4	1.2	-65	+191
Highest Availability					Highest Availability				
Iceland	319.0	209.9	-34	+ 52	Gabon	328.0	258.3	-21	+ 27
Asia					*Central America*				
Cyprus	0.06	0.05	-17	+ 22	Jamaica	1.1	0.6	-45	+ 93
Arabian Peninsula	0.7	0.3	-57	+106	Haiti	1.4	0.8	-43	+ 81
Pakistan	1.1	0.5	-55	+125	Dominican Republic	2.8	1.2	-57	+142
Bangladesh	1.8	0.9	-50	+102	Cuba	3.1	1.8	-42	+ 74
South Korea	1.9	1.0	-47	+ 93	El Salvador	4.2	1.7	-60	+146
India	2.9	1.5	-48	+ 92	Highest Availability				
Syria	3.0	1.0	-67	+165	Belize	123.0	63.7	-48	+ 93
Iraq	3.6	1.3	-64	+173					
Japan	3.8	2.7	-29	+ 43	North America, South America, Australia and New Zealand				
China	3.8	2.7	-29	+ 42	have a high water availability.				
Highest Availability									
Laos	77.0	37.0	-52	+108					

The numbers have the following approximate meaning: A value of 1.0 or under is a very low availability; 1.0 - 5.0 is low; 5.0 - 10.0 is medium; 10.0 or above is high.

Adapted from; Global 2000 (1980).

The diversion to the south of large rivers in the USSR and Canada for the irrigation of large agricultural areas has also been considered for some time (Roberts and Lansford, 1979). This would lead to an increase in the salt content of the Arctic Ocean, so that the pack ice could form less quickly in winter, which would have an influence on the energy budget and atmospheric circulation.

The provision of sufficient water is already a problem and will be even more so in the future because of the high energy requirement and the possibly unfavourable side-effects on climate. A CO_2-induced climatic change that could lead to a disappearance of the Arctic pack ice in summer and also to a shift of the climate zones (see also V.7.2) would have serious consequences for the water supply of the industrial countries in the temperate latitudes (35-50°N). The already precarious water shortage would become worse along the southern flanks of this zone, especially in California, the whole Mediterranean area and the Near East (Flohn, 1980). Detailed regional studies are not yet available so that, at present, it must suffice to merely list the problem.

V.5.3 Global water problems

The most fertile soil and densest settlements mostly lie along large rivers that are shared by several nations. Examples are the Nile (Uganda, Ethiopia, Sudan and Egypt), the Jordan (Israel and Jordan), the Euphrates (Turkey, Syria and Iraq), the Indus (Pakistan and India), the Ganges (India and Bangladesh), the Rio de la Plata (Brazil, Paraguay and Argentina), the Columbia (USA and Canada) and, finally, the Colorado and the Rio Grande (USA and Mexico). The priorities for use and questions of **water rights** have been regulated fairly satisfactorily between the USA and Mexico only, while in other areas this represents a long-term potential for conflict that can become more serious as water shortages increase (Global 2000, 1980).

Fig. V.7 shows the various water problems of individual regions (Lindh, 1981). Droughts and floods influence agriculture in particular and lead to harvest failures. Low water levels lead to suffocation of the fishes as a result of warming and pollutant enrichment. Water shortages restrict energy conversion (too little cooling water and strongly heated rivers; increase of river evaporation at low water; empty reservoirs); restrict the mining and processing of various fuels (e.g. in synthetic fuels production) and hinder the transport of goods as well as traffic on inland waters. Finally, changing water levels lead to many different health and hygiene problems that are considered in more detail in section 8.

Some of these problems can be at least mitigated by far-sighted planning. For example, the effects of floods and

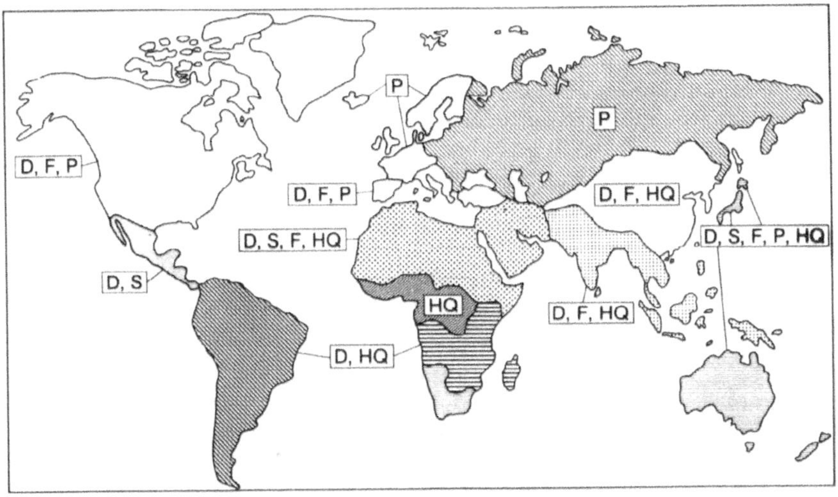

Fig. V.7 : Global water problems: D (drought), S (water shortage), F (floods), P (industrial pollution), HQ (health and water quality problems).
Source: Lindh (1981).

droughts can be weakened by the building of reservoirs and canals. Multiple use and cleaning mean that water can be recycled and the potential use can be extended. The potential climatic developments play a decisive role in the provision of future water supplies. The information about temporal and spatial changes of water availability, necessary for the planning of water supplies, cannot yet be derived from present climate model calculations (US NAS, 1977b).

V.6 Fisheries

Fish is an important component of human nutrition and animal feed. Fish have a low proportion of polysaturated fats and a high proportion of essential minerals, and its protein content is equal to that of meat (US NAS, 1977c). The global fish catch has more than trebled between 1950 and 1978 from 21 to 72 million tons (Brown, 1978; Global 2000, 1980). Of this, 60 million tons are sea fish and 12 million tons are freshwater fish. The yields from Antarctic krill and fish farming are presently still small.

Over the years enormous fleets of fishing boats with the most modern techniques (echo sounders) and processing methods (freezer factory ships) have fished the traditional catching grounds almost empty (Bardach, 1974; Uthott, 1978). Overfishing, like overgrazing in the semiarid belts, disrupts the ecological equilibrium (Eckholm, 1976). The result of years of overfishing has now become apparent to everyone through the disagreements within the European Community. In the case of fish, which have become scarce and are only renewable to a limited extent, it is apparently difficult to agree on reasonable, i.e. stock-maintaining, catch quotas. Together with ecological problems, economic problems come more and more into the foreground, since in the case of overfishing the catching costs are much higher than with a reasonable stock.

In this tense situation, effects that could further cut fish yields are of special importance. The following selected examples give a review of the possible influences of climate on fisheries.

In the mid-1960s the **herring catch** in Icelandic waters decreased from about 750,000 tons/yr to about 50,000 tons in 1970 and this had catastrophic effects on the Icelandic economy. The cause was probably the shift in the atmospheric circulation that led, via strengthened northwesterly winds, to a stronger southward movement of the Arctic pack ice and thus to a shift of the ocean currents, and the herring migration paths moved out of the Icelandic fishing grounds (Johnson, 1976). A comparison of the herring yields in northern Japanese waters with sunspot numbers in the last hundred years shows that when the sunspot numbers were high the herring yields were always low. High sunspot numbers signify strong solar activity with a corresponding temperature increase, particularly in polar latitudes. This leads to a poleward shift of the circumpolar frontal zone and a corresponding shift of the Arctic Ocean currents to the north. Since herring prefer cold ocean currents, they move out of the Japanese fishing grounds when the sunspot numbers are high with the concomitant warming (Takahashi and Nemoto, 1978).

The **cod** supplies another good example for the close interactions between fish yields and climate. Parallel to the warming trend in the 1930s and 1940s, in which the mean temperature over the area 72.5–87.5°N was 1.4°C above the long-term mean, the cod catch of West Greenland increased strongly. Since the temperature decrease in the 1950s (see Fig. IV.17), the cod have almost vanished (Cushing, 1979).

The influence on the **anchovy catch** off the Peruvian coast has become associated with the term **El Niño** (Spanish for the Christ child, since this phenomenon generally begins around Christmas-time). In this case, there is a weakening of the trade winds for a few months, which leads to a breakdown of the entire normal ocean current system. The weakening of the Humboldt Current and the inflow of warm tropical water prohibit the **upwelling** of the cool and nutrient-rich water that normally comes from several hundred metres depth. Therefore, in El Niño years the anchovy catch is much smaller. For example, the catch was 0.4 million tons in 1957, 7.2 million tons in 1965, 8.5 million tons in 1969, and 4.6 million tons in 1972, compared with the long-term average catch of about 10 million tons (Idyll, 1973). Excessive fish catches and the occurrence of the El Niño phenomenon between 1977 and 1979 cut the anchovy catch to about 1 million tons (Walsh, 1981).

The guano-birds that depend on the anchovies are also strongly influenced by El Niño. The El Niño of 1957 reduced their numbers from more than 27 million to 5 million (Idyll, 1973). Today they still have not got back to their former numbers, probably because the fish food for the birds has become scarce as a result of increased commercial fishing. As a result, the amount of commercially-used guano fertilizer is reduced. The coast of Peru is only one of many upwelling areas. Other important upwelling areas that are very significant for fisheries, are along the coasts of California, Mauritius, Namibia and Somalia.

These examples show that the food production in the ocean can be influenced by a number of interacting factors (see Fig. V.1) such as solar energy, atmospheric pressure, the ocean circulation, and temperature, cloud and precipitation fields. The survival of the fish larvae, the migration to spawning grounds, and pubescence are strongly dependent on these climatic influences (Bardach and Santerre, 1981). A CO_2 doubling in the atmosphere might reduce the pH of surface sea water from the usual 8.1 to 7.6 (Holm-Hansen, 1982). It has been shown that a reduction of the pH to 7.6 would increase the copper ion activity by 10-fold, a change to which the marine phytoplankton would dramatically react, and that a pH of 7.7 could cause death of fish larvae. A CO_2-induced climate change could become a critical factor if the demand for foodstuffs continues to increase and the yields of global fisheries decrease. Efforts have, therefore, been made for some time to establish commercial fish cultivation that is independent of climate in the form of aquaculture (Hodges et al., 1981). Whether this can be carried out on a significant scale depends on economic and other considerations and is not yet known.

V.7 Population and settlement

In the history of mankind, climate has always played a decisive role, both in the growth and decline of major civilisations. This is shown here using examples from the past from the Mediterranean area, a subarctic region, the Midwest of the USA, and from central Europe. Whatever happened in the past is also possible in the future, even if the boundary conditions are different. As a result of a CO_2-induced climatic change the following scenarios are plausible in the near future, namely a sea level rise and the concomitant impact on coastal regions, as well as an ice-free Arctic Ocean causing a shift in climatic zones, thereby affecting Man's basis for growing food. All of this has happened before throughout the Earth's history; what is new is that Man's activities can greatly accelerate the process and that an ever increasing number of people will be affected by it.

V.7.1 Examples from the past: Decline of civilisations - migrations - abandonments - climate as a causal factor?

In Plato's "Timaeus", Solon and an Egyptian priest discuss droughts and floods, whose effects, the priest claims, could be so serious that a nation even loses the art of writing. The sudden **decline** of the **Mycenaean culture** in Greece around 1200-1300 BC could be an example of this, since there is no evidence for a conquest by the Dorians (Bryson, 1975). Real palaeoclimatological detective work by Bryson et al. (1974) established that about 1200 BC whole regions of Greece were depopulated. These regions, according to present-day synoptic experience, could have undergone drought periods, while regions with precipitation increases experienced a large population growth. The explanation, that a climatic change with a mean temperature change in the Northern Hemisphere of about 1.5^{o}C (see Fig. II.1) and regional precipitation anomalies could have brought about the decline of the Mycenaean civilisation, does not seem to be implausible.

The influence of climate on settlement can be particularly well followed in northern regions. The naming of **Iceland** and **Greenland** by the Vikings is of interest. About 1000 AD it was relatively warm with a 1-2°C higher mean temperature than today (Flohn, 1978b), so that Iceland was suitable for agriculture with a luxurious vegetation, while Greenland, which was not particularly green even then, was too far west for the Gulf Stream to guarantee productive agriculture (Bryson and Murray, 1977). The christening of Greenland appears to have been a piece of clever salesmanship of Erik the Red, who was banned to Greenland as a murderer and wanted to lure settlers into the inhospitable land by giving it a promising name.

A further example is the decline of an **Indian culture** called **Mill Creek** in the Midwest of the USA between 1300 and 1500 AD (Bryson, 1975; Bryson and Murray, 1977). Reconstructions based on pollen analyses, counting of animal bone remains at fireplaces and of pot-sherds, showed that the occupants, who were hunters and farmers, had less and less venison, bison and corn to satisfy the demand for foodstuffs. Thus they were forced to migrate to the south. The marked temperature decrease that began about 1200 AD (see Fig. II.1) probably led to a change from a forest vegetation to a treeless steppe, since the meridional circulation was strengthened, the cyclone tracks were forced southward and this caused a precipitation deficit of about 50%. In addition to the climatic change, the heavy overhunting by the Indians also contributed to the extinction of many types of mammal. This example is even more striking because the region, from which the Indian tribes were driven between 1300 and 1500 AD as a result of the drought (which persisted for 200 years!), is today the main area in the USA for the cultivation of spring wheat, corn and soybeans.

In the well-known events involving **abandonment of settlements and agricultural areas**, that occurred not only in the German-speaking area but in the whole of central, west and north Europe, 20-60% of all villages were abandoned (Flohn, 1949/50). Several causes can be listed for these occurrences, including the decimation of the population by plagues, falling agricultural prices, and increasing prices for trade products, that eventually led to the peasant revolts. An important contributing cause was certainly the climatic deterioration (see the "Little Ice Age" in Fig. II.1 and Appendix II.2), that led to a series of harvest failures as a result of a succession of several damaging cold winters and summers that were too wet.

V.7.2 Possible scenarios for the future: partial disintegration of the West Antarctic ice sheet - rise of sea level - impact on coastal regions; an ice-free Arctic Ocean - shift of agroclimatic zones.

The melting of the Arctic ice masses and the rise of sea level are processes that the public are most often made aware of with regard to the effects of a CO_2-induced warming. Here we want to investigate which of the events are more likely to occur and could most influence mankind.

The components of the cryosphere that could be influenced by a warming include the Arctic pack ice, the ice sheets of Greenland, the West Antarctic and the East Antarctic, as well as the continental glaciers and snowcover (see also Fig. II.2). On the basis of the present mass distribution, a hypothetical complete melting would give the following sea level increases (Hollin, 1980; Budyko, 1982): for the ice sheet of East Antarctica about 60m, for the ice sheets of West Antarctica and Greenland 6 m each, for the glaciers and snow cover about 0.3 m.

Which effects are possible for the expected CO_2-induced climate change in the near future, i.e. in about the next hundred years? A complete melting of the Arctic pack ice would not affect the sea level, since it is in floating equilibrium (Flohn, 1980). However, it could have a strong influence on climate by reducing the albedo (Parkinson and Kellogg, 1979; Hollin and Barry, 1979). Radiation balance considerations suggest that the Greenland ice sheet would melt slowly over thousands of years. In this case, sea level variations comparable with those of today (1.2 mm/yr) could occur. With a surface temperature of $-20^{o}C$ in summer and an annual mean temperature of $-50^{o}C$ to $-60^{o}C$, the ice sheet of East Antarctica would be relatively insensitive to the expected warming.

The situation is different for the smaller West Antarctic ice that represents only 10% of the total ice mass; this is the origin of the fears of a coming sea level rise that must be taken seriously (Mercer, 1978). As Fig. V.8 shows, about 70% of the West Antarctic ice sheet (cross-hatched area) rests on a rock base below sea level, held in place by the ice shelves (dotted area) and sea ice. If, according to the argumentation, the expected polar temperature increase of 5-10°C (see Chapter IV.1.7.1) leads to a weakening of the protective ice barriers, then about 2-2.5 million km^{3} of ice could surge into the ocean. Such a surge, accompanied by calving processes, could then result in a relatively rapid disintegration of the interior of the ice sheet (Hughes, 1977). According to Thomas et al. (1979) this process may have already started whereby the smaller ice shelves of the Pine Island and Thwaites glaciers may be already collapsing. Sugden and Clapperton (1980) do not agree, however, because their geomorphological evidence does not support this hypothesis.

There is ample evidence of a partial disintegration of the West Antarctic ice sheet during the last interglacial, the Eem, some 120,000 yBP (Hollin, 1980). This was accompanied by a mean global sea level rise of 5-7 m as is shown by the raised coral reefs in New Guinea (Aharon et al., 1980) and similar evidence at many other places such as Timor, Hawaii, Barbados, Mallorca and the Thames basin (Flohn, 1983). What has happened in the past could happen again. The major question that concerns mankind is about the likely timing of a possible collapse of the marine-based West Antarctic ice sheet. A calving mechanism has been proposed by Denton and Hughes (1981) where large crevasses disintegrate the glacier rather rapidly. While such a mechanism has not been reported for present conditions, it is, however, thought to have occurred some 8000 yBP, when the Laurentide ice sheet near what is now Hudson Bay disintegrated within a period of 200 years or less (Andrews et al., 1972). The Pine Island and Thwaites glaciers which may already be collapsing, as indicated above, could take as little as 40 years to disintegrate. Hughes (1982) thinks that there is now enough field evidence for a dramatic confirmation of the Mercer hypothesis that the marine West Antarctic ice sheet is inherently unstable. Bentley (1982a, b), on the other hand, contends that the West Antarctic ice sheet is relatively invulnerable to a CO_2-induced warming because it would take hundreds of years for a temperature rise to penetrate deep enough into the ice. He concludes that the minimum time span for the ice to disintegrate would be about 500 years. Many uncertainties remain, and at the present state of knowledge the disintegration of the West Antarctic ice sheet and a concomitant sea level rise could take anywhere from a few decades to a few centuries. It is important to clarify these conflicting issues by a combined programme of field observations and computer modelling.

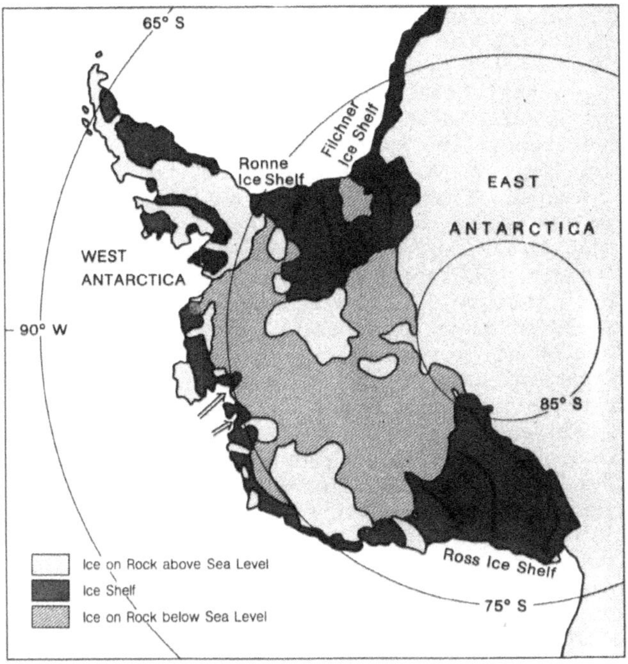

Fig. V.8 : Ice shield and ice shelves in West Antarctica. Double arrows point to critical glacier flows.
Adapted from: Mercer (1978) and Flohn (1980).
(reprinted with permission from *Nature*, Vol. 271, p. 322, © 1978 Macmillan Journals Ltd.)

Past research into a potential sea level rise has almost exclusively focussed on the evidence of a catastrophic ice sheet disintegration. More recently research has also begun to look into the causes of the ongoing sea level rise (Gornitz et al. 1982; Etkins and Epstein, 1982). The sea level is determined by the mass of water in the ocean basins, the volume of that mass, and the topography of the ocean basins. The volume of the ocean depends on the average temperature and mass of each of the ocean layers. The accumulation of the volumes of the individual layers makes up the sea level. A sea level rise may then be caused by a global warming and the resulting thermal expansion of the ocean volume plus the melting of the polar ice.

Depending on the data set used, the mean northern hemispheric temperature increase was between 0.3 and 0.6°C from 1890 to 1940 (see also Fig. IV.17). Over the same time period the global mean sea level rose by about 45 mm. Calculations with a 1-D coupled atmosphere-ocean radiation balance model show that a temperature increase of 0.3°C results in a sea level rise of 24 mm, which is about half the actual sea level rise. Besides thermal expansion there must have been additional forces at work. Moreover, during the recent 40-year period, the sea level has risen at an average rate of 3 mm/yr, while during the same time period the global mean temperature decreased by about 0.2°C. A reasonable hypothesis is that the rapid recent sea level rise is due to the accelerated melting of polar ice. Calculations show that during the past 40 years some 50,000 km^3 (a 50th of the West Antarctic ice sheet) of polar ice have melted and that the heat of fusion necessary to melt the ice has reduced the surface warming due to CO_2 and other factors by about a factor of two. Gornitz et al. (1982) have calculated, using reasonable energy projections, that by 2050 the mean global sea level should rise another 40-60 mm above the present, due to thermal expansion and the melting of polar ice.

Measurements from the global tide-gauge network clearly show an upward trend in global sea level and the above hypothesis and calculations point toward a continuation of this trend. There are also opposing views, such as those of Oerlemans (1982), who argues that an atmospheric warming would not raise the sea level at all, on the contrary, it would reduce it by tenths of centimetres over the next centuries. The reason for this, according to Oerlemans, is that during a warming period, precipitation and, hence, snow accumulation, would exceed the rate of melting of the polar ice. There are conflicting forces at work, and the question, which of these will dominate and by how much, is far from clear. Furthermore, due to the isostatic and eustatic processes involved, a global sea level rise may be quite unevenly distributed so that the mean sea level rise in some locales may be substantially different from the average global change in mean sea level.

In such an uncertain situation it is instructive to look at the potential effects of a sea level rise using the results of a detailed study by Schneider and Chen (1980). As Table V.6 shows, a sea level increase of about 5 m on the coast of the USA would mean that 12 million people would become homeless (about 6% of the total population) and a property loss of about $110 billion would ensue (prices in 1971). A sea level rise of about 8 m would increase these numbers to 16 million people (8% of the total population) and about $150 billion of property losses. Since in the US a third of the landowners pay no land tax, and were consequently not included in the study, the corresponding number increases to $160-220 billion. Further, the secondary costs, economic and population developments, as well as the expected strong migration to the southern coastal regions, have not been considered. The costs would thus be somewhat higher, especially in a few decades, when the losses start. Overall, about 2% of the total area of the USA would be flooded. Regional considerations are, however, more important, since, for example, one third of Florida would be covered by sea water (Table V.6). In the remaining area, the land prices could increase immeasurably.

Table V.6 : Summary of estimated geographic, demographic and economic impacts of 15- and 25-foot (4.6 and 7.6 m) rises in sea level for the continental United States.

Region/State	15-Foot Case			25-Foot Case		
	Percentage Flooded	Population (millions)	EMV[2] (bill. $)	Percentage Flooded	Population (millions)	EMV[2] (bill. $)
North Atlantic	0.9	3.6	33.3	1.3	5.0	47.6
Mid-Atlantic	5.3	1.8	11.6	7.6	2.5	16.6
Florida	24.1	2.9	33.4	35.5	3.8	41.7
Gulf Coast	4.7	2.7	21.3	5.8	3.3	26.9
West Coast	0.6	0.8	7.8	1.2	1.4	14.2
All regions[1]	-	11.8	107.4	-	16.0	147.0
Percentage of Continental USA	1.5	5.7	6.2	2.1	7.8	8.4

[1] Totals may not add exactly due to rounding errors.

[2] Estimated Market Value derived by dividing the "locally assessed taxable real property" in each country by the "aggregate assessment sales price ratio", which is based on a sample of market values.

Adapted from: Schneider and Chen (1980).

A comparison of the damages with other catastrophes shows that coastal erosion in the USA causes annual damages costing about $300 million, and tornadoes, lightning and earthquakes cost hundreds of millions of dollars each year. The possible sea level increase would involve billions of dollars. The estimate of the damage depends strongly on the speed of the sea level rise. Assuming that the sea level rise would take a decade or more, then the 16 million people involved could be moved out at a rate of 1-1.5 million per year. With a total construction volume of 2 million units per year in the entire USA this might be manageable. Schneider and Chen (1980) think that a direct loss of $15-20 billion would be serious but not catastrophic for the whole of the US economy.

A sea level rise could also affect large coastal areas in Europe. For the German North Sea and Baltic Sea coasts, the areas below the lines marking 5 m and 10 m levels above sea level (Fig. V.9) were planimetred using a map with scale 1:500,000. The corresponding population counts were determined*. The results show that for a sea level rise of 5 or 10 m the states of Lower Saxony, Schleswig-Holstein, Hamburg and Bremen would have to reckon with losses of 17% or 23% of the land area, respectively. The affected population amounts to between 2 and 3 million people, which corresponds to 1/6-1/4 of the total population of these states.

In view of the potentially far-reaching consequences more detailed studies are warranted. Since a small vertical sea level rise can result in the erosion of large horizontal areas, an increase in coastal erosion and shoreline shifts can be expected. This may lead to a change in the extent and severity of storm surge patterns with enhanced flooding of coastal areas. The intrusion of saltwater into the marshland and the groundwater aquifers may lead to a loss of surface freshwater and groundwater supplies. All of this can have a significant economic and environmental impact by affecting harbours and holiday resorts, residential areas and industrial centres, as well as agricultural land and wildlife refuges.

The other scenario for a CO_2-induced climatic change considers a complete melting of the Arctic sea ice with far-reaching consequences. Budyko (see Chapter IV.1.7.1) calculated that a 4^0C temperature increase would lead to a complete disappearance of the Arctic pack ice in summer, while the Antarctic ice cap would remain. The main question is, can an **ice-free Arctic Ocean** exist simultaneously with a glaciated Antarctic?

Flohn (e.g. 1978b, 1980) has repeatedly investigated this question of the **unipolar warm period** and the resulting **asymmetry of the climate zones** of both hemispheres. We know from the Deep

---

* This work was carried out by Dr. Rauschelbach of Dornier System in Friedrichshafen, Federal Republic of Germany.

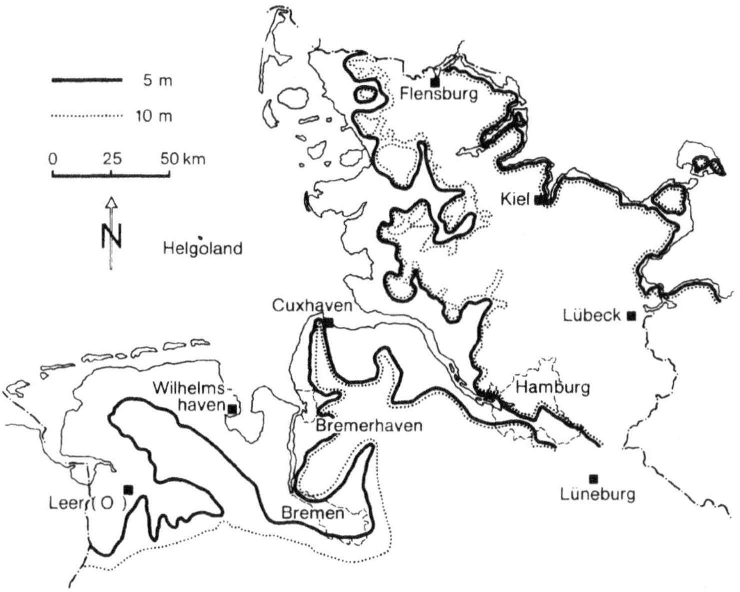

Fig. V.9 : German coastal areas affected by a potential sea level rise. Heavy line denotes a 5 m and dotted line a 10 m rise.

Sea Drilling Program with the research ship Glomar Challenger that such an asymmetry of the polar areas existed continuously for 10 million years directly before the Pleistocene (Kennett, 1977). At that time the Antarctic troposphere was about 18–20°C (today only 11°C) colder than the Arctic troposphere (Flohn, 1980). The meteorological equator was shifted to 9–10°N (today at 6°N on an annual average), whereby the Hadley cell in the Northern Hemisphere was shifted by about 2–3° latitude in summer and by about 4–6° latitude in winter (Flohn, 1983). The extension of the dry area into southern central Europe and the central part of North America is especially significant, since a similar future constellation would, in particular, have serious effects for water and food supply.

Using empirical data from observed circulation parameters, Flohn (1981b) derives the temperature (Ts) and precipitation (P) changes to be expected with an ice-free Arctic (Fig. V.10). In agreement with climate model calculations (see Chapter IV.1.7.1), we find a very strong increase of Ts and P in Arctic and subarctic latitudes. The displacement of the tropical rain belt is responsible for the relative increase of precipitation at about 15°N, and the slight displacement of the intertropical convergence zone is together with the equatorial upwelling responsible for the precipitation decrease at about 10°S. The precipitation deficit at about 38°N, that is confirmed both by climate model calculations and palaeoclimatic reconstructions, is of particular economic significance given the simultaneous temperature increase and resulting increased evaporation. In this area, water supply and agricultural production would be most at risk, as discussed above. The smaller temperature and precipitation changes in the Antarctic, as compared to the Arctic, are a result of the very low water vapour content of the air and the smaller warming with the CO_2 increase.

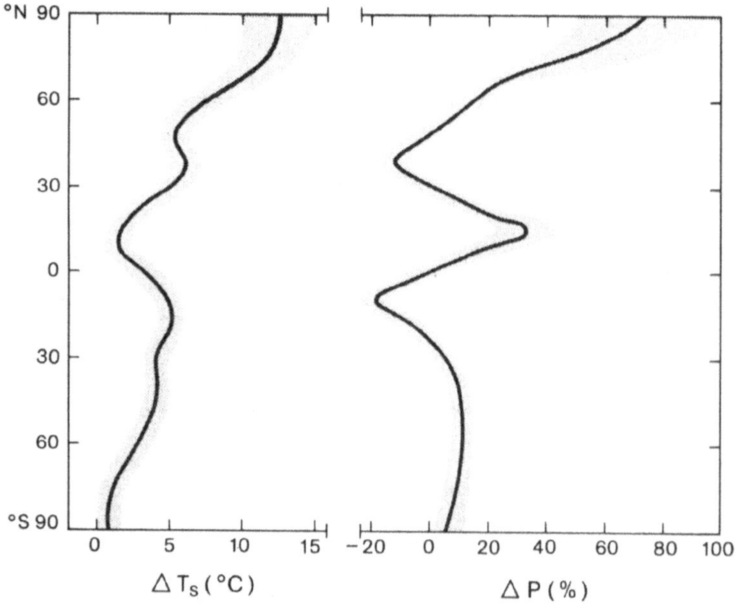

Fig. V.10 : Projected changes of annual surface air temperature (left, °C) and annual precipitation (right, % of present) for an ice free Arctic Ocean.
Source: Flohn (1981b).

The important question, what is most likely to occur first, the collapse of the West Antarctic ice sheet with a 5 m high sea level increase, or the disappearance of the Arctic sea ice and the corresponding shift of the climate zones, can be answered as follows using the presently available knowledge. On the basis of the involved geophysical processes, the melting of the Arctic pack ice should be faster than that of the West Antarctic ice sheet. In addition, the high variability of the extent of the thin sea ice, during the last 1,000 years, suggests a much higher sensitivity to climatic change compared with the enormous ice sheets that hardly change. From all this Flohn concludes that a disappearance of the Arctic sea ice could occur faster (perhaps after a transition period of a few decades already in the middle of the next century) and probably sooner than the disintegration of the West Antarctic ice sheet (Flohn, 1980, 1981b). There is also, however, an alternative view by Clark (1982) who has argued that the geologic evidence for unipolar glaciation is not compelling.

These uncertainties, of course, affect research priorities. It would be important to continue the remote-sensing programmes with the aid of GEOS-3, SEASAT and LANDSAT satellite pictures and radio-echo investigations with airplanes to determine the melting and movement of the West Antarctic ice sheet (Hughes et al., 1980). For a more systematic understanding of the polar climate, the Polar Group (1980) proposed the establishment of a data bank. In addition, the existing climate models must be so modified that they can simulate better the seasonal behaviour of the sea ice. The influence of the diversion of the large Siberian rivers to the south and the related reduction of the freshwater input on the growth of Arctic sea ice should be investigated in more detail. Finally, an intensive palaeoclimatic research programme could provide data necessary for the measures that must be taken.

V.8 Health, disease and welfare

Human health and sickness are strongly influenced by weather and climate. Nobody is immune to the power of influence of the physical environment, in which climate plays a special role. Already in 400 BC the Greek doctor Hippocrates described the physiological and psychological responses of humans to weather and climate effects in his book "On air, water and places". Physiologically, humans are adapted to a relatively narrowly limited set of climatic conditions, and even with the application of the most advanced technology this comfort zone could only be significantly extended with unreasonable effort. A CO_2-induced climate change could lead to regionally differing adaptation and redistribution difficulties in this area, too.

V.8.1 Climate and comfort

Humans can only survive in a relatively limited climatic zone called the **comfort zone**, which is characterised by an upper and lower critical temperature. If temperatures are too low, humans must try to increase the metabolic or body heat through muscular activity. If temperatures are too high the thermal equilibrium can be reestablished through perspiration. If temperature and humidity are simultaneously too high, the cooling mechanism through perspiration cannot operate, and this can lead to a **heat-stroke** that can be fatal. Clothing, heating and cooling can extend the comfort zone somewhat.

Fig. V.11 shows the limited area where humans can live as a function of temperature (Hoffmann, 1960). The white area encloses the entire continental surface; the lightly shaded area shows the extent of existing settlements; it is bounded by an annual maximum temperature $\bar{T}_{max}$ = 55°C and an annual minimum temperature $\bar{T}_{min}$ = −60°C. The heavily dotted area includes 60% of the global population. Generally, for self-sufficient food production, the vegetation period in the living zone must be at least 3–4 months (Weihe, 1979).

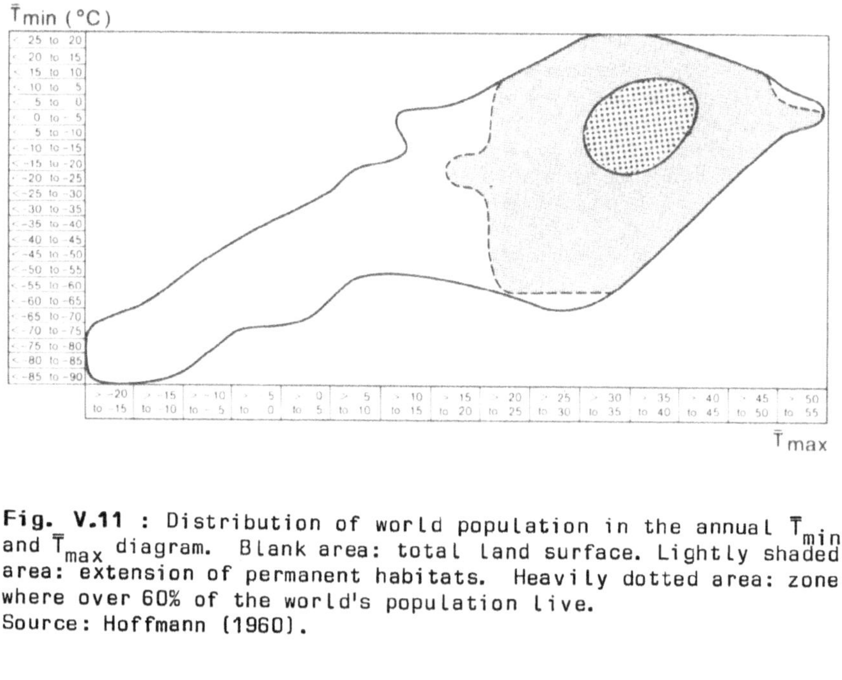

Fig. V.11 : Distribution of world population in the annual $\bar{T}_{min}$ and $\bar{T}_{max}$ diagram. Blank area: total land surface. Lightly shaded area: extension of permanent habitats. Heavily dotted area: zone where over 60% of the world's population live.
Source: Hoffmann (1960).

V.8.2 Climate and disease

Diseases carried by insects are found, above all, in moist-warm climates. However, under certain conditions epidemics can also occur in cool-temperate climates. For example, the louse that thrives well with the human body heat causes **typhus and spotted fever epidemics** in places where the cold climate means that heavy clothing is seldom changed and bodily hygiene leaves much to be desired (Weihe, 1979).

Yellow fever is a frequently occurring virus infection that is passed on to humans by the mosquito vector (disease transmitter) (Brown, 1977). Studies have shown that the virus requires a temperature of at least 24°C and high humidity in order to develop in the mosquito. Yellow fever epidemics were therefore limited in the past to a zone between 40°N and 35°S with a mean annual temperature of about 20°C. **Malaria** is likewise transmitted by the mosquito, in which case a minimum temperature of 16-18°C is necessary (Learmonth, 1977). The disease is found between 65°N and 32°S and up to 3,000 m altitude.

In the tropics and subtropics it is estimated that half a billion people suffer from **hookworm disease** that leads to anaemia and apathy and to mental and physical retardation of children (Kellogg and Schware, 1981). The larvae of the hookworm develop in the soil best when the soil moisture is high and the temperature is 25-30°C.

Frequently, more than one vector is involved. **Bubonic plague** is transmitted to humans by fleas that have been infected by rats. Studies in Vietnam have shown that during the warm and dry season the bubonic plague reaches a maximum and in the rainy period a minimum, while in the latter period the number of fleas is drastically reduced (Olson, 1970).

These few examples show clearly that the individual disease agents respond variously to the different climate factors. Generally, however, it can be stated that a temperature increase benefits most disease transmission and promotes a larger areal distribution.

V.8.3 Climate change and welfare

The CO_2 increase and related temperature increase are the most direct effects on health and welfare that result from human activities. The **CO_2 limit** determined by the US Environmental Protection Agency is a continual exposure at 1200 ppmv (US DOE, 1980). If the present trend in CO_2 growth were to continue, this limit could be reached in 2100.

A temperature increase, even if it were not so large, would be significant because it would strengthen the **heat-waves** that have occurred more frequently in the recent past. The heat wave that occurred in the USA in 1980 led to about 1,300 deaths of humans during the four months from June to September (US DOC, 1980; Karl and Quayle, 1981) and caused losses of about $20 billion (EDIS, 1981).

Fig. V.12 shows how a mean global temperature increase could affect the welfare of the population in the various climate zones. Accordingly, life in the cooler zones would be more pleasant but in the temperate and warm zones **heat stress**, in particular, would increase. Studies have shown that **labour productivity** is reduced by 2-4% for every degree of temperature increase (Harrison, 1979). With a simultaneous high relative humidity, the output would decrease even more drastically.

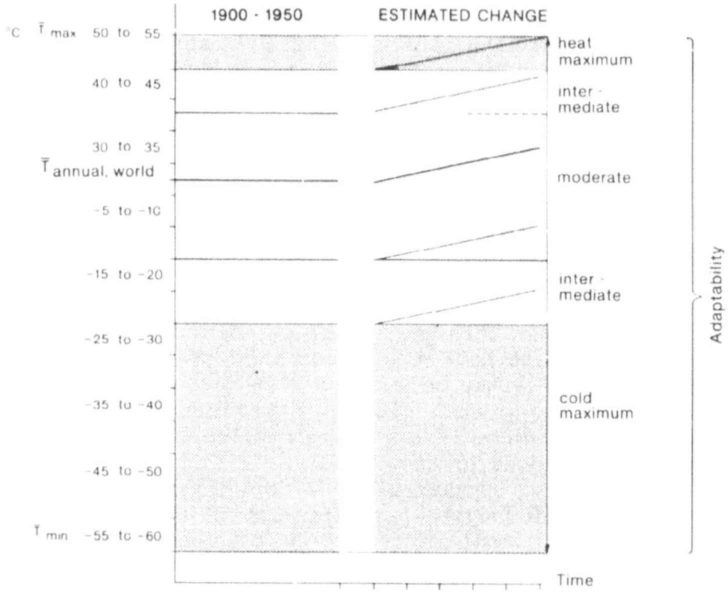

Fig. V.12 : Effect of estimated change of mean world temperature in the human population-climate interaction diagram.
Source: Weihe (1979).

V.9 Recreation

The right to recreation and the need for a sensible use of leisure time are generally recognised today. The need for change and distraction draws the working person to recreation centres, discotheques and pleasure parks, and to satisfy the need for recreation they head from the crowds to the recreational areas near towns, to game parks and nature parks (Schnell and Weber, 1980). Sports, travels, holidays, as well as visits to health resorts are significant components of the organisation of leisure time. To satisfy the need for leisure activities, enormous undertakings are involved. These include the tourist trade (travel agants, train services, airlines and shipping companies), businesses (hotels and guest houses, restaurants, shops) and industries (game and sport article production, clothing and food industry, souvenirs etc.) as well as the insurance companies.

In total, this is an industry with **sales of billions** of dollars which is very strongly dependent on the prevailing weather and climate. If the weather is too rainy, too cool or too hot at the camping area, if there is too little snow or it is too warm with avalanche danger in the winter sports areas, or if there is a drought or too little water in the water sports areas - there are many more examples (Landsberg, 1976) - both those who are seeking leisure and the leisure industries suffer.

In a Canadian sports centre more than $100,000 per day were lost as a result of avalanches (Nowicki, 1980). Other studies in Alaska suggest that the early winter of 1981, that was too warm and too dry, led to a daily loss of income of about $2,500 in the winter sports areas (US DOE, 1980). In conclusion, it is clear that it would be important to make climate impact studies also for the leisure industries since these could show ways in which enormous losses could be avoided.

You must dig the well
before you are thirsty.

Oriental proverb

VI. STRATEGIES FOR AVERTING A CO_2/CLIMATE PROBLEM

VI.1 The scope of alternatives and concepts for action

The previous chapters have shown that Man is still much influenced by weather extremes and climate anomalies, in spite of remarkable technological advances. Food, energy and water supplies appear to be particularly sensitive and their disturbance by climatic influences together with the growing population pressure would have serious social and economic impacts on human welfare. Because of the sensitivity of these supply systems, the long-term influence of climate becomes more significant, especially that resulting from CO_2, whose effects will probably not be differentiable from natural climatic changes much before the year 2000 and consequently not provable till then. With the aid of climate modelling and the study of the climatic history of the Earth, we have been able to provide some plausible answers. At the moment, however, we are not yet able to make reliable predictions of regional temperature and precipitation changes, or the speed and magnitude of possible climatic changes and their effects on mankind.

The decision-makers see themselves confronted by these and a series of other environmental problems that could indeed have possibly catastrophic consequences for mankind, but that can also be easily pushed aside, since the information is at present not sufficiently certain and the problem threatens to become acute only in the distant future. For the introduction of measures, general statements about possible climate changes are not enough. Detailed regional predictions about the magnitude and timing of the occurrence of climatic changes and how these influence socio-economic and political sectors are required. When this kind of information will be available, is, however, uncertain.

This **course of action**, first of all to collect sufficient information (who decides how much information is sufficient?) and only then to make decisions, sounds reasonable but brings us into a **dilemma**. This is because the present energy policies that select particular energy carriers, thereby determine the CO_2 concentration and related climate variations for the coming decades. This means that **relief measures**, if they are to be at all effective, must already be introduced **now**. If we wait until the global climatic effects can be proved, it would be too late for countervailing measures.

Fortunately, the lack of information does not completely bind us to one course, for the **scope of the alternatives** is not so narrow and the choice does not only lie between sufficient energy supply and a high standard of living on the one hand and the risk of a threat to climate on the other. The scope for action can be widened if the **flexibility**, even to go back on false decisions, is maintained and if new knowledge can be brought into and used in the **decision-making process** at any time.

The scope of actions depends on our **objectives**. The **aim** is to guarantee energy and food supplies, to reduce dependence on external energy sources (especially on oil), to use available energy sources more sensibly, and to build up alternative energy sources as quickly as possible to maintain the quality of living standards and to reduce the risk of an irreversible environmental and climatic catastrophe.

This can be achieved with a **low-risk climate, energy and land use policy**. This is characterised by a series of **precautionary measures** that all serve to reduce the CO_2-climate effects. Fig. VI.1 shows, in an overview of the interactions, the most important parts of such a **catalogue of measures**, that ranges from technological fixes and biological methods, to energy conservation through improved energy use and thereby to a reduced fossil fuel use, to the introduction of alternative energy sources. This is supplemented by flanking measures and additional decision aids.

The **precautionary strategy** proposed here extends the scope of alternatives for the decision-makers.

VI.2 Technical fixes

For the reduction of the atmospheric CO_2 concentration, the methods that have been seriously considered so far have been almost entirely technical fixes. This is typical for the kind of thinking in our technocratical era. The methods include, as Fig. VI.1 shows, CO_2 removal from the concentrated stack gases, the atmosphere and the ocean, CO_2 storage in the oceans and in caverns, a few techniques for conversion of CO_2 to useful products, and finally, methods involving indirect effects, for example, through albedo modification.

VI.2.1 CO_2 removal

One problem is the enormous amounts to be removed and disposed of. Presently, about 20,000 million tons CO_2/yr are released by fossil fuel consumption (compared with the more modest amounts of about 150 million tons SO_2/yr and about 300 million tons particles/yr). Another problem is the economic practicability, which

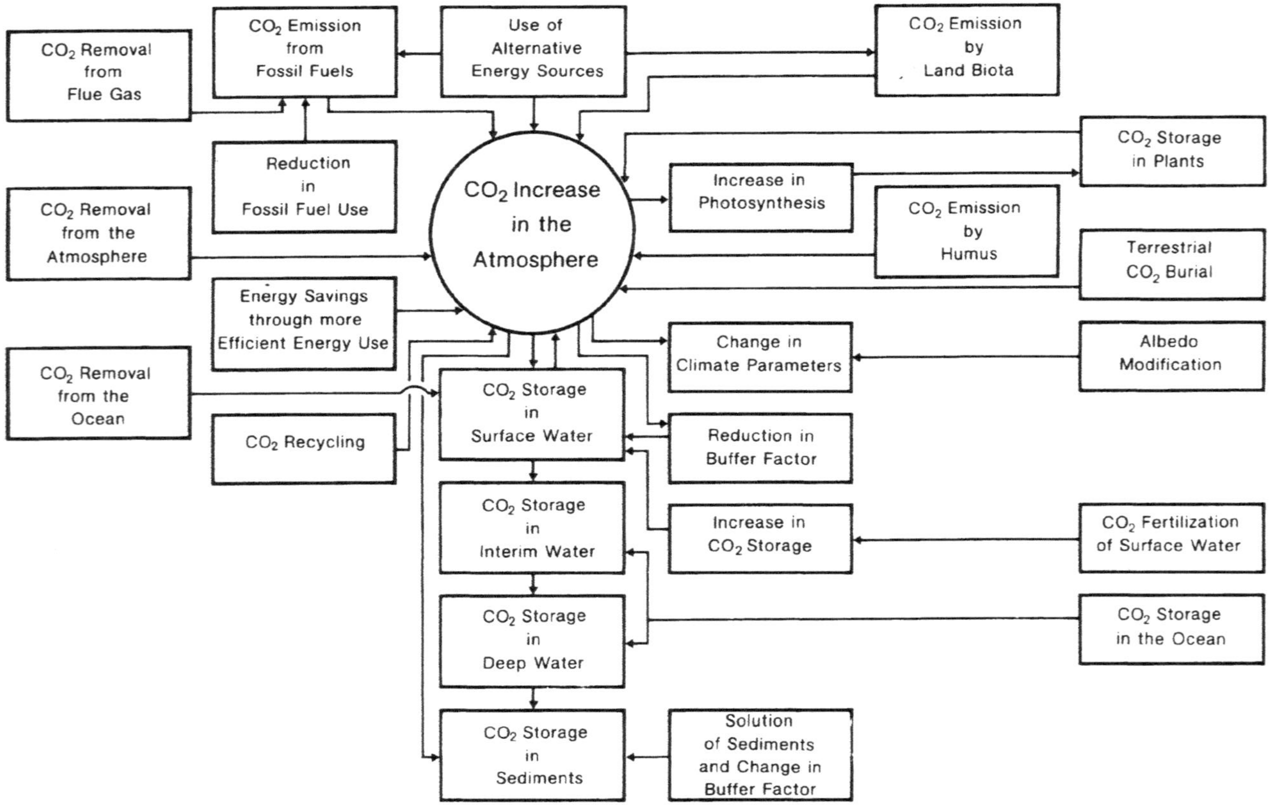

Fig. VI.1 : Possibilities to influence the CO_2 increase in the atmosphere – an interaction scheme.

depends, above all, on the additionally required energy. Since a CO_2 filter system reduces the total efficiency of a power plant, the amount of CO_2 release into the atmosphere per kilowatt-hour is increased. A point can thus be reached at which the only effect of the CO_2 control system is the reduction of the total efficiency of the power plant without a net reduction of the CO_2 emissions (Bach and Jung, 1981).

Of the three potential CO_2 removal sites, the atmosphere and ocean are not as good as the concentrated stack gas, since the latter requires only a third of the CO_2 separation energy (Albanese and Steinberg, 1980). If a cheap non-fossil energy source were available, the CO_2 removal from the atmosphere for methanol fuel production (recycling) could provide an important precautionary measure for the reduction of the CO_2/climate problem. Up till now developments suggest that the introduction of cheap energy sources should not be expected in the foreseeable future.

Even with the application of the presently most effective method, the removal with monoethanolamine (MEA), with an efficiency of 90% CO_2 removal from the stack gas, about 43% of the combustion energy from the fossil fuel is consumed (Table VI.1).

Table VI.1 : Cost of collecting and processing carbon dioxide from stack gas.

	Energy Cost Gcal/t CO_2	Per Cent of Combustion Energy of coal*	Monetary Cost \$/t CO_2
Capture CO_2**	1.31	43	20
Compress to 150 bar and transport 40 km**	0.10	3	4
Liquify CO_2	0.12	4	4
Solidify CO_2			
to Dry Ice	0.36	13	14
to $CO_2 \cdot 6H_2O$	0.47	17	17

* Assuming one ton of coal yields 2.55 t of CO_2 and 7.0 Gcal of heat, the boiler efficiency for steam is 90%, and the plant efficiency for electricity is 38%.

** Refers to 290 t/hr monoethanolamine (MEA) process that captures 90% of the carbon dioxide.

Source: Baes et al. (1980).

With an energy price of \$7.14/Gcal* the CO_2 removal costs about \$20/ton CO_2. The considerable costs for compression, liquefaction or solidification of the CO_2 and the transport costs must also be added (Baes et al., 1980).

The simultaneous removal of SO_2/CO_2 appears to be impracticable because of the large differences in concentration (Steinberg and Albanese, 1980). More noteworthy is the proposal of Marchetti (1976, 1979) to burn the fossil fuel in pure oxygen rather than air, since the entire stack gas then consists of CO_2 and H_2O and could be entirely removed without pretreatment. Also, the energy costs for CO_2 removal by combustion in pure oxygen are about 30% smaller than by combustion in air (Baes et al., 1980).

Given the sizes and isolated locations of many power plants, CO_2 removal is feasible, if at all, probably at half of the power plants, so that in 1978 about 2.7 billion tons, or about 14% of the total global CO_2 emission (see Chapter IV.1) would have been affected (Baes et al., 1980). This would have cost \$54 billion/yr for removal costs alone and would therefore not be practicable.

VI.2.2 CO_2 storage

For the disposal of the enormous amounts of CO_2 there are oceanic, terrestrial and biogenic stores (Fig. VI.1). The latter are considered in section 3. The extraterrestrial disposal and storage in Antarctic ice that have been proposed, are unfeasible from the start because of the amounts involved and the related transport costs.

CO_2 can be introduced into the ocean, which is the most important net sink for atmospheric CO_2, as a concentrated CO_2/seawater solution, liquid CO_2 or as hydrate or dry ice. For each form of injection and for various temperatures and pressures, one can determine the minimum depth for the CO_2 input, at which the CO_2 concentrate sinks into the deep layers of the ocean without larger bubbles forming and related CO_2 losses. Dry ice appears to be most suited, since of all the injection types it has the largest specific density (1.5 g/cm^3), and at a pressure of 1 bar it has a much lower temperature (-78.5^{o}C) than the surrounding water, so that it is most likely to reach the ocean floor without much dissolution (Steinberg and Albanese, 1980).

In order to save the energy for the injection, proposals have also been made to use the naturally sinking thermohaline currents, in which seawater with a higher salt concentration sinks to deeper levels, e.g. in the Straits of Gibraltar, the Red Sea, or in the downwelling waters of the Norwegian and Weddell

---

* Gcal = 10^9 or one billion cal.

Seas (Marchetti, 1976; Mustacchi et al., 1979). One disadvantage is, of course, that they are far-removed from the industrial areas. It has therefore been suggested that fossil energy power plants should be built on enormous swimming platforms and anchored in suitable locations like the oil rigs. In addition to the advantages for CO_2 disposal, the sites would have unlimited availability of cooling water and would lead to lower air pollution in the populated continental areas.

Hoffert et al. (1979) have used a box-diffusion model (see Chapter IV.1.4.4) to investigate what effects the 5 injection strategies, shown in Fig. VI.2, could have on the future atmospheric CO_2 concentration. They assumed that the fossil fuel consumption would increase from 1970 onwards by 5%/yr according to a logistic function until the reservoir was exhausted (about 7.09×10^{12} tons C). The results are very instructive. If all of the CO_2 is emitted into the atmosphere, the maximum of about 2,800 ppmv is reached in 2100 (curve a). In contrast, if 50% of the CO_2 production is diverted into the deep sea (1,500 m depth), then the atmospheric CO_2 increase is reduced almost by half (curve b). If 100% of the CO_2 is injected into the deep sea (1,500 m depth), the atmospheric CO_2 concentration could be held at its present level till at least 2100 (curve d), and a 100% injection at ocean floor level could reduce the CO_2 concentration to the preindustrial value by 2300 (curve e). If this method

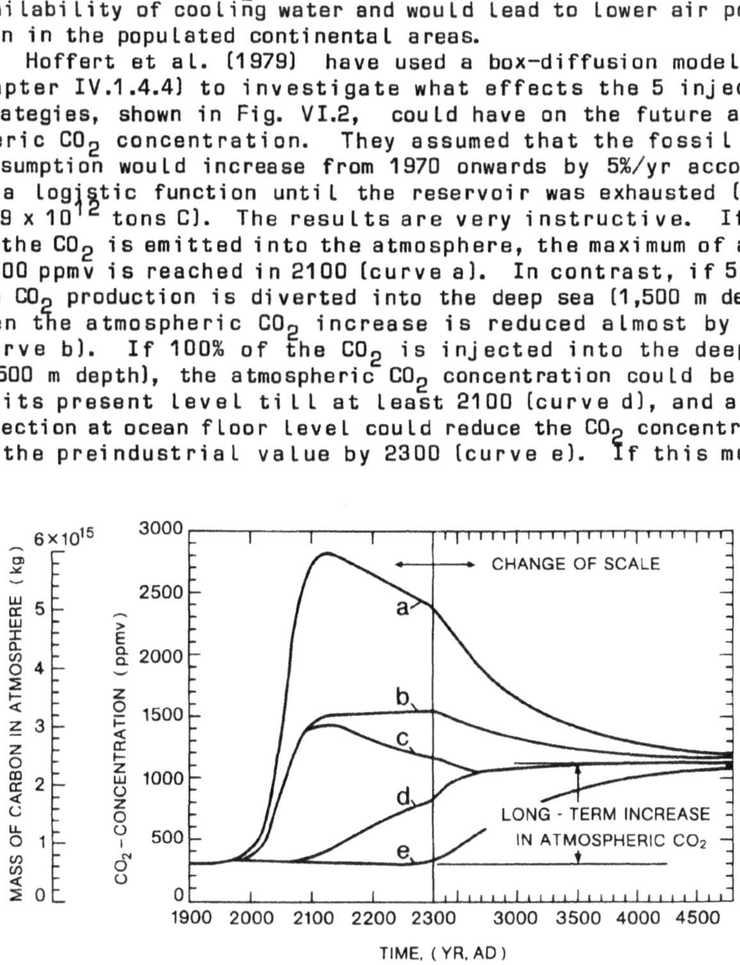

Fig. VI.2 : Model projections of atmospheric carbon dioxide variations assuming the entire fossil fuel reserve is depleted as a logistic function of time. All oceanic injections are initiated in 1985. Note fivefold compression of time-scale beyond the year 2300.
Source: Hoffert et al. (1979).

were economically and technically feasible, we would gain the necessary time for the reorganisation of the energy system from the fossil to the non-fossil energy carriers.

Another important result is that in all projections, regardless of the depth and amount of injection, a new equilibrium concentration of CO_2 in the atmosphere is reached after several thousand years. The concentration levels off at about 1,150 ppmv or about four times the preindustrial value. Of course, these model calculations ignore possibly important effects, such as the dissolution of the $CaCO_3$ in deep sea sediments. Normally, the ocean floor sediments that are saturated with $CaCO_3$ would become undersaturated as a result of the CO_2 injections and begin to dissolve. This could lead to the change of the buffer factor referred to in Fig. VI.1, such that for every mole of $CaCO_3$ that dissolves, the capacity of the ocean increases so that it can absorb one additional mole of CO_2.

The storage of CO_2 in abandoned natural gas and oil wells, as well as CO_2 fixation in natural clays have also been considered (Steinberg and Albanese, 1980). The latter is, however, impracticable because of the enormous amounts of clay that would be required. On the other hand, the caverns of depleted natural gas fields could take up about 2.6 billion tons CO_2/yr and depleted oil fields about 0.6 billion tons CO_2/yr, which in theory would correspond to 16% of the present global annual production of CO_2 (Baes et al., 1980).

VI.2.3 CO_2 conversion to useful products

As pointed out in section 2.1, the production of methanol, a high quality fuel, from atmospheric CO_2 and H_2 could become interesting in the future. For the CO_2 removal from the atmosphere 30 kcal/mole are necessary, which corresponds to about 15% of the required energy for the entire methanol production (Häfele, 1978). An inexpensive, non-fossil fuel source would, of course, be needed for this **recycling** which helps to preserve the environment and climate.

In another method, Williams et al. (cited in Brennst.-Wärme-Kraft, 1978) suggest that excess electricity can be used to electrolytically split water into H_2 and O, then CO_2 can be bubbled in and this combines with the H_2 to make formic acid. This could be used to replace gasoline in combustion engines or to produce electricity in fuel cells.

Since CO_2 reduces the viscosity of crude oil, it is often pumped into oil wells to increase the production (Shah et al., 1978). Krickenberger and Lubore (1980) suggest that CO_2 should be taken from synthetic gas and used for so-called tertiary (increased) oil mining, since it only disturbs further synthesis, for example, in the production of ammonia. They estimate that with an additional cost of $37/ton, about one billion tons of oil that would otherwise remain in the well, could be mined using this method.

Pure chemistry offers multiple possibilities for producing higher hydrocarbons, such as gasoline, heating oil and diesel oil, through CO_2 hydration. All of these methods require very large amounts of energy and could only be implemented if the hydrogen is produced without the use of carbon, so that the CO_2 burden of the atmosphere is not increased (Janssen-Hering and Janssen, 1981).

VI.2.4 Albedo modification

The possibilities to compensate for the warming of the troposphere, resulting from the CO_2 greenhouse effect, by purposeful anthropogenic albedo modification have been reported in various papers (Bach, 1978 a, b; Bach and Schwanhäusser, 1978). For example, the distribution of small reflecting discs, like confetti made of latex foil of 0.01 mm thickness, on the surface of the ocean or in the stratosphere have been suggested. For an amount of 10 tons/km^2 per year, the costs would be astronomical and there would be a risk of considerable environmental damage, even if only 1% of the Earth's surface were treated. Ignoring economic and environmental costs, Budyko (1982) has estimated that the release of some 40,000 tons of sulphur per year into the lower stratosphere would form a layer of sulphuric aerosols there which could reduce total solar radiation by 0.3%, sufficient to compensate for the expected warming.

Investigations with the aid of climate model calculations show that the modification of the annual mean of the zonal albedo between 20°N and 10°S, where the largest albedo changes are to be expected in the foreseeable future, as a result of deforestation and desertification, could lead to a temperature decrease of 0.3-0.7°C (Bach and Jung, 1981). An albedo change through the construction of areally extensive solar farms is not to be expected, since the exclusive installation of this energy conversion system in the necessary magnitude is unrealistic. Albedo modifications in higher latitudes, as a result of industrialisation, urbanisation, and other land use changes (see also Chapter IV.4) are also limited. It is therefore unlikely that purposeful albedo modification can make a significant contribution to the compensation of the expected CO_2 warming that could amount to 2-3°C by the middle of the next century (see Chapter IV.1). Within a system so complex as the climate system, a compensation of global effects by individual regional effects cannot be expected.

In conclusion, we find that **all technical fixes** for avoiding a CO_2 burden on the atmosphere need too much energy and are therefore too expensive (Baes et al., 1980; Steinberg and Albanese, 1980).

VI.3 Biological methods

Biogenic CO_2 storage (Fig. VI.1) includes both the carbon fixation in plants and forests and in organic substances in the soil. If a simultaneous loss through litter and humus does not occur, plants accumulate carbon. The accumulation period extends from the initial stage of plant formation to the mature stage, which is mostly reached after centuries in forests (Hampicke and Bach, 1979/1980). Such a succession can be brought about by a direct human action in the form of **afforestation** or also through **spontaneous reforestation** of an unforested area and thereby remove CO_2 from the atmosphere.

VI.3.1 CO_2 storage in land plants

Dyson (1977) was the first to suggest the forming of a biogenic **"carbon bank"** (similar to a blood bank) of fast-growing trees or water plants, that bind the carbon in wood, humus and peat. For example, a species of American sycamore stores 750 t $C/km^2/yr$. To remove the approximately 20,000 million t CO_2 presently added to the atmosphere each year, an area of 7 million km^2 (roughly the size of Australia) would have to be planted (Dyson and Marland, 1979).

Assuming that within the next 20-30 years at least 2% of the land area can be afforested, Breuer (1979) suggests that a reforestation of 1.5 million km^2 area should be possible in each of the regions where deforestation was largest up to now: the Mediterranean area and the tropical rain forests. Under the further assumption of a net primary productivity of 585 t/km^2 for evergreen forests in temperate latitudes and 900 t/km^2 for tropical rain forests and with a 60% carbon storage in the new forests, about 5 billion t CO_2/yr could be withdrawn from the atmosphere. That would correspond to about 1/4 of the present total emission and would thus be considerable. This sink would, however, only persist for several decades, i.e. for as long as the trees grow, and as maturity is reached the sink would become smaller. Thus, it is argued, one would have several decades for the development of alternative, CO_2-free technologies, and, in any case, reforestation is, in its own right, an ecological necessity.

The success of this enormous effort could, however, be nullified because the planting of new enormous forest areas would reduce the albedo over large areas, and this could add to the warming effect of the CO_2, due to the increased absorption of solar radiation (Marchetti, 1979). With the aid of model calculations that consider the interactions of atmosphere, biosphere and ocean, this supposition should be investigated in more detail. Such models are, however, presently only in the development stage. The contention, that due to the enormous demand for fuelwood in the developing countries, a larger reforestation might never take place, is certainly more realistic than these hypothetical considerations.

VI.3.2 CO_2 storage in water plants

Intensive cultivation of water plants has been proposed as an alternative to purposeful afforestation (Dyson and Marland, 1979). The water hyacinth, a globally distributed freshwater plant that originated in South America, can, for example, store about 6,000 t C/km^2/yr. That is, eight times as much as the American sycamore. To remove the 20,000 million t CO_2/yr from the atmosphere, one would need an area of about 800,000 km^2 for the hyacinths. The hyacinths only grow in alluvial land with enough freshwater, in ponds, lakes and canals. The total alluvial land in the tropics suitable for hyacinth growth amounts to about 2.5 million km^2. If this plant were to be used as a carbon sink, it must be stored after it is harvested so that the conversion to methane is not possible. To do this one must artificially copy the conditions under which the natural peat deposits become peat and not methane.

In addition to hyacinths, a whole series of useful plants, such as sugar cane have been proposed as carbon sinks. It would, of course, be difficult to justify the use of plants necessary for food as carbon sinks or for energy production (Brown, 1980).

The surface layers of the ocean could take up more CO_2 from the atmosphere if the rate of sedimentation of biological and organic substances from the surface layer into the deep sea could be accelerated. This could be done by stimulating the photosynthetic production of organic mass in the surface water, in which case large amounts of phosphorus (as Na_3PO_4) and nitrogen (as $NaNO_3$) would be needed for "fertilization" (see Fig. VI.1). A first guess suggests that a fertilization of the surface waters with 10 million tons of phosphorus would give a deposition of about 300 million tons of organic carbon (Dyson and Marland, 1979). It is therefore very questionable whether such projects can be carried out at all.

VI.3.3 CO_2 storage in soil humus

As a result of changing anthropogenic stress, soil humus is a very variable reservoir for organic carbon (Hampicke and Bach, 1979/1980). The carbon storage in the soil depends on climatological and pedological conditions of the location (e.g. temperature, precipitation, groundwater budget and original rock type), the annual supply of perishable organic substance through litterfall and dying of roots, as well as the type of soil preparation (e.g. depth and frequency of ploughing). Apparently less humus is at present formed than destroyed.

Within the next few years only a policy of energy saving can be quickly enough introduced to bring a decisive change.

Chauncey Starr
Director of the
Electric Power Research Institute,
USA.

VI.4 Energy saving through better energy use

Apart from the technical fixes that need too much energy and are too expensive, there are promising strategies that guarantee an extensive CO_2 reduction. These include energy conservation through rational energy use, whereby the fossil fuel use and, hence, the CO_2 emission, can be simultaneously reduced, and the introduction of alternative energy sources that release little or no CO_2 into the atmosphere (see Fig. VI.1).

Most studies of the CO_2 problem begin with the assumption that the use of fossil fuels will increase quickly and continually, as it has been in the past. It is assumed that there will be steady economic and energy growth and a stabilisation of energy demand is held to be impossible. The knowledge gathered in very careful and convincing studies in more than a dozen countries have, however, seriously questioned these assumptions (Lovins et al., 1981). The results clearly show that the most economic strategy for energy supply does not require an increasing use of fossil fuels but much more a sinking use. The reason for these interesting results is that the same or even higher **energy services** can be provided with less energy solely by raising **energy productivity**. The much reduced energy demand then means that the energy supply that was secured up to now by fossil fuels can be quickly replaced by CO_2-free, renewable energy sources. This **two-pronged strategy** relies on technologies that are already developed, highly economical, and indistinguishable with respect to reliability and convenience from current energy systems.

Every national energy policy, also that of the F.R. Germany, propagates the improved use and thus the conservation of energy as its main goal (Energiediskussion, 1979). The primary aim of this policy is to reduce the dependence on external energy sources, especially on oil, and to prevent a further increase of the deficit in the balance of payments. The oil consumption of the F.R. Germany fell by 30 million tons between 1979 and 1981 (consumption in 1981: 115 million tons). Schiffer (1981) estimated that 70% of the reduction was due to conservation and substitution of oil and 30% due to economic stagnation and advantageous weather conditions. In 1980, compared with the previous year, the average price for a ton of crude oil increased from 278 DM to 456 DM - in spite of a temporary oversupply (Voss, 1981). Although the crude oil imports decreased by 9% during the same period, the bill increased by about 50% to 44.7 billion DM and was, therefore, the main cause of the deficit in the balance of payments of about 28 billion DM.

In as much as the energy carriers are preserved or substituted as a result of energy-saving measures, **rational energy use** is factually another **energy source** (Meyer-Abich, 1979). The more careful use of energy in all sectors is therefore an additional energy source that is independent from foreign supplies, that could, according to the estimate of von Weizsäcker (1979), supply 20-25% of the expected energy demand by the year 2000 and would thereby be equivalent in size to other major energy carriers. The decisive advantages of more efficient energy use as opposed to the traditional energy wastage are an increasing security of supply, a lower growth of energy demand and positive employment effects through investments for the purpose of energy saving. A **special bonus** of such an energy policy is that the environment in general is preserved and, crucial for the question we are looking at, it can thereby reduce or even eliminate the CO_2/climate problem.

It is not possible to give a comprehensive description of all possibilities for the efficient use of energy within the frame of this book. Rather, I would like to show, using some developments in projections and selected studies, that the present CO_2 increase can be decidedly influenced by a more rational energy use without undue changes of the energy supply and demand system.

VI.4.1 Future energy use

Projections of fossil fuel use are the starting point for the estimation of the future CO_2 development, as we have seen in Chapter IV.1. Actually, the energy consumption of a country cannot be projected (Herz and Jochem, 1980) because the future is not predictable (Voss, 1981). It is, therefore, not surprising that the reliability of the projections is hardly more than 50%. Therefore, it is preferable to approach the future through scenario analyses (Voss, 1982) and to examine various energy paths that are to be understood as reflections of plausible perspectives of the future rather than forecasts (Enquete-Kommission, 1980).

In spite of this, we shall use the word projection in the following sections because it has become generally accepted. The possibilities of a more effective energy use can be seen, for instance, in the temporal development of energy demand projections and their wide spread. The actual energy demand apparently

Table VI.2 : Projections of primary energy demand (mtce) of the Federal Republic of Germany and the Institute of Applied Systems Research and Prognosis (ISP) (Consumption 1982: 362 mtce).

	Projections of the German Government for 1985			for 1995	Reference Scenario of ISP from 1975	
	1973[1]	1974[2]	1977[3]	1981[4]	for 1985	for 2000
Oil	330	245	226	160-169	205	240
Hard Coal	50	79	73	103-109	83	85
Gas	92	101	87	75-80	71	104
Brown Coal	38	38	35	37-40	36	37
Nuclear	90	81	62	80-84	37	113
Other[5]	10	11	13	14-15	5	5
Total	610	555	496	469-497	437	584

* 1 mtce = 9.26×10^8W ≃ 1 GW.

[1] Energy Programme of the Federal Republic of Germany (1973).
[2] 1st continuation of the Energy Programme (1974).
[3] 2nd continuation of the Energy Programme (1977).
[4] 3rd continuation of the Energy Programme (1981).
[5] Various groups of other energy carriers.

Source: Pestel et al. (1978) and Die Zeit (1981).

continually forces drastic reductions of the projections being made. Some examples of this are now presented. Table VI.2 shows projections of the primary energy demand for the F.R. Germany. In 1973, just before the energy crisis, the projection of the Federal Government for 1985 was 610 million tce (consumption in 1982: 362 million tce). The oil price increases and consequent reductions of energy consumption meant that in the continuation of its Energy Programme, the Government was forced to drastically reduce the numbers to 496 million tce in 1977, corresponding to a reduction by 19% in four years. The projections of the Institute of Applied Systems Research and Prognosis for 1985 are considerably lower. The most recent (third) continuation of the Energy Programme projects a range of energy demand for 1995 that is lower than that originally considered possible for 1985. The projections of the Federal Government are produced by the three cooperating economic research institutes referred to in Table VI.4. The Government gives them an economic growth rate which they aim to achieve within the framework of their economic policies (Geberth, 1980). With this kind of handicap a projection that is halfway realistic cannot be expected.

In the USA the energy projections of recent years have also been drastically corrected downwards (Table VI.3). The present official forecasts of the US Department of Energy are lower than

Table VI.3 : Evolution of approximate estimates of U.S. primary energy demand (TW) in the year 2000 (q/yr $\simeq 10^{15}$ BTU/yr $\simeq$ EJ/yr $\simeq$ 36 mtce/yr $\simeq 3.35 \times 10^{10}$ W) (1980: 76 q/yr $\simeq$ 2.6 TW).

Year of forecast	Magnitude and source of forecast: Beyond the pale	Heresy	Conventional wisdom	Superstition
1972	4.19[1]	4.69[2]	*5.36[3]	*6.37[4]
1974	3.35[5]	4.15[6]	*4.69[7]	*5.36[8]
1976	2.51[1]	2.98[9] - 3.18[10]	4.15[7]	*4.69[8]
1978	*1.11[11] - 1.84[1]	*2.11[12] - 2.58[12]	3.18[13] - 3.22[12] - 3.38[14]	4.12[12] - 4.15[16]
1980	0.50[1]	*1.64[17] - *1.81[18]	2.08[19] - 2.14[20] - 2.21[21] - 2.71[22]	3.25[23] - 3.42[24]

1 Lovins speeches
2 Sierra Club
3 U.S. Atomic Energy Commission
4 Other Federal Agencies
5 Ford Foundation (Zero Energy Growth)
6 Ford Foundation (Technical Fix)
7 U.S. Energy R&D Admin. (ERDA)
8 Edison Electric Institute
9 Von Hippel and Williams (Princeton)
10 Lovins, Foreign Affairs (1976)
11 Steinhart (Wisconsin) for 2050
12 CONAES-Study (1978) for 2010
13 U.S. DOE $32/bbl (1977 $)
14 Weinberg (IEA - Oak Ridge), low scenario
15 USDOE, $18 & 25/bbl av.
16 Lapp (believes E/GNP fixed)
17 Stanford III for 2010 (USDOE, 1980)
18 CONAES (1980) for 2010
19 SERI (1981) Sawhill Report, low scenario
20 Ross and Williams (1981)
21 SERI (1981), high scenario
22 Sant (1981), least-cost
23 Exxon (1980)
24 Norman (1981)

* With lifestyle changes; **bbl = barrel $\simeq$ 159 l

Source: Lovins (1980) and Lovins et al. (1981).

the projections that Lovins made eight years ago and which were, at that time, criticised in official places as being completely unrealistic. The range of the more recent global energy demand projections for the year 2000 is likewise very large (Table VI.4). It appears that more recent projections find that lower energy amounts are more realistic. The magnitude of the projected primary energy demand appears, however, to depend above all on the private interests at the time the projections are made. In section VI.6 we shall come back to the energy scenarios of Häfele et al., Colombo/Bernardini and Lovins et al. in order to investigate their influence on the CO_2 and temperature increase.

Table VI.4 : Range of recent world energy projections for global primary energy use (TW) in 2000.

	Date of Projection	Primary Energy Demand (TW)
Knop-Quaas (GDR)*	1975	24.07 - 27.78
Frisch (French Electricity Board)*	1977	25.51
Rotty (Oak Ridge Assoc. Universities) (USA)	1976	23.12
German Shell AG*	1978/79	16.67 - 21.85
WEC	1978	16.46 - 21.84
WAES	1977	17.69 - 21.04
German Esso AG*	1978	19.44
Häfele et al. (IIASA)	1981	13.58 - 16.83
DIW/EWI/RWI*	1978	16.61
Colombo/Bernardini (Ital. Atomic Energy Commission)	1979	12.17
Marchetti (IIASA)	1977	8.80
Lovins et al.	1981	7.07

WEC = World Energy Conference
WAES = Workshop on Alternative Energy Sources
IIASA = International Institute for Applied Systems Analysis, Laxenburg, Austria
DIW = German Institute for Economic Research, Berlin
EWI = Energy Institute, Cologne
RWI = Institute for Economic Research, Essen

*All data with * extracted from Bienewitz et al. (1981); for all other data see references.*

The causes of the general trend towards lower energy demand projections - which is observed in all western industrialised countries - are, according to Rotty and Marland (1980), not only due to the increasing energy prices and the general slowing of economic growth, but also due to the fact that the projections and those who make them are also prisoners of the prevailing attitudes. Detailed studies of Lovins et al. (1981) have shown that the various high energy projections arose through the neglect of a whole series of important factors. These include, for instance, the facts that the high prices increase the energy efficiency, that the high growth rates of energy consumption in the past were based on falling real energy prices, that the demand for energy services reaches saturation, that frequently (as a result of historical conditions) a distorted picture of the energy markets was used in the projections, that the structural changes in the individual national economies and in the global economy have not been sufficiently taken into account, and that excessively aggregated analyses often conceal the additive effect of many small contributions to saving.

It is certainly correct that a better energy utilisation provides the possibilities to get by with less energy in the future and this has implications for the future CO_2 development. Therefore, in the following paragraphs we want to look more closely at the potential for saving energy in a number of countries.

VI.4.2 Some selected studies of rational energy use

Only after the shock of the first oil crisis in 1973/74 did people begin to think more seriously about a better utilization of the available energy sources. The first comprehensive study that considered the multiple possibilities of a rational energy use and, amongst others, a zero growth, was the classic Energy Policy Project of the Ford Foundation (1974). Then came a long series of detailed studies. The best-known include those of Lovins (1975, 1978a, b, 1980); Bossel et al. (1976); Bossel and Denton (1977); Council on Environmental Quality (1979); Meyer-Abich (1979); Leach et al. (1979); Stobaugh and Yergin (1979); Norgard (1979, 1982); CONAES (1980); Krause et al. (1980); Enquete-Kommission (1980); SERI (1981); Lovins et al. (1981); Krause (1982); Bzserghi et al. (1982); Sorensen (1982a, b); Olivier et al. (1983) and Bach (1983a, b).

VI.4.2.1 German studies

The study of the Öko-Institut - The basic idea of this study by Krause et al. (1980) is that energy and environmental problems can be best solved if one uses the available energy carriers better and introduces more environmentally favourable, renewable energy sources quickly. The consumer would then receive more energy services (e.g. warm rooms, transportation in a car, goods production, etc.) with less energy consumption. The positive consequences are lower resource consumption, reduced dependence on energy imports, lower energy costs, less environmental stress, and additional jobs.

For the improved conversion of primary energy into energy services, they see three main starting points:

a) improvement of the technical efficiency of conversion, distribution and use;
b) improvement of the application of energy to services; and
c) use, above all, of those end-use energies that require the least amount of primary energy for their production and distribution.

On the basis of their investigations, they come to the conclusion that the present energy productivity, in comparison to that of 1973, could, for a weighted mean, be improved by at least a factor of 2. With technologies that could be introduced already today, the specific end-use energy could be reduced by the year 2030 by the following percentages:

- 70% for space heating (e.g. by the use of Swedish insulation techniques, and improvements in heating efficiencies through electronic sensors and regulators);
- 65% for electrical household equipment (e.g. through better insulation of refrigerators and ovens, more efficient heat pumps and motors, heat recovery from waste water in washing machines and dishwashers, or direct heat recovery from driers etc.);
- 60% for car fuel (this corresponds, for example, to the improvement of the new VW Rabbit Diesel that is to be introduced in the 1980s, compared with the VW Beetle);
- 30% for industrial process heat (e.g. waste heat recovery, insulation, cogeneration, heat recuperators, electronic process controls);
- 30% for electric drives (e.g. improved sizing of motors, use of improved torque converters, operating of motors from alternating-current synthesizers).

After the studies of individual sectors, the development of the primary energy demand until the year 2030 is estimated using three scenarios, in order to illustrate the range for future energy policies (Table VI.5). In the scenario "Business as Usual", substitution processes between energy carriers are neglected and it is assumed that coal, oil, gas and other sources are used in about the same proportions as today. This scenario is felt to be unrealistic but it shows what amounts of conventional energy carriers will be required if renewable energy sources are not introduced.

In the scenario "Coal and Gas" it is assumed that coal and natural gas can fully replace oil by the year 2030 without resort to other energy sources. This scenario should give an impression of the environmental problems to those who believe that renewable energy sources cannot be introduced quickly to any significant extent.

In the scenario "Sun and Coal" it is assumed that non-renewable energy carriers are quickly substituted by renewable

Table VI.5 : Primary energy use (mtce/yr) in three different scenarios for the Federal Republic of Germany.

	1973	1980	1990	2000	2010	2020	2030
1. *Example "Business as Usual"*							
Primary energy, total	354	413	376	293	254	215	192
index	100	117	106	83	72	61	54
- coal	115.0	110	105	85	75	70	60
- oil	184.2	210	200	150	130	110	100
- gas	35.5	65	60	50	40	30	27
- uranium	3.9	15	-	-	-	-	-
- hydro and others*	15.4	13	11	8	9	5	5
2. *Example "Coal and Gas"*							
Primary energy, total	354	413	375	300	267	241	232
index	100	117	106	85	75	68	65
- coal	115.0	110	115	200	130	140	155
- oil	184.2	210	180	110	60	25	-
- gas	35.5	65	70	70	70	70	70
- uranium	3.9	15	-	-	-	-	-
- hydro and others*	15.4	13	10	10	7	6	7
3. *Example "Sun and Coal"*							
Primary energy, total	354	413	381	298	260	224	207
index	100	117	108	83	73	63	58
- coal	115	110	115	110	115	115	110
- oil	184.2	210	200	100	60	15	-
- gas	35.5	65	60	40	20	-	-
- uranium	3.9	15	-	-	-	-	-
- wind	-	-	0.5	3.5	7.0	12.0	13.2
- hydro	2.2	2.2	2.8	3.0	3.0	3.0	3.0
- solar (direct)	-	-	7.5	17.9	23.4	25.5	26.6
- biomass	1.7	2.0	5.0	20.0	30.0	50.0	50.0
- others*	11.5	9	8	4	2	3	4
fossil share							
Index	100	117	104	72	56	38	32
%	99	99	96	85	76	60	55
renewable share							
mtce	2.8	2.8	16.0	44.4	63.4	90.5	91.5
%	1	1	4	15	24	40	44

* Includes imports of electricity and rounding errors.

Extracted from: Krause et al. (1980).

energy sources (especially solar and wind energy, biomass, biogas, biofuels, etc.) and by indigenous coal. Of all the scenarios this is felt to be most realistic because in the short-term there is already a tendency to substitute for oil, and in the long-term we must reckon with substitution for natural gas. Thus, the remaining alternatives are abundant indigenous coal and increasingly used renewable energy sources.

As shown in Table VI.5 in the row for Primary Energy Index for all three scenarios, the primary energy use in 2030 is reduced to 54-65% compared with 1973, i.e. an average reduction of 40%. In the scenario "Sun and Coal" the fossil fuel share declines from 99% in 1973, to 55% in 2030, with a corresponding increase of the renewable share. It is concluded that this would drastically reduce the environmental burden, e.g. through sulphur and nitrogen oxides. The numbers suggest that the CO_2 emissions would also be significantly reduced, by about 45% in the "Business as Usual" scenario, about 29% in the "Coal and Gas" scenario and about 53% in the scenario "Sun and Coal".

As a result, the authors emphasize that the low energy demand values are realistic because with a more efficient energy utilization and a possibly lower population (between 40 and 60 million) one must expect a considerably lower energy requirement.

The Report of the Enquete Kommission (1980) - The Kommission, set up by the 8th Federal German Parliament, was asked to estimate the amount and type of the primary energy carriers to be used in the next 50 years, taking into account ecological, economic, societal and security policy aspects. To do this, four energy paths were defined and their characteristics are presented in Table VI.6.

Paths 1 and 2 follow the conventional practice of determining the energy demand from the supply side. In Paths 3 and 4 the starting point is the demand and the extent to which this can be reduced by rational energy use is investigated, so that it can primarily be covered by indigenous coal and renewable energy sources. In comparison to the base year of 1978 Path 1 assumes a very strong, and Path 2 a strong increase of the primary energy demand; Paths 3 and 4, in contrast, reckon with a demand reduction. Consequently, the latter cannot only cut the fossil fuel demand, but also, by simultaneous expeditious build-up of the renewable energy sources from 2000 onwards, have no need for the use of nuclear energy and synthetic fuels.

Assuming that the shares of bituminous coal (65.8%) and brown coal (34.2%) of 1978 are also valid for 2000 and 2030; that the oil share declines from 77% (1978) to 60% (2000) and 50% (2030); that the natural gas share increases from 23% (1978) to 40% (2000) and 50% (2030); and, finally, that synthetic fuels production is done autothermally (i.e. with coal providing the energy) until 2000 and allothermally (i.e. without a carbon-containing energy source) after 2000, the following changes of the CO_2 emission from 1978 to 2030 arise: for Path 1 an increase of about 35%, and, in contrast, decreases for Paths 2-4 of about 9%, 15% and 36%. The decrease for Path 2 can be explained by the halving of the consumption of oil and natural gas (Table VI.6).

In 1980 a majority of the first Enquete Kommission (the second has been meeting since 1981) came to the conclusion that in about 1990 one should examine whether the development of nuclear energy is necessary or whether the use of nuclear energy can be abandoned in the future. Until then great efforts should be made to support the introduction of energy efficiency measures

Table VI.6 : Results of the four energy paths (mtce) of the Enquete Kommission of the German Parliament.

		Path 1		Path 2		Path 3		Path 4	
Characteristics									
Economic growth									
- before 2000		3.3%		2.0%		2.0%		2.0%	
- after 2000		1.4%		1.1%		1.1%		1.1%	
Structural changes in the economy		medium		medium		strong		strong	
Growth of basic industry		like GNP/2		like GNP/2		zero		zero	
Energy savings		trend		strong		very strong		extreme	
	1978	2000	2030	2000	2030	2000	2030	2000	2030
Demand									
Primary energy	390	600	800	445	550	375	360	345	310
End-use energy	260	365	446	298	317	265	250	245	210
Electricity[1]	36	92	124	47	57	39	42	36	37
Non-energy use	32	50	67	43	52	34	34	34	34
Supply									
Coal and lignite	105	175	210	145	160	145	160	130	145
Oil and gas	265	250	250	190	130	190	130	165	65
Nuclear in GWe	10	77	165	40	120	0	0	0	0
- Breeder reactors	-	-	84	-	54	-	-	-	-
Renewable sources	8	40	50	40	50	40	70	50	100
Other									
Electricity from coal	65	80	80	29	22	76	77	52	33
Coal gasification	-	18	50	18	56	-	-	-	-
Electricity share in %									
- residential heating	3	14	17	8	7	3	2	2	0
- process heat	7	19	17	8	8	8	8	7	6
Demand of uranium in									
1000 t accumulated		until 2030		until 2030					
- without reprocessing		650		425					
- with breeder reactors		390		255					

[1] Electricity use refers to end-use and not to gross electricity production given here in mtce, where 1 mtce electricity use ≃ 8.13 TWh.

Source: Enquete-Kommission (1980).

and the use of renewable energy sources. If this occurs, the Federal Republic of Germany (with a primary energy consumption of fossil fuels of about 367 million tce in 1980) could provide a substantial contribution to the reduction of the CO_2 risk.

VI.4.2.2 European studies

A strategy of low energy consumption for Great Britain – A significant aspect of this study by Leach et al. (1979) is the assumption that through the introduction of energy-saving technologies, a zero-growth in the energy sector can be achieved with a simultaneous economic growth. To prove that energy and economic growth are not necessarily coupled, the fuel consumption in England was used as an example showing that it was lower in 1977 than in 1970, although during the same period the Gross National Product rose by more than 10%.

As is common in energy analyses these days, two scenarios, a low one and a high one, are considered. These differ due to differing assumptions about the future population and economic growth. In the low scenario for England it is assumed that in comparison to 1976 the primary energy consumption would be reduced by 7% by 2000 and by 22% by 2025*. This means that the oil consumption is reduced by about half, coal consumption remains constant and the shares of nuclear power and hydropower continually decrease after 1990. In the high scenario, the total primary energy consumption in 2000 is somewhat higher than that of today, but falls to 8% below the comparable value for 1976 by 2025. In this scenario it is assumed that the oil consumption decreases by 37% by 2025, the coal consumption increases by 25% in the same time period and nuclear power, as in the low scenario, continually decreases after 1990.

If these energy scenarios were to become official British energy policy, then over the period 1976–2025 the CO_2-emission could be reduced by about 24% in the low scenario and by 10% even in the high scenario. Thus Great Britain (with a primary energy consumption of fossil fuels of about 337 million tce in 1976) could also make an important contribution to the diminishing of the CO_2-problem.

The energy development of Sweden – In 1979 the energy demand in Sweden was covered by oil (67%), hydro- and nuclear power (19%),

---

* In an even more elaborate study Olivier et al. (1983) have shown that it would be very economical for the U.K. to quadruple its energy productivity by 2025. They also present a very detailed comparison of costs of energy efficiency improvements and energy supply systems which is outside the scope of this book. The reader is referred to the original work.

coal (12%) and wood wastes (2%) (Dampier, 1981). Since they have no reserves of their own, the Swedes must import all of their oil. The Swedish Parliament therefore passed an energy law in 1981 that aims to halve the oil imports within 10 years. Compared with the annual total consumption of about 40 million tons of oil it is expected to save some 3 million tons of oil through more efficient energy use and some 9 million tons of oil through substitution (i.e. 3 million tons by wood, peat or coal; and 1 million each by solar energy, hydro- and nuclear power, plus heat pumps and district heating). On the basis of these numbers, the CO_2 emission into the atmosphere in 1990 would be reduced by about 3.9% through oil substitution and about 7.5% through improved energy use, compared with 1975 values.

VI.4.2.3 US studies

The Harvard Study - This refers to the well-known report of the Energy Project carried out by a team at the highly reputed Harvard School of Business (Stobaugh and Yergin, 1979). The authors doubt that the deregulation of fuel prices and the fast expansion of coal production and nuclear energy are politically enforceable in the USA. They also believe that synthetic fuel production cannot make an important contribution in this century because of technical problems and long market penetration times. Further, they doubt that Saudi Arabia, that is presently the only reliable oil producing land for the western world, can maintain its oil production, let alone expand it.

They come rather to the conclusion that it would be better during a transition period to concentrate on the more rational use of energy sources. The effort required for this is much smaller than the investments that would be required for the production of additional energy carriers. Thereby, about 40% of the energy used in the USA in 1973 could be saved. The development of renewable energy sources, especially all forms of solar energy, should be strengthened, then they could reach a share of 20% of the total primary energy consumption by 2000. This concept, in which more efficient energy use and the speedy introduction of renewable energy sources have highest priority, was also propagated at that time by the Carter Administration (Council on Environmental Quality, 1981). If such an energy policy were to be carried out in the future in the world's most important industrialised nation, this would be a clear signal for other countries and could diminish the CO_2 problem as well as other environmental problems.

The CONAES Study - Because of the difficulties in finding a strong concensus, this report of the Committee on Nuclear and Alternative Energy Systems (CONAES, 1980) of the US National Academy of Sciences appeared only after a delay of several years.

The main results of the study can be summarised as follows. In national and international energy policies, energy-saving measures for reducing energy growth should have highest priority. Thus, over a period of 30–40 years, measures for better energy utilisation could halve the present ratio between energy consumption and Gross National Product (GNP), which means that energy growth and GNP are only loosely coupled. The former could readily be reduced without endangering the economic welfare. However, the reduction of energy growth with the aid of energy price increases is considered a slow process.

Coal and nuclear power are seen to be necessary for energy supply in the near- and medium-term future, and a large share of coal and nuclear power is recommended for electricity production. The expeditious development of synthetic fuels (above all, coal liquefaction) should be strongly supported for supplying the indigenous market with liquid fuels and to reduce the dependence on imported oil. The Committee finds that the renewable energy sources could only make a significant contribution in the near future, if they were supported by strong federal subsidies during the market penetration period. If the synthetic fuel industry were to be strongly developed, as recommended here, and the development of the renewable energy sources were to proceed hesitantly because of lack of capital investment, then this could lead to a noticeable increase of the atmospheric CO_2 concentration (see section VI.5.2).

VI.4.3 Efficient energy use in industrialised and developing countries

Above we have only looked at the rational energy use in industrialised countries, because detailed studies are only available for these so far. This should not, however, lead to the assumption that efficient energy use is not important in the developing countries, because the per capita energy consumption is so small there. The opposite is true. If very little capital is available to develop indigenous resources or to import external resources, then also in the developing countries, as in the industrialised countries, more **effective energy use** becomes the **most important energy source** (Hauff, 1981). The efficiency of energy use is poor in industrialised countries; it is even worse in the developing countries, despite the much lower energy consumption (e.g. because of cooking over open wood fires and using obsolete machines). If we really aim for a **fair energy distribution** between industrialised and developing countries, then this cannot be achieved by an increase of energy production and energy demand, but rather by a better utilisation of the available energy sources and, indeed, many opportunities exist in all sectors to save energy even at low consumption levels (Dunkerley and Ramsay, 1982). This measure produces fastest relief, is most economic, and is most likely to reduce the risk of a deterioration of environment and climate as a result of deforestation and fossil fuel consumption.

VI.5 Introduction of renewable energy resources

Measures for reducing energy waste can be supplemented in their effectiveness by the introduction of renewable energy sources that emit no CO_2 (e.g. solar energy) or CO_2 as part of a recycling (e.g. bioenergy) into the atmosphere (see Fig. VI.1). The chances of covering a significant part of the demand by renewable energy sources are small as long as there is a strong exponential energy growth. The chances increase rapidly if the energy demand is reduced by efficient energy use.

Renewable energy sources are often associated exclusively with so-called small and decentralised applications. We are not concerned, however, with a particular size but rather with providing the most sensible energy service for a particular energy demand, taking account of ecological and economic conditions, i.e. we are concerned with matching the respective demand with the most **appropriate form of energy** (Krause, 1978). Reason demands that the exhaustible and irreplaceable fuels, such as oil, natural gas, coal and uranium, must be preserved and the satisfying of our energy demand must be increasingly shifted to renewable, and thus inexhaustible, energy sources. An ecologically sound future can only be sustained on the basis of the renewable energy resources.

VI.5.1 Substitution of non-renewable by renewable energy sources

A common feature of the energy plans presented in Fig. VI.3 is the complete substitution of non-renewable by renewable energy sources within the given time period. The exception is the scenario for the F.R. Germany that is based on the assumptions for the "Sun and Coal" scenario shown in Table VI.5 and included here for comparative purposes. According to this scenario that appears rather conservative in comparison with the other proposals, the share of the renewable energy sources should only be 44% in 2030. It is also interesting that the per capita energy consumption decreases only slightly from 5.2 kW/cap. in 1973 to 3.6 kW/cap. in 2020, as a result of the better energy utilisation. This is because of the projected population decrease from 62 million (1973) to about 39 million (2030), which means that even with a further reduction of the primary energy consumption, the share of energy available per capita can increase.

Also, almost all other scenarios assume a continual increase of the energy efficiency, so that the per capita consumption decreases strongly over the given time period, especially in those countries and regions where it is very high, such as the

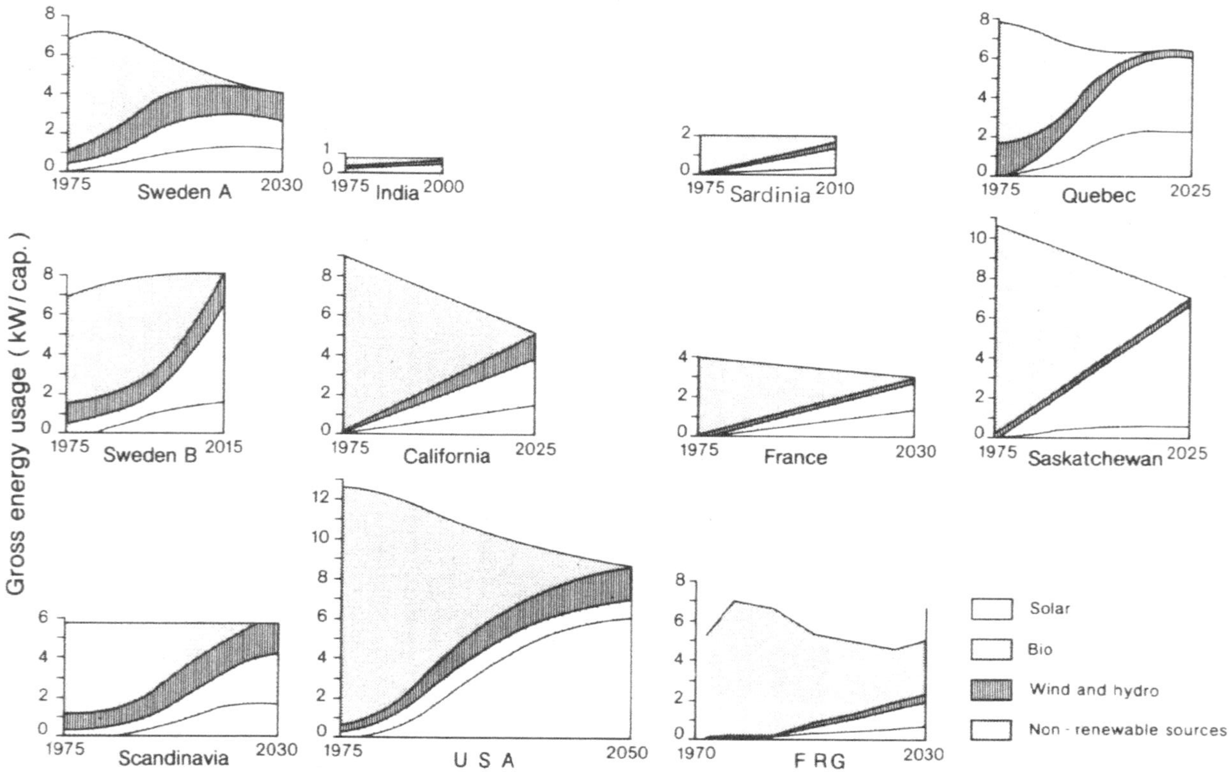

Fig. VI.3 : Per capita primary energy use for selected regions according to renewable energy models.
Source: Sorensen (1981) and supplemented.

USA, California, and Saskatchewan. Energy plans on the basis of renewable energy sources thus show, in contrast to the conventional energy plans, a progress towards a reduced energy consumption. Both strategies, namely the low energy consumption through increased efficiency and the introduction of renewable energy sources, could prevent a CO_2/climate problem.

The Canadian scenarios for Quebec and Saskatchewan (Fig. VI.3) are distinguished by a high share of bioenergy, which is not unrealistic in view of the availability of large forests. Quebec also envisages a large share for solar and wind energy and wants to gradually phase out hydropower because of environmental factors (Sorensen, 1981).

The energy scenario for California is based on direct solar energy collection for the low and high temperature applications, wind energy and bioconversion (wood and wood waste products, bioplantations on land and in coastal waters) but only little hydropower and photovoltaic conversion (Craig et al., 1978). The energy model for the USA puts more emphasis on solar and wind energy both for high and low temperature applications and electricity production by photovoltaic conversion and gives bioenergy less consideration (Kendall and Nadis, 1980).

Both of the Swedish models depend primarily on bioenergy (wood, wood waste products, peat), water, wind and solar power (Sorensen, 1981). Scenario B from 1978 shows a constant high energy consumption based on the view that Sweden has a low population density and large supply and thus does not need a more efficient energy use. This opinion has changed drastically in favour of a more careful approach to energy use with the energy law of 1981. The regional scenario which considers the whole of Scandinavia as a uniform energy market, emphasises energy exchange between the individual countries similar to the exports and imports of other goods.

The French model assumes, in contrast to the present energy policy, a lower demand for electricity (Sorensen, 1981). The remainder could then be provided by hydro- and windpower, solar energy and bioconversion. The energy plan for the island of Sardinia puts the use of bioenergy in the foreground.

VI.5.2 Global potential of renewable energy sources

If one considers the development of renewable energy sources, not only in a few countries but in the whole world, then it is appropriate to make an estimate of this largely untapped potential. Table VI.7 gives a survey of the estimated potential of renewable energy sources in 2030. The opinions on what can be implemented in practice are still very variable at the International Institute for Applied Systems Analysis (IIASA). Thus, the two estimates of what is implementable, i.e. 6.3-7.3 TW (Häfele, 1980), or 9.7 TW (Häfele and Rogner, 1980) were made only one year apart. The statements about the maximum potential

Table VI.7 : Estimated global potential of renewable energy sources for 2030.

Source	Potential (TW) theoretical (1)	practicable (1)	theoretical (2)	practicable (2)	maximum (3)	maximum (4)	System effects and constraints (5)
Solar (hard)	5.0	1.0	5.0	1.0	6	65-270	Large areas needed with hard solar, changes in heat budget (?) technology, economy
Biomass	7.5	2.5	6.0	5.1	3	10	CO_2 only recycled, changes in albedo and water budget, large areas, competition with food production
Organic wastes	0.1	0.1	0.1	0.1			None
Wind	3.0	1.0	3.0	1.0	3	2	Economy
Hydro	2.9	1.5	2.9	1.5	3		Environment and safety problems
Glacier	0.1	0	0.1	0			Environment, technology, economy
Ocean temperature gradient (OTEC)	1.0	0.1	1.0	0.5	10	1	CO_2-release from deep layers, albedo changes, effect on ocean currents, technology, economy
waves and ocean currents					1		Ocean currents, technology, economy
tidal	0.04	0	0.04	0			Only a few suitable sites
Geothermal	0.4	0.2	2.0	0.6			Emission of gases and aerosols, earthquake (?), economy
Total	20	6.3-7.3	20	9.7	26	78-283	

Sources: (1) Häfele (1980); (2) Häfele and Rogner (1980); (3) Weingart (1980); (4) Caputo (1980); (5) Matthöfer (1976); Bach and Schwanhäusser (1978); Bach (1979b).

made by Caputo (1980), who worked for some time at IIASA, cover a very wide range from 78-283 TW. The high values arise above all from the large-scale technological development of solar energy (solar towers, solar farms) in the sunny desert and fallow lands with bounding ocean regions (see also Chapter V.3.2).

At a UN conference about the prospects for renewable energy sources, there was a general consensus that renewable energy sources could make a considerable contribution to the energy supply, in the short term on the local scale and in the medium and long term on the global scale (Bach et al., 1980). The maximum potential compiled by Weingart (1980) (Table VI.7) is based on estimates made by experts at this conference. The estimates are based on the assumptions that the development and installation of renewable energy sources are purposefully supported by official bodies, that they become economically competitive and that unforeseen barriers in the social, political and environmental sectors do not arise. Given these preconditions, a maximum potential for renewable energy sources of about 26 TW is felt to be possible, which does not differ significantly from the theoretical value of IIASA.

Table VI.7 further gives a compilation of the system effects and limitations that could arise during the development of renewable energy sources. From our point of view, the possible effects of ocean energy use using the temperature gradients in tropical oceans (OTEC) and the use of biological and geothermal energy are of interest (Bach and Matthews, 1980). With OTEC systems, enormous amounts of water must be pumped from the deep ocean layers to be used for cooling purposes at the ocean surface. Since the deep water is enriched with CO_2 and nutrients, a CO_2 emission into the atmosphere and a stimulation of phytoplankton growth with a possible change of the albedo of the ocean surface could occur. The extent to which the positive feedback of the CO_2 emission would be compensated by the negative feedback of the albedo increase is as yet unclear. In addition, a large-scale OTEC development could influence the ocean currents and thus the atmospheric circulation and climate. Changes of albedo, roughness factor and the hydrological cycle etc. would also be expected with the large-scale development of bioplantations or firewood plantations. Of course, if reforestation occurred immediately, the burning of the recent biomass, in contrast to fossil biomass, would only recycle the CO_2. During the production of geothermal energy the emission of CO_2 into the atmosphere can be prevented if the so-called "closed-cycle" method were applied, in which pollutants are trapped and reinjected into the earth.

Finally, it should be noted that with the possible exception of OTEC none of the renewable energies on solar and related basis contributes to a CO_2 increase in the atmosphere. The problem with the use of renewable energy sources is apparently not the offer of too few attractive possibilities but which combination one should choose from the multitude of possibilities. Frequently, the question of costs is brought up as being a major problem for the speedy introduction of renewable energy sources. A detailed treatment of this problem is beyond the scope of this book. The economic aspects are, however, apparently quite positive, as the interested reader can see in the literature on the subject (e.g. Lovins, 1978a, b, 1979; Bruckmann, 1978; Sorensen, 1979, 1982a, b; Sant, 1979; Caputo, 1980; Zraket and Scholl, 1980; Seri, 1981; Holdren, 1981; Lovins et al., 1981; Olivier et al., 1983; Bach, 1983b).

VI.6 Effects of various energy strategies

In the following, we select a variety of energy scenarios for the purpose of assessing the wide-ranging spectrum of the differing energy strategies from fossil fuel consumption with synthetic fuel production, through high and low energy scenarios, to zero growth and a more efficient energy use. Using the most important characteristics of these scenarios, we then estimate the possible effects on the future CO_2 and temperature development.

VI.6.1 Characteristics of selected scenarios

VI.6.1.1 Synthetic fuel production scenario

The oil crisis of 1973/74 contributed to the renewed interest in coal. Following the energy-political principle of "away from oil", the development of the synthetic fuel industry (coal gasification and coal liquefaction) has generally a high priority. At the beginning of 1980 the government of the F.R. Germany passed a synthetic fuel production programme which considers 14 projects for coal gasification and liquefaction (BMFT, 1980). For these, annual amounts of 12 million tons of bituminous coal (consumption in 1980: about 77 million tons) and 10 million tons of lignite (consumption in 1980: about 38 million tons) are planned. These are to be supplied from indigenous sources (Schilling and Krauss, 1981). The development of the synthetic fuels programme is, however, slow, probably because of the high investment costs and the low competitiveness.

In the USA a strong development of the synthetic fuels industry is envisaged in the scope of the national energy programme passed in 1980. As the programme became known, a group of scientists pointed out that in the production and consumption of synthetic fuels 2-3 times more CO_2 is added to the atmosphere than with the direct combustion of coal and therefore warned against a stronger development. To clarify this question a US Senate Committee called a conference of experts (US Senate, 1979).

For this hearing, Rotty (1979) estimated the additional effect of synthetic fuels with the aid of three different energy scenarios. The input data for the first scenario came from the continuation of the second US National Energy Programme. It considers annual growth rates of 2.1% till 1985, 1.8% till 2000, 2.1% till 2020 and 2% till 2050. The synthetic fuel production should likewise increase at 2%/yr till 2000 and afterwards, however, at 5%/yr. The plan envisages a daily production of synthetic fuels of 0.14 million tons by 1985, a doubling of this value by 1990, an increase by a factor of 5 by 2000 and an increase by a factor of 60 up to 8.4 million tons in 2050. The first scenario only considers a synthetic fuels programme for the USA.

In the second scenario it is assumed that the rest of the world also has synthetic fuels production with the same growth rates as the US and proportional in size to the available fossil fuel reserves in each country. In the third scenario distinct alternatives are considered. With a projected global energy demand of 26.9 TW in 2025, 8 TW should be provided by solar and wind energy, biomass and hydropower and 4 TW by nuclear energy. A market introduction rate of 9%/yr is assumed for the non-fossil energy systems. The effects of these different scenarios are considered in VI.6.2.

VI.6.1.2 IIASA scenarios

The IIASA study investigated on the basis of various scenarios (high energy demand, low energy demand, increased deployment of nuclear energy; nuclear moratorium, zero-growth of the per capita energy consumption) the global primary energy demand and ways of satisfying it (Häfele et al., 1981). The goal of the study is to develop strategies with which the transition from dependence on oil and gas to a less vulnerable energy supply can be achieved in the next 50 years. According to the study, the transition will probably proceed more slowly than originally assumed and take place in two stages. The first stage is the transition from the conventional fuels, oil and gas, to the more environmentally damaging and more expensive oil production from oil shales or tar sands, as well as coal mining and synthetic fuels production. This process should be finished by 2030. Then, the transition to solar and nuclear fuels should follow until the end of the next century.

Table VI.8 shows the projections of the global primary energy demand for 2000 and 2030, for both a high and a low scenario. With an assumed doubling of the global population from 4 to 8 billion people, the high scenario assumes an energy demand increase of a factor of 4.4 in 2030 in comparison to the base year 1975. The scenario referred to as low has a factor of 2.7 energy increase over the same time period.

Table VI.8 : Global primary energy supply (TWyr/yr) for different energy sources and a variety of energy scenarios, 1975–2030.

Source	Reference Year		IIASA - Scenarios (1981) High Scenario				IIASA - Scenarios (1981) Low Scenario				Colombo/ Bernardini (1979) 16 TW (2030) Scenario[1]				Lovins et al. (1981) Efficiency Scenario			
	1975		2000		2030		2000		2030		2000		2030		2000		2030	
	TW	%	TW	%	TW	%	TW	%	TW	%	TW	%	TW	%	TW	%	TW	%
Oil	3.83		5.89		6.83		4.75		5.02		4.26		3.58		1.77		0.24	
Gas	1.51		3.11		5.97		2.53		3.47		2.27		2.48		1.51		0.34	
Coal	2.26		4.94		11.98		3.92		6.45		3.51		4.60		1.77		0.38	
Subtotal	7.60	92	13.94	83	24.78	69	11.20	82	14.94	67	10.04	82	10.66	67	5.05	71	0.96	18
Nuclear (LWR)	0.12		1.70		3.21		1.27		1.89		1.13		1.36		-		-	
Nuclear (Breeder)	0		0.04		4.88		0.02		3.28		0.02		2.35		-		-	
Subtotal	0.12	2	1.74	10	8.09	23	1.29	10	5.17	23	1.15	10	3.71	23	-	0	-	0
Hydro	0.50		0.83		1.46		0.83		1.46		0.74		1.04					
Solar[2]	0		0.10		0.49		0.09		0.30		0.08		0.20		2.02		4.27	
Other[3]	0		0.22		0.81		0.17		0.52		0.16		0.36					
Subtotal	0.50	6	1.15	7	2.76	8	1.09	8	2.28	10	0.98	8	1.60	10	2.02	29	4.27	82
Total	8.22		16.83		35.63		13.58		22.39		12.17		15.97		7.07		5.23	
Index	100		205		435		166		273		146		195		86		63	

1 Using the fuel shares of the low IIASA scenario; the 16 TW scenario is also known as the "zero growth scenario" because it assumes that the present per capita energy consumption will be the same in 2030 but that a doubling in world population from 4 to 8 billion will lead to a doubling in total energy consumption.
2 Mostly soft solar (e.g. rooftop collectors) but also small amounts of centralized solar electricity.
3 Includes biogas, geothermal, commercial wood use.

Sources: Colombo and Bernardini (1979), Häfele et al. (1981) and Lovins et al. (1981).

For the question we are considering, the development of the absolute amounts of the fossil fuels and the changes of their relative shares are important. In both scenarios the absolute amounts of oil, gas and coal increase for both 2000 and 2030. The increase is greatest for coal production. According to the arguments of Chapter III.4, however, there is good reason to doubt the factor of 5.3 increase of the coal supply in the high scenario for 2030, in view of the limited export capacities and the cost developments, etc.

The fossil fuel shares of oil (50%), gas (20%) and coal (30%) in 1975, shift in the direction of coal in both scenarios. In the high scenario in 2030, the distribution is coal (48%), oil (28%) and gas (24%). Both IIASA scenarios show not only a shift towards coal but also the greatest absolute increase in the coal production sector. This is of no small importance, since coal has a much higher CO_2 emission per unit energy production than oil and gas. Clearly, in both scenarios the fossil energy share remains dominant up till 2030. If the future energy supply were to develop along the lines of the IIASA scenarios, then one must reckon with a considerable increase of the CO_2 emission in comparison with energy scenarios that envisage a lower fossil fuel use or a more favourable energy mixture.

VI.6.1.3 "Zero growth" scenario

In contrast to the IIASA scenarios, whose purpose is to satisfy the energy demand by adjusting the supply, the so-called "zero growth" scenario of Colombo and Bernardini (1979) and Colombo (1982) assumes a lower demand. "Zero growth" refers here to the mean global per capita energy consumption of 2 kW in 1975, which is maintained until 2030. With the assumed population growth from 4 billion today, to 8 billion people in 2030, the present commercial total energy consumption is doubled, from about 8 TW to about 16 TW (Table VI.8). The expression "zero growth" is misleading, particularly for the questions being asked here, because in the estimation of effects the total amount and individual shares of the fossil fuel amounts are of prime importance. For 2000 and 2030, the mixture of fossil fuels corresponds to that in the low IIASA scenario.

This 16 TW scenario, that was prepared for the Commission of the European Communities, considers no significant changes in the economic and social sectors in the model calculations. A major characteristic of this scenario is the emphasis of the rural in contrast to the urban development, especially in developing countries. The average urban population fraction of about 30% in 1975 would, according to this study, also be valid in 2030 (see also Chapter III.2.2 and Table III.3). Since, on average, the energy service demand per unit gross national product is 30%

lower in rural areas, this scenario arrives at the relatively low global energy demand of 16 TW in comparison with the IIASA scenarios.

VI.6.1.4 Efficiency scenario

This scenario, developed by Lovins et al. (1981) for the German Federal Environmental Agency is based on a two-pronged strategy: by increasing the energy productivity the desired energy services can be provided with less energy, and the strongly cut demand for fossil fuels can then be relatively quickly supplied by renewable and CO_2-free energy sources.

The starting point of the study is the energy-economic situation of the F.R. Germany in 1973 (before the energy crisis). It investigates, using 15 consumer sectors, by how much the specific energy intensity (i.e. the energy amount needed to produce a unit of good or service) can be realistically reduced. For the extrapolation of these findings to the global scale, the same population increase to 8 billion people in 2030, as in the IIASA and "zero growth" scenarios is assumed, and the economic growth rates of the low IIASA scenario for industrialised countries (2.04%/yr from 1975-2000 and 1.19%/yr from 2000-2030) and developing countries (1.86%/yr and 1.53%/yr for the corresponding time periods) are taken. It is further assumed that the present global average energy consumption of 2 kW will reach the energy consumption level of the F.R. Germany of 1973, i.e. 5.2 kW/cap. On the basis of detailed national and international studies it is shown that with the aid of energy technologies that are already economic and that are already in use or will soon be applied, the energy consumption can be reduced considerably in all sectors.

Table VI.9 shows the large energy savings both due to structural changes and better end-use technology in the individual regions by 2000 and 2030. The interesting result is that, despite the assumed doubling of the global population and the 2.6 fold increase in global per capita energy consumption, the primary energy demand can be reduced on the basis of efficiency improvements by 14% in 2000 and 37% in 2030, compared with the 1975 value. This means, if nations adopt an economically efficient energy strategy, their energy needs can be expected to go down not up.

For the CO_2/climate question, the critical variables include the absolute amounts of fossil fuels burned, the change in their relative shares, and the change in the absolute contribution of CO_2-free energy sources. We see from Table VI.8 that efficiency improvements can effect a reduction in fossil fuel demand from 7.6 TW in 1975 to less than 1 TW in 2030 with a concomitant increase of CO_2-free renewables from 0.5 TW to 4.3 TW over the same time period. The 1975 fossil fuel shares of oil (50%), coal (30%) and gas (20%) shift by 2030 to coal (40%), gas (35%) and oil (25%). Since the CO_2 emission of coal is 1.85 times and that

Table VI.9 : Possible world primary energy demand (excluding feedstocks) with efficient use, strong economic growth, and constant urban fraction.

Region	1 primary energy TW	2 GDP / 1975 GDP		3 ES/GDP[1] / 1975 ES/GDP		4 E/ES[2] / 1975 E/ES		1×2×3×4 primary energy demand TW[3]	
	1975	2000	2030	2000	2030	2000	2030	2000	2030
1. North America (U.S.A. and Canada)	2.65	1.68	2.37	0.8	0.65	0.5	0.26	1.78	1.06
2. U.S.S.R. and East Europe	1.84	2.57	4.98	0.8	0.65	0.5	0.26	1.89	1.55
3. West Europe, Japan, Australia, New Zealand, South Africa, Israel	2.26	1.67	2.46	0.8	0.65	0.5	0.26	1.51	0.94
Developed countries	6.75	1.97	3.27	0.8	0.65	0.5	0.26	5.18	3.55
Developing countries	1.46	2.35	4.92	1.1	0.90	0.5	0.26	1.89	1.68
World	8.22	2.13	3.69					7.07	5.23
Index	100							86	63

[1] ES/GDP = Energy service intensity per unit of gross domestic product (*structural changes*, e.g. from energy-intensive to less energy-intensive sectors with more emphasis on research and development).
[2] E/ES = Primary energy intensity of services delivered to end-users (better *end-use technology*, e.g. heat from the radiator or motive power from the power train).
[3] May not exactly equal column products due to rounding.

After: Lovins et al. (1981).

of oil 1.50 times greater than that of gas, these shifts are important. The effects of these shifts on CO_2 and temperature are investigated next and they are compared with those obtained from other energy strategies.

VI.6.2 Effects of various energy strategies on CO_2 and temperature changes

VI.6.2.1 Synthetic fuel strategy

For the three scenarios described in section 6.1.1, we obtain the carbon emissions and atmospheric CO_2 concentrations for 1978–2050 given in Table VI.10. In all three scenarios the development is about the same until 2000. Up till 2050 the carbon emission in scenarios 1 and 2 increases by a factor of 8 to 9, which leads almost to a doubling of the atmospheric CO_2 content. The fact that the atmospheric CO_2 concentration doubles by 2050, compared with today, regardless of whether the USA or the whole world develops a synthetic fuels industry, is surprising. From this Rotty (1979) has concluded that it is not so much the kind of fossil fuel source but more the magnitude of the fossil energy consumption that is decisive for the high CO_2 concentration.

An atmospheric CO_2 concentration of 600–700 ppmv represents a critical threshold range, as we have seen in Chapter IV.6.2 and Fig. IV.27. Only with the transition to the expeditious develop-

Table VI.10 : Projection of synfuels industry growth and effect on atmospheric carbon dioxide.

	Energy Scenarios					
	Use of synthetic fuels in U.S. only (1)		Use of synthetic fuels also in other countries (2)		Use of non-fossil energy (3)	
Year	C-emission Gt	CO_2-concen. ppmv	C-emission Gt	CO_2-concen. ppmv	C-emission Gt	CO_2-concen. ppmv
1978	7.2	335	7.2	335	7.2	335
1990	10.70	354.9	10.89	355.1	10.34	353.2
2000	13.98	377.4	14.32	378.1	13.08	374.5
2010	17.52	406.3	18.06	407.8	15.33	400.7
2020	23.65	444.7	24.57	447.5	17.67	432.0
2030	31.92	496.5	33.43	501.6	18.68	467.9
2040	43.09	566.4	45.59	575.3	10.1	496.7
2050	58.17	660.8	62.29	675.9	-	501.4

Adapted from: Rotty (1979).

ment of non-fossil energy carriers in scenario 3 is the increase of the atmospheric CO_2 concentration drastically reduced in comparison with the other two scenarios. This underlines the assumption that together with more rational energy use, the forced development of non-fossil energy sources is an effective measure for reducing the CO_2 increase. The extent to which allothermal processes (in which the synthetic fuel production is carried out using non-fossil energy sources; see Chapter IV.1.4.2) affect the CO_2 content of the atmosphere has not yet been investigated (Niehaus, 1981).

VI.6.2.2 Pure fossil fuel strategy

With the aid of three models, Niehaus (1979) investigated the influence of the hypothetical case of a pure fossil fuel strategy. With a global energy model, similar to the IIASA energy models, the future energy demand is estimated (Voss and Niehaus, 1977); using this information, a carbon cycle model calculates the future development of the atmospheric CO_2 (Niehaus, 1977); and the temperature increase resulting from the CO_2 increase is estimated using a one-dimensional radiative-convective model (Augustsson and Ramanathan, 1977).

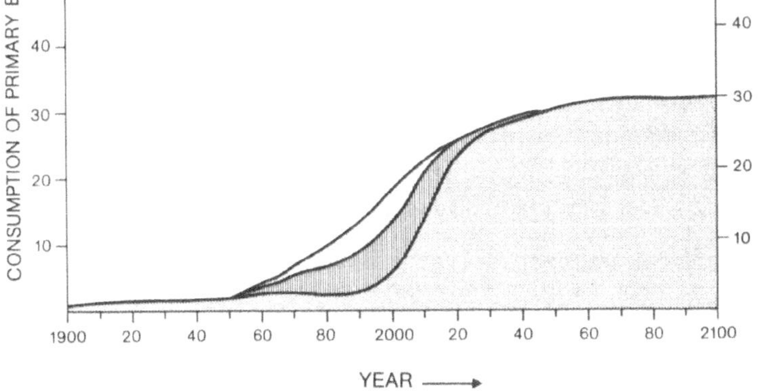

Fig. VI.4 : The 30 TW fossil fuel strategy.
Source: Niehaus (1979).

The simulations assume a global energy demand that, beginning in the middle of the 21st century, asymptotically approaches a value of 30 TW by the end of the century (Fig.VI.4) and is thus in the range of many conventional projections (see section VI.4.1). With the exclusive use of fossil fuels (coal, oil, gas) the CO_2 emission would increase by about a factor of 5 from about 5 billion tons C/yr today to 24 billion tons C/yr, and the CO_2 concentration would increase by almost a factor of 3 from about 335 ppmv to more than 1,000 ppmv (Fig. VI.5). The resulting mean global temperature increase would be about 4°C above the base value of 1967, which would have serious climatic consequences as shown in the results of Chapter IV.6.2.

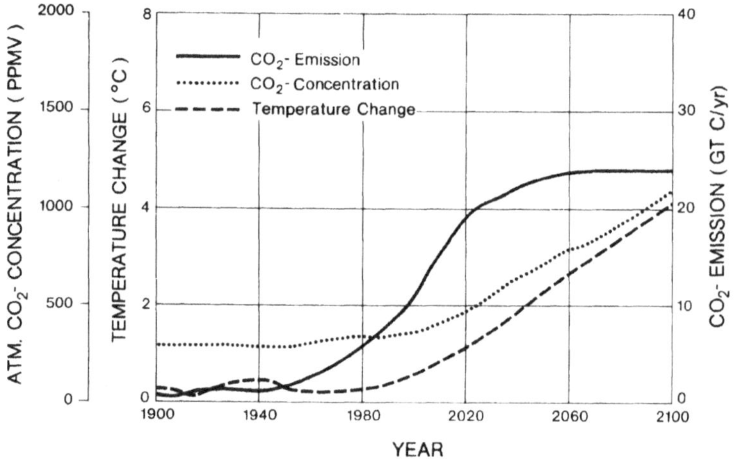

Fig. VI.5 : Simulated CO_2 and temperature change for the 30 TW fossil fuel strategy.
Source: Niehaus (1979).

VI.6.2.3 Spectrum of various energy strategies

The basis for a more realistic estimate of the potential climate risk can only be provided by a comparison of energy scenarios that deal with an order of magnitude and energy mix that are held by many to be possible. The information gained in such an exercise establishes the starting point for precautionary measures.

Using the current approach that was described in section VI.6.2.2, the future CO_2 and temperature developments are computed (Table VI.11) with the aid of the energy data for the

Table VI.11 : Development of CO_2 emission, CO_2 concentration and temperature change from 1975–2030 for the energy scenarios in Table VI.8.

Item	Reference Year		IIASA - Scenarios (1981) High Scenario				IIASA - Scenarios (1981) Low Scenario				Colombo/ Bernardini (1979) 16 TW (2030) Scenario				Lovins et al. (1981) Efficiency Scenario			
	1975	%	2000	%	2030	%	2000	%	2030	%	2000	%	2030	%	2000	%	2030	%
CO_2-Emission (GtC/yr)																		
Oil	2.32	48	3.77	41	4.37	26	3.04	41	3.21	32	2.73	41	2.29	32	1.13	35	0.15	24
Gas	0.65	13	1.34	14	2.57	15	1.09	15	1.49	15	0.98	15	1.07	15	0.65	20	0.15	24
Coal	1.72	35	3.77	41	9.14	55	2.99	40	4.92	49	2.68	40	3.51	49	1.35	41	0.29	48
Cement & Gas Flaring	0.19	4	0.35	4	0.62	4	0.29	4	0.38	4	0.25	4	0.27	4	0.11	4	0.02	4
Total	4.88	100	9.23	100	16.70	100	7.41	100	10.00	100	6.64	100	7.14	100	3.24	100	0.61	100
Index	100		189		342		152		205		136		146		60		13	
CO_2-Concentration (ppmv)	329		379		495		372		445		368		424		354		364	
Index	100		116		150		113		135		112		129		107		111	
Temperature Change (°C)																		
with Ocean Effect	0.4		0.8		1.6		0.7		1.3		0.7		1.2		0.6		0.8	
Index	100		200		400		175		325		175		300		150		200	
Temperature Change (°C)																		
without Ocean Effect	0.7		1.3		2.5		1.2		2.1		1.2		1.8		1.0		1.1	
Index	100		186		357		171		300		171		257		143		157	

various scenarios given in Table VI.8. The historic development of the CO_2 emission from 1860 to 1975 (Fig. VI.6) is based on UN consumption statistics (see Chapter IV.1.4.1), and the future development from 1975 to 2030 on the data in Table VI.8. The calculations of the historical curve of the CO_2 concentration in the atmosphere from 1860–1957 in Fig. VI.7 was carried out using the carbon cycle model described in Chapter IV.1.6.2. The measured values from 1957–1975 come from the longest available series of observations at Mauna Loa, Hawaii (see Chapter IV.1.3.1). The estimate of the future development of the atmospheric CO_2 content was made using the carbon cycle model referred to above and the information in Table VI.8. The temperature values shown for 1880–1970 in Fig. VI.8 are based on observed annual average temperatures in the latitude band 0°–80^oN (Mitchell, Jr., 1975). The base point is the annual mean value from 1880–1884. With the future development of the CO_2 content as input, the resulting temperature curve can be calculated, e.g. with the aid of an energy balance model (see Chapter II.3.2 and Appendix II.3), coupled to an ocean model (Cess and Goldenberg, 1981). The ocean model used here consists of a mixed layer about 70 m deep and the deep sea of 3–4 km depth. The calculation of the global temperature change is a result of the averaging of the sea surface temperature and continental temperature taking into account their proportional area.

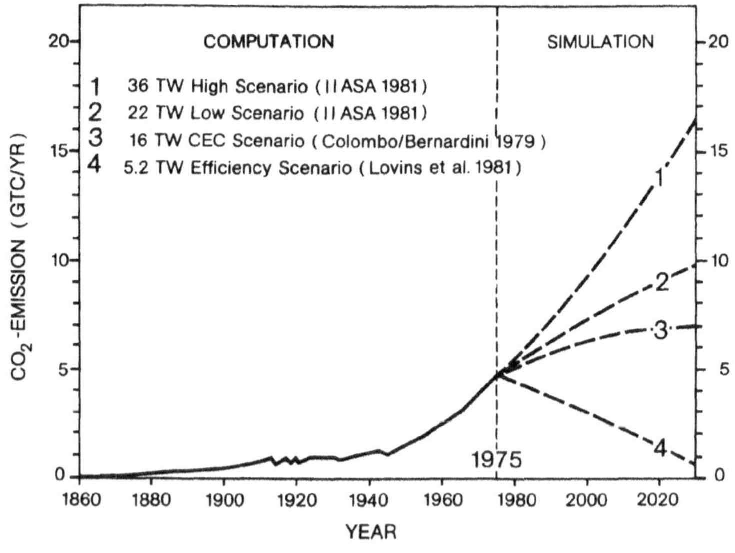

Fig. VI.6 : CO_2 emission for a variety of energy scenarios. Computation: 1860–1975. Simulation: 1975–2030.

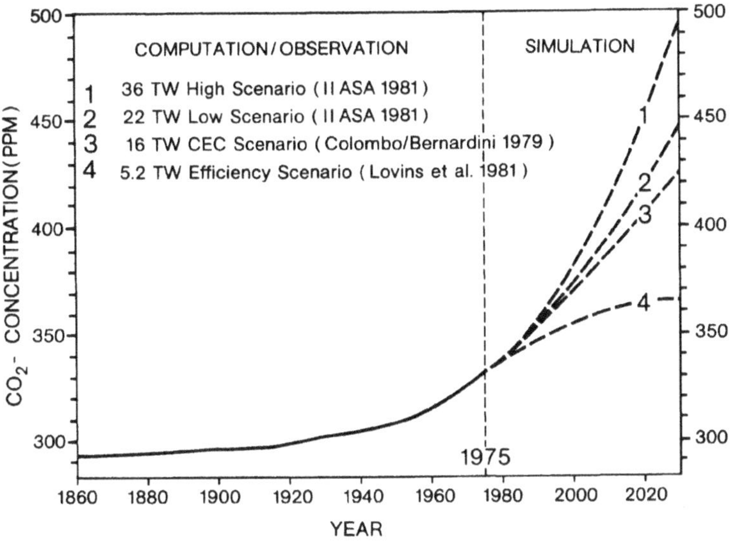

Fig. VI.7 : CO_2 concentration for a variety of energy scenarios. Computation: 1860–1957. Observation 1957–1975. Simulation: 1975–2030.

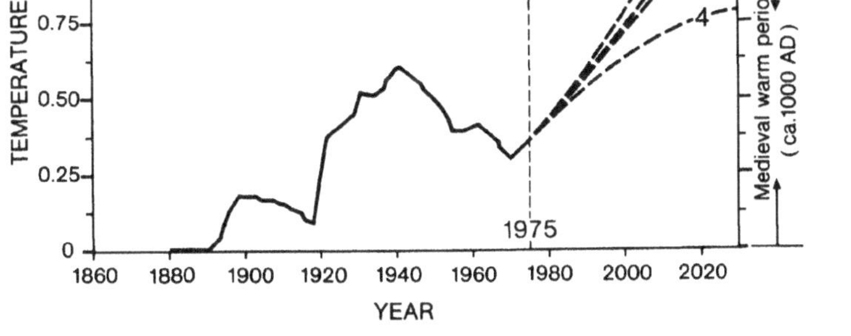

Fig. VI.8 : Temperature change for a variety of energy scenarios. Observation: 1880–1975. Simulation: 1975–2030.

The results of the model calculations give valuable indications for the choice of precautionary measures. For example, in comparison to 1975, in the high IIASA scenario, the CO_2 emission is a factor of 3.4 larger and the CO_2 concentration about 50% higher resulting in a 4 fold increase in excess temperature over the 1880 value from 0.4°C to 1.6°C (Table VI.11 and Figs. VI.6–VI.8). In the efficiency scenario, on the other hand, the CO_2 emissions decrease by a factor of 8 over the same time period, so that the CO_2 concentration only increases by 11%. The resulting temperature increase of 0.8°C is only half as large as that in the high IIASA scenario.

Beside the actual magnitude of these numbers, the inertia of the atmosphere-ocean system is important. In the efficiency scenario the CO_2 emission is notably reduced straight away in 1975 (Fig. VI.6), but only after about 54 years in 2029 does the atmospheric CO_2 concentration respond with a slight decrease (Fig. VI.7). The temperature, on the other hand, continues to increase slightly (Fig. VI.8). The thermal inertia of the ocean reduces the warming effect (see in Table VI.11 the numbers with and without ocean effects), and thereby delays the point at which a temperature increase can be detected by some decades (see Chapter IV.1.7.2).

The pure numerical results of the simulations must be interpreted with care in view of the many remaining uncertainties, especially in the carbon cycle and climate models. Given the fact that economic and energy policy decisions have long-term effects that could influence environment and climate in the next 30–50 years drastically or even irreversibly, the results should, however, be appraised despite their preliminary character.

According to our present knowledge the first critical threshold area is for a CO_2 concentration of 400–450 ppmv with a corresponding mean global temperature increase of about 1–1.5°C (see Chapter IV.6.2 and Fig. IV.27). Catastrophic climatic changes would be expected with an increase of the CO_2 concentration to 600–700 ppmv in combination with other influencing factors and the resulting mean global temperature change of 4–5°C. According to Fig. VI.8 the temperature increase for the high IIASA scenario is somewhat above the first critical threshold; the low IIASA scenario and the "zero growth" scenario are within the so-called Holocene Optimum with a warming that has not been experienced in the last 6,000 years. Also the efficiency scenario leads to a warming which has not been equalled since the Middle Ages, about 1000 AD. Moreover, it is well to realise that the above temperature simulations are not based on any forcing functions other than carbon dioxide.

The most important result is, however, that apparently only the measures of the efficiency scenario (section 6.1.4) lead to a significant reduction of the temperature increase. Thus, something very important would be gained, namely **time** for the continual transition to a **sustainable energy future**, which does not change the environmental and climate system in an inadmissable way. The measures introduced in the efficiency scenario are immediately effective through the more efficient energy use, since they contribute directly to the reduction of the fossil fuel combustion. This provides the time that is necessary for the longer-term transition to CO_2-free renewable energy sources. Both measures thus complement one another perfectly (Bach, 1983a).

VI.7 Flanking measures

Up till now we have treated, above all, measures which are purposefully introduced to reduce or prevent the CO_2/climate problem. In addition there is a multitude of other activities that have a completely different purpose than the CO_2 reduction, but by reducing the vulnerability of our environmental system to a CO_2-induced climatic change could have positive side-effects.

The flanking measures that can make us more resilient to climatic changes include, in particular (Roberts et al., 1980; Kellogg and Schware, 1981, 1982):

The protection of the top soil - Every year enormous areas of agricultural land are affected by erosion, salinisation, desertification or turn to steppe to a large extent as a result of poor soil management (Glantz, 1977; Kuntze et al., 1981; Hallsworth, 1981; Brown, 1981). A possible increase of climatic variability can strengthen this process, which would have serious consequences for the security of global food supply. The **protection** of the **soil** is one of the most important tasks of mankind, that is unfortunately most badly neglected. Apparently, however, organic cultivation methods that protect the soil can be competitively applied both in developing countries (Glaeser, 1980) and the highly specialised agro-industrial countries, such as the USA (Lockeretz et al., 1981).

Application of agricultural technology - Agricultural technology, which is exclusively devoted to the application of enormous amounts of artificial fertilizers and the cultivation of highly specialised but disease-vulnerable plant types, has apparently brought the "Green Revolution" in disrepute (Lappe and Collins, 1977). Agricultural technology, which, on the other hand, strives to improve the irrigation systems, to cultivate plants that tolerate saline soils or bind nitrogen from the air and thus require no artificial fertilizers, makes plants more resistant to climate variability and maintains genetic diversity, is to be welcomed (Lindh, 1981; Hodges et al., 1981; Hekstra, 1981; Cooper, 1982).

Maintaining a food reserve - This should primarily serve to bridge emergency situations, such as droughts (Schneider and Bach, 1981). In recent years, the reserves have been almost used up through the frequent occurrence of harvest failures (see Chapter V.4). In particular, the lands with sufficient foreign exchange holdings, such as the USSR and the People's Republic of China control the grain imports, so that hardly anything is left for poor Third World countries.

Improvement of water management - The fertility of about 40-60% of all irrigated lands is much reduced already by salinisation; and no less than 20 million hectares are lost annually as a result of it (Kovda, 1979). In the planning of irrigation systems, dams and reservoirs, as well as straightening and diversion of rivers, all imaginable factors and side-effects must be considered, otherwise more damages can be caused than are prevented in times of drought and floods.

Enhanced coastal protection - Storms and high-water catastrophes afflict whole coastal regions over and over again (Burton et al., 1978). They often cause large property damages and frequently they cost many lives. In the course of time, the appropriate precautions can be taken against a CO_2-induced warming with a possible sea level rise (see Chapter V.7.2) by strengthening and raising dykes.

Development of support programmes - With the increasing population (by 2000 about 2 billion more people) the effects of climatic influences will increase strongly in magnitude, especially in the food sector. Aid programmes must be developed appropriately in order to be prepared for all unforeseen events. In the past, supporting activities did not always meet with undivided agreement, since the aid, rightly or wrongly, was not always given in the best interest of those seeking help (Garcia, 1981).

VI.8 Decision aids

Objective decisions are based on a good information system. Timely decisions require a warning and precaution system. It is, however, not enough only to measure climatic events or only to calculate possible CO_2-induced climatic changes. If the whole system is to make sense, the knowledge must flow into the decision-making process. Sensible **decision aids** are characterised by the following chain of actions: Observe/calculate → inform → decide. In the following paragraphs I sketch some of these activities as part of a **total system for precautionary measures.**

World Climate Programme - The worry about possible climatic changes on a global scale and the understanding that for the future planning and handling, mankind should have a better knowledge of the functioning of the climate system, eventually led the 8th Congress of the World Meteorological Organisation to approve the following programme (WMO, 1980; Kellogg, 1981):

- **Climate Data Programme, CDP** - The main purpose of this programme is the more reliable observation of climate through a more complete monitoring network with beter data collection and data processing.
- **Climate Applications Programme, CAP.** - The information from this programme should be profitably applied, especially in the climatically vulnerable human sectors, such as food security, availability of water reserves, energy supply, settlement and health, etc. A better application of the climate information should especially benefit the more vulnerable developing countries.
- **Climate Impact Study Programme, CIP** - In this case the effects on the vital interests of humanity, referred to in the previous programme, are considered so that they can be avoided through appropriate measures. The responsibility for this programme is shared by the WMO and the UN Environmental Programme (UNEP).
- **Climate Research Programme, CRP** - This is to provide an improved understanding of the physical processes that are the basis of the climate system and climatic changes and is a continuation of the former Global Atmospheric Research Programme (GARP). The responsibility is shared by the WMO and the International Council of Scientific Unions (ICSU).

Informing the public - Here, public refers to both the individual citizens and the decision-maker. It is simpler to interest the public in a local problem because it is more readily appreciated by those affected. The CO_2/climate problem has, on the other hand, a global dimension, whose effects remain incomprehensible to most people. In addition, the CO_2 effect is not yet detectable. Because of this some conclude - perhaps correctly - that the CO_2 problem is not taken seriously enough and is even suppressed in public. Others, it is claimed, use it as an alibi, to divert attention from their own difficulties. In addition, the public is confused and unsure because of the reports in the media of apparently contradictory statements of scientists about glacial and interglacial periods. The result is a **lack of credibility**, which must first be rectified if the remaining problems are to be mastered.

The CO_2/climate problem involves almost all disciplines, from the natural sciences to the social and economic sciences. Unfortunately, many are not prepared to do interdisciplinary work because they are prejudiced through a lack of knowledge about the capabilities of the other sciences. Given such an attitude, the public can hardly expect realistic recommendations for the solution of their complex problems. The problems must be attacked, however, with all of the scientific methods available to us. But it is certainly not enough just to write scientific articles. The knowledge must also be passed on in appropriate form to the user. Science has an **obligation to advise**, but the public also has an **obligation to consult.**

As we have seen in the previous chapters, the CO_2/climate problem is the result of complicated interactions within an extremely complex dynamic system with non-linear feedbacks, whose interlinked structures are part of a **cybernetic system** (Vester, 1980). The information about it is indeed available to the public but in view of the difficult material, it is not surprising that the absorption of the information remains limited. A relatively new educational tool could provide assistance, since it is especially suited for making difficult, interlinked processes more transparent.

This new approach is called **gaming,** in which the players assume the roles of the decision-makers and either, as in monopoly, can represent their own interests or those of a firm or a town, or, as in the case of the CO_2/climate problem, can represent the interests of the global public (Ausubel et al., 1980; Robinson and Ausubel, 1981). This means that a number of **goals** can be achieved simultaneously. In order to be able to play successfully, the participants must prepare carefully and have an understanding of the entire CO_2/climate problem. Such things as model development and scenario analysis are thereby practiced. The participants gain a deepened understanding, not only of the physical/chemical/biological aspects, but also the social/economic/institutional/strategic behavioural aspects of this problem. The greatest **educational value** comes from the fact that scientists, decision-makers and citizens work together on the solution of the problem and thus the understanding of the other's problems, and also the ability to judge what is sensible at any given time can increase.

International cooperation - There is no country in the world that is not somehow involved in the CO_2/climate problem. The type and magnitude of the effects are, however, quite variable. As we have seen in Chapter IV.1.4.1 and Fig. IV.5, the industrialised countries of the middle latitudes in the northern hemisphere are primarily responsible for the CO_2 emissions from the consumption of fossil fuels. On the other hand, the CO_2 emissions due to deforestation and land use changes largely come from the developing countries in the tropical regions. But the industrialised countries are also involved in the destruction of the latter ecosystems, directly as a result of timber and pasturage and indirectly as a result of neglected aid, thus perpetuating the subsistence agriculture in the developing countries.

It is true that a global climatic change, set off by the CO_2 increase in the atmosphere, would affect all countries, since we are literally all sitting in one boat. Of course, there will be differences in degree of the effects. Already, the developing countries in the belts between steppe and desert are most vulnerable. The industrialised countries are probably less susceptible since, with a lower population pressure than in the developing countries, they can resist possible effects using their larger technical potential. The fact that they have greater economic and technical possibilities obliges the industrialised countries to take the lead in solving the CO_2/climate problem (Bach, 1979a).

A whole series of supranational organisations are already working on the CO_2/climate question. In addition, a number of committees, within the organisations referred to above, such as WMO, UNEP and ICSU and also the UN organisation for Education, Science and Culture (UNESCO) and the Food and Agriculture Organisation (FAO) are involved. In particular, the Scientific Committee for Problems of the Environment (SCOPE), the Man and the Biosphere (MAB) programme, the Committee for Climate Change and the Oceans (CCCO) and finally the Interparliamentary Oceanographic Commission (IOC) are the most important (USNAS, 1980).

The work of the scientists and the coordination by the scientific organisations result in a broad information flow, which must be globally directed to the decision-makers in order to help them reach decisions. This would close the circle of information acquisition → information transmission → information use → introduction of corrective measures for the solution of the CO_2/climate problem.

Not blind opposition to progress,
but opposition to blind progress.

Sierra Club

VII. OPPORTUNITIES FOR THE FUTURE

What realistic possibilities do we have to avoid a serious threat to climate due to CO_2 and other influences? Up till now we have concentrated on the development of energy, carbon and climate models, in order to recognise the danger in due time, and we have built a global observation network, in order to prove a possible influence on climate. That is, however, not enough. In addition, measures for reducing the CO_2/climate risk or avoiding the CO_2/climate problem must be introduced now, and these clearly fall in the socio-economic-institutional field.

It is necessary to see the CO_2/climate problem as a part of the other important human problems. If the historical trends continue we run the risk of a vicious circle, namely: population increase → energy growth → climatic impact → environmental destruction → threat to food security → arms race → international conflicts → increased deployment of energy-intensive technologies → resource scarcity, etc. (Bach, 1983b).

In order to break out of this vicious circle, we must learn to **think differently**. Above all we must recognise that **climate**, as part of the environment, is a **common good**, which must be treated cautiously and with consideration. Man has the means to steer an influence on climate. The key to the avoidance of a CO_2/climate problem is to place **new emphasis** on a sensible energy and land use policy.

VII.1 The old credence in progress – the new credibility

The generation that experienced the Second World War and that carried the main burden of the reconstruction is rightly proud of the high **living standard** that has been achieved. In order to maintain this level, which has, so far, only been reached in the major industrialised countries, mostly complex technological means are introduced. This has led to a plundering of our non-renewable resources and to a noticeable influence on the environment. Instead of analysing the environmental problems ahead of time, and thus deriving necessary precautionary measures, most governments and their clumsy bureaucracies tend to wait until the difficulties are visible to all. If this is followed, mostly as a result of public pressure, by some reaction, it is ill-conceived, hasty and is often limited to the subsequent observational confirmation of the problem, without introducing simultaneous measures for hindering a repetition.

Particularly in the younger generation this has led to the feeling that "with regard to the **blind faith in progress**, they are no longer confident that this opaque complexity of technological development is going in the right direction. In place of this lost confidence, there is, on the contrary, a fear that the technical complexity leads unavoidably to a precipice. Confidence in technical progress can therefore no longer be the starting point for a responsible environmental protection and security policy in the area of technological developments" (Baum, 1981).

Basically, a **fading credibility** can only be regained, if those responsible either change their attitude or make it more understandable (Meyer-Abich, 1981). Therefore, one of the main duties of the politicians must be to reduce, step-by-step, the immediate worries about survival and the fear of a future inimical to life and to replace this by a policy of mutual trust and assistance for a more humane life-style. Thus, the question arises, to what extent is man still free to shape his own future?

VII.2 Man: Shaping his own future?

A look at the past can help to answer this question. About 1,000 years ago, there was a quite sudden collapse of the Mayan Civilisation. With the help of palaeoecological detective work, the events that could have led to the collapse can be reconstructed (Deevey et al., 1979; Wilhelmy, 1981).

The core of the Mayan Civilisation was situated in an area of about 20,000 km^2 with lakes and tropical rain forests in the northern part of present-day Guatemala. The first agricultural settlement is shown to have existed about 800-900 BC, i.e. during the time of Homer. For the entire 17 centuries of Mayan history there was a uniform, low population growth rate of only 0.17%/yr (the present globally averaged rate is 2%/yr), which corresponds to a doubling time of 408 years. Such a low growth rate must have seemed like a state of equilibrium to the Mayan leaders; and yet the Mayan population reached about 5 million people in about 850 AD, which roughly corresponds to the high settlement density of the major river valley settlements. Lake deposits show that, in about 250 AD, almost all of the jungle had been cut down and that soil erosion and loss of nutrients no longer permitted intensive cultivation: overstressing of the natural resources, population growth and socio-revolutionary agitation – and, to some extent, warlike disputes with the highland tribes – apparently led to the collapse of the Mayan Culture. It really looks as if the Mayan leaders and their economic advisers, as those taking part in the events, did not recognise the problems ahead of time to be able to introduce effective countermeasures. Those observing from the present-day perspective and with the present state of knowledge, find it easier to recognise the difficulties of that period. The question is whether we, as participants in the present-day events, are capable of realistically assessing the potential dangers and to respond appropriately.

Let us consider an example from the present. In the 1950s and 1960s, the savanna and steppe regions on the southern borders of the Sahara, known as the Sahel region, were favoured by an unusually long period without droughts (Mensching, 1978; Schneider and Bach, 1980). The agronomical boundary that separates the areal cultivation from traditional grazing lands, lies where there are 4 to 5 months with sufficient rainfall. In the climatically favourable period, this boundary shifted as much as 200 km northwards, so that arable farming could extend at the expense of grazing areas. When a high rainfall variability in the 1970s gave rise to frequent dry periods, this zone, with a mixture of cultivation and grazing, was particularly affected. The extension of cultivation had led to a contraction of the grazing area and more and more animals were kept on an ever-decreasing grazing area. The water holes used in common did not supply sufficient drinking water for the large numbers of cattle and most of the animals had to be slaughtered or they perished. One main reason for the famines in the Sahel area is certainly that the **carrying capacity of the land** was exceeded, thereby destroying the ecological base. Another important reason might lie in the large extension of cash crops at the expense of traditional, indigenous cultivation and grazing methods that are more resistant to climatic stress factors.

In an essay "The Tragedy of the Commons", Hardin (1968) gives an account of what happens if the interests of the individual are maximised when there is limited availability of common goods: it leads to collapse of the system, to a tragedy. The events that Hardin has portrayed on the local scale for the commons, i.e. for the agricultural area jointly cultivated by a village community, can also occur on the global scale, if the **overstressing** of the **atmosphere**, the **biosphere** and the **ocean**, which are man's most valuable **common goods**, continues to the same extent as till now.

Are we to expect an irreversible change of the **common good, climate**, as a result of the CO_2-induced effects described in the previous chapters, or can we bring about a change in due time? I believe that we have the chance, but only for a short time, i.e. during the next 15-20 years, to successfully intervene in the entire course of events. In contrast to the Mayas, we have the ability today, not only to analyse our problems, but also to estimate their consequences, and thus to avert a threatening catastrophe through countermeasures, if they are introduced in time. This requires, however, a **change of mind**, that does not aim for short-term maximisation of profits of influential interest groups, as was often the case up to now, but considers much more the **global commons** necessary for our survival and adopts the **general well-being** as the main operating principle.

Constraints are the result of ill-conceived actions. Once they have been set in motion, they lead, if they are too far advanced, a fateful life of their own. We still have the chance to choose between various paths into the future. But soon we have to decide upon a particular future (Tinbergen, 1977).

VII.3 Prospects: Extrapolating old trends

This potential path is based on the philosophy that continual growth is possible and also necessary for the survival of the human race. The satisfaction of all demands is seen as the main task. The supporters of this "supply-side" economics, that is particularly fashionable at the present, believe that only an increase of the energy supply can continue the strong economic growth of the past into the future.

We all know that economic growth cannot go on forever and that we could also live quite happily with half of the energy that we consume now.

Heinz Meier-Leibnitz,
former President of the
Federal German Research Foundation

VII.3.1 Growth

While many economics experts find it difficult to imagine a future without economic growth, it is clear to systems analysts that in a closed system, such as that of the Earth, an unlimited growth with a constant rate is quite simply impossible (Vester, 1976). It is not only questionable whether it is physically possible but, above all, how far undifferentiated growth is socially acceptable and thus at all desirable.

Dürr (1977) answered the question by stating that a continual increase of the consumption of raw materials and energy is not possible in the long run and that it will presumably bring the human race into major difficulties in the next 50-100 years. Even with very optimistic assumptions about the global reserves of raw materials, and future technological advances, Mishan (1973) finds that the present population and economic growth rates can be kept up for 100 years at most. The question of the social desirability of unlimited growth is answered by him with a clear "No". This type of economic growth and technological progress had led, after all, to problems which are symptomatic of today's cities (debts, slums, waste disposal, drugs, criminality, unemployment, lack of housing, loneliness, reduced living standard, etc.), and, in general, to increasing destruction of the environment. The economic growth of the past and the enormous accumulation of material goods has not contributed to the expected increase of the living standard in all countries. Hirsch (1980), therefore, seriously questions the doctrine that has prevailed up to now, that **social justice** is not possible without growth.

However, **growth** is still taken to be a symbol of progress and **consumption** a sign of living standard. It is, therefore, most important to convince the public that unlimited growth is not possible (Flohn, 1980), and that the continuation of past behaviour will certainly lead to catastrophe (Eppler, 1980). This work of conviction can only be successful, if we succeed in bringing about a **change of behaviour** through a **change of consciousness.**

A society that has primarily quantities in mind, thereby establishes usually very questionable qualities (Dürr and Hähnle, 1980). At the beginning of a process, the quality increases with the quantity. However, as Ginsburg (1979) has shown, beyond a respective culmination point, a further increase in quantity is accompanied by a decrease in quality. More quality can then only be achieved through less quantity and more knowledge. In the entire **controversy about growth**, the question is not whether there should be growth or not. It is more important to discuss **how** and in **what direction growth** should proceed.

In order to get out of the dead end, to which the old quantitative growth ideology has led, a **new understanding of growth** is required. This should be characterised by an emphasis on quality and not on quantity, in which the irreplaceable resources are preserved, environmental and climate risks reduced and a lasting survival of mankind with a high standard of living is guaranteed (Polunin, 1980). **Full employment** is generally held to be an important index for the living standard in a land. As we have shown in Chapter VI, on the basis of a whole series of examples, it is not tied to the extension of raw materials consumption and economic growth, but can be achieved much sooner through an improved utilisation of materials and energy, i.e. through a reduced raw materials and energy consumption per unit product.

The **inequality of opportunity** that has existed up to now, between north and south, represents one of the largest **potentials for conflict** for the future. The question is, can both be reduced with the continuation of the past trends?

If a free society cannot save the many poor,
it cannot also save the few rich.

John F. Kennedy

VII.3.2 North-south conflict

Population growth and distribution of resources are the critical sensitive points in the north-south dialogue. Fig. VII.1 symbolises the relevant points of view. The existing disparities between developed and developing countries are a result of the differing historical, cultural, economic and political developments. A maintenance of the present condition would be serious enough, but an increase of the inequities, as has been noted for some time, gives rise to a highly explosive situation and represents a **real threat to world peace.** Although acute east-west tensions have at present overshadowed the north-south problems "one should see the north-south relations as they are, namely a new historical dimension for the active protection of peace" (Brandt, 1980). Without question, north-south problems will play an even greater role in international relations in the future.

The disparities between developed and developing countries are large in all areas. In comparison to the developing countries, the developed countries with 28% of the world population receive 70% of the total world income, use 7 times as much fibre and 11 times as much metal, consume 2 1/2 times as much basic foodstuffs per capita and have a 5 times higher total energy consumption and a 12 times higher per capita energy consumption (Bach, 1980). What are the chances that these inequalities can be reduced? Let us consider the energy sector.

In 1978 the developing countries paid 8 billion dollars for oil imports alone, two years later the sum was almost 10 times larger, and an end to the price increases in the various trade sectors is not in sight (Venzky, 1981). It is naturally questionable whether an unadjusted development aid makes sense with such a cost explosion. The 119 developing countries, that are presently included in the Group 77, are trying to persuade the developed countries to increase their development aid to at least 1% of the Gross National Product (GNP) by 1990 (Der Spiegel, 1980). For development aid, the eastern developed countries divert only 0.04% of their GNP, the USA only 0.19% and the F.R. Germany only 0.44%. The Reagan Government thinks that even this share is wasted and has cut development aid by a third in 1981. Thus the USA takes an unglorious last place among the 17 western developed countries. The debts of the developing countries to the developed countries were, in 1980, already 450 billion dollars and will increase immeasurably if this development continues.

Some energy projections (e.g. Rotty and Marland, 1980; Häfele et al., 1981) propagate a strong energy growth, both in the developing countries and the developed countries, because

Fig. VII.1 : What's your problem?
Source: Asian Action (1980).

this would be the only way to reduce the large disparities. One can of course ask why, in the past, with the strong energy growth and comparatively low energy prices, the discrepancy between the developing and developed countries could continually increase, and who, in the future, with the tendency to cut development aid rather than to increase it, should pay the ever-increasing energy bills.

Interestingly, the actual numerical values of some of the projections, referred to above, show exactly the opposite of what they propagate, namely a reduction of the differences between developed and developing countries and thus an improvement of the opportunities in the case of lower energy growth. As the results in Chapter III.5.2 and Table III.9 have clearly shown, a quite considerable share of the north-south tensions in the energy sector could be best reduced by a more efficient energy consumption and a consequently diminished energy demand in the developed countries, with a simultaneous preservation of climate and environment. In contrast, with the continuation of the present trends, one should reckon with an even faster exhaustion of resources, leading to wars of distribution and more price increases, which further endanger the security of raw materials, energy and foodstuff supplies, and lead to an increase of the global potential for conflict.

To see clearly, only takes clearing your mind.

Antoine de Saint Exupéry

VII.4 Prospects: Shifting emphases

If the continuation of the previous economic and energy policies, described above, only deepens the societal crisis of which the first signs are already visible in individual countries, and increases the potential for conflict between the various countries, then a **reorientation** is urgently called for.

VII.4.1 Reorientation

The process of a reorientation has already begun, even if it is still much too faint-hearted. It started after the energy crisis of 1973/74, when the previous official economic and energy growth projections were very far from the actual events (see Chapter VI.4.1) and the former hard rule that energy and economic growth are coupled, turned out to be a myth on the basis of detailed international investigations (see Chapter VI.4.2). In addition, the observed reduction of the growth rate of CO_2 emissions, from the long-term average of 3.4%/yr to about 1.8%/yr after 1973/74 (see Chapter IV.1.4.1 and Fig. IV.4), indicates a reduction of the fossil energy consumption that results from such factors as the sharply increased energy prices, a slowing-down of economic activities and real energy saving. Energy growth that was previously declared unavoidable is a myth.

An absolute decline in energy consumption is therefore not impossible in the future (Rat der Sachverständigen für Umweltfragen, 1981). Indeed, it is in the interest of the maintenance of a healthy global energy, as the investigations in Chapter VI have shown, since the most economic way to provide energy does not require an increasing use of fossil fuels, it requires a sinking use (Lovins et al., 1981). In order to achieve this, the energy sources used in the past must be used more sensibly, i.e. more effectively. Thereby the total energy demand is reduced, which means that the non-renewable energy carriers are preserved and, the renewable energy carriers that initially can only be introduced on a modest scale, can already make a notable contribution. This **double strategy** has a decisive advantage over all other procedures. The more efficient use of energy can be introduced without any large build-up time into active energy policies with a success that is immediately visible. For the introduction of inexhaustible energy sources there is then enough time to be able to integrate them slowly and thoughtfully into the new economic and energy system. The relatively cheap conventional fossil fuels that are still available have an important job. With their help, the still attainable capital for investment in a stable future with a life-supporting environment must be obtained during this historically unique **transition period** of only a few decades.

This policy requires a reorientation especially in the deployment of the various technologies and in the introduction of innovative measures that are sketched in the following sections.

VII.4.2 Using available technological know-how

VII.4.2.1 Cogeneration

In a conventional power plant about 36%, on average, of the primary energy in the form of coal, oil, gas or uranium is converted into electricity and the remaining 64% is released as waste heat into water bodies or the air. Plants with cogeneration, i.e. the combination of two different processes to produce power (electricity) and heat, increase the total efficiency from about 36% to 60-85%. Not only are considerable amounts of energy saved, but the gas, particle and waste heat emissions are significantly reduced (Rat der Sachverständigen für Umweltfragen, 1981). In cogeneration, the waste heat is reduced by increasing the useful heat.

Basically, cogeneration can be applied in power plants, combined-heat-and-power-plants, neighbourhood cogeneration systems, in combination with fluidised bed technology and district heating (Hein, 1981). Cogeneration is also promising in various heat pump systems. Goetzberger (1981) shows that solar cells and cogeneration, when combined correctly, can make electricity available all of the year. The use of cogeneration in power plants, which have a size of 1,000 MW and more and thus supply a large area and have large transmission losses, on the other hand, only makes sense for a few locations near industry or towns with high energy densities (Ruske and Teufel, 1980).

VII.4.2.2 Combined-heat-and-power-plants

The first combined-heat-and-power-plants, in which both electricity and heat are produced, were installed in Germany as early as 1919 (Hein, 1981). In 1979 there were combined-heat-and-power plants in Germany with an electrical supply of about 6,600 MW and a thermal energy of about 35 TWth. These could, however, only cover about 1/10 of the total household demand for heat. Speedy development is hindered by institutional barriers, as discussed in section 4.4.

The best-known combined-heat-and-power-cycle system in the F.R. Germany has been operating in Flensburg since 1969. With a capacity of 120 MW, 20.8% of the primary energy input was used for electricity and 40.5% for district heating in 1978, so that about 35,000 tce/yr were saved, in comparison with a conventional power plant (Ruske and Teufel, 1980). The total efficiency was 61% in 1978 and should be increased to 71% in 1985, as a result of technical improvements, in which case it would be about twice as efficient as a power plant producing only electricity. Using the price level for oil of 1979, the unconventional power plant was about 25% cheaper than conventional heating in both single and multiple family housing with individual systems.

VII.4.2.3 Neighbourhood cogeneration systems

As the distance of energy transport increases, the losses increase correspondingly and so do the investment costs for the power lines. In order to minimise the energy losses and costs, mini power plants or neighbourhood cogeneration systems have been developed. There are already 50 such plants in the F.R. Germany (Hein, 1981). The drive assembly is not a turbine, as in the conventional combined cycle power plants, but a piston-stroke engine driven by natural gas or diesel oil. This drives a generator which delivers electricity. From the cooling water of the motor and the waste gases, heat for combined cycle plants is removed using a cooling basin. From 100% of the primary energy input, 32% are converted to electricity and 54% to heat, i.e. in addition to the 32 parts electricity of a conventional power plant, this type of power plant gives 54% heat. This system mostly consists of a series of motors, and, depending on demand, motors can be switched on or off so that they can always be run at full load with the highest efficiency. The system size varies between 250 kWe and 10,000 kWe electrical capacity (large power plants have, in comparison, capacities of 300,000 and 1,000,000 kWe), which can provide heat for 50-3,000 dwellings. The cheap series motors are practically everywhere usable with natural gas, sewage gas, pyrolysis gas, gasified coal, oil, etc. (Spreer, 1982). The amount of waste gases is controllable for a reasonable price.

VII.4.2.4 Fluidised-bed combustion

In this new combustion technology, finely ground coal is burned in suspension over a bed of moving air, and not burned in large pieces on a grate, as in conventional power plants. The movement of the coal particles gives a larger heat transfer, making a boiler of half the usual size possible. Under pressure, up to about 16 bar, a 10% higher efficiency is achieved (Peters, 1978).

Fluidised-bed combustion has a whole series of other advantages compared with conventional energy conversion (Ruske and Teufel, 1980): as a result of improved efficiency and cogeneration, the waste heat amount is about 80% smaller; with the addition of about 8% ground chalk, a desulphurisation of more than 95% is achieved; as a result of the lower combustion temperatures (in this case between 800 and 900°C, usually 1,600^{o}C) less nitrogen oxides form; and, finally, as a result of higher efficiency and, consequently, reduced consumption per unit energy output, a smaller amount of CO_2 is released into the atmosphere. The increased production of suspended particulates can be trapped by electrostatic precipitators with a removal efficiency of about 99%. Thus, qualitatively poor coal with a high SO_2 and ash content can be used without disadvantage. Fluidised-bed technology and cogeneration are not only extremely energy saving and environmentally sound, they also fit very well into the energy concept of the industrialised nations, since they reduce dependence on imported oil and simultaneously make a more intensive use of indigenous coal possible (Energiediskussion, 1980a; Langhoff and Krischke, 1981).

VII.4.2.5 Central station district heating

In 1978, all public power plants of the F.R. Germany produced a waste heat amount of 420 TWh, in addition to the 280 TWh of electricity (Hein, 1981). The heating demand of all 24 million dwellings was, on the other hand, only 360 TWh, so that, theoretically, the heat that was uselessly released into the atmosphere and water bodies could have heated all dwellings. The import of about 30 million tons of oil could be saved annually in this way and the balance of trade could have been cut by about 4.3 billion DM (1980 prices) (Hoffmann, 1981).

In 1979, in the F.R. Germany, the total length of the district heating system was only 4,886 km (Deuster, 1980), while in comparison the 380 kV electric circuit was 7,250 km, the 220 kV electric circuit was 16,600 km (in 1978) and the natural gas network was about 17,000 km (in 1977) (Bischoff and Gocht, 1979). The district heating sales amounted to about 42.9 TWh in 1979, of which 72.9% were produced by combined-heat-and-power-plants. The share of district heating in the end-use energy consumption in 1978 amounted to only 2.1% (Deuster, 1980). The utility companies have so far shown no serious interest in the use of waste heat.

The Federal Land Nordrhein-Westfalen has presented a 5-year programme, according to which many decentralised district heating islands should be installed on the basis of small environmentally sound coal-based combined-heat-and-power-plants (Hoffmann, 1981).

For the total investment of 50 billion DM, of which one half each should go into the construction of decentralised cogeneration plants and the extension of the district heating network, an employment effect of about 215,000 jobs per year is expected. A joint programme for district heating, planned by the Federal government and the states, fell through due to opposition from Schleswig-Holstein. No state will be able, in the future, to ignore waste heat usage, i.e. an energy source that cannot be influenced by international energy and price cartels.

VII.4.2.6 Heat pumps

Heat pumps operate according to the same principle as the refrigerator. While, in the case of a refrigerator, the heat is removed from the interior, the heat pump takes the heat from the surroundings. Thus, the heat that is available with lower temperatures, for instance, in air, water or in the ground, can be converted to heat with higher temperatures. Gas, diesel and electric heat pumps can be used, for example, for space heating, heating of swimming pools, air conditioners, heat retrieval or waste heat recovery (Ruske and Teufel, 1980). Advanced heat pump systems are already in operation, especially in industry and business and make considerable energy savings possible (Pfeiffenberger, 1981). A well-known example of the successful use of such technology are the tandem units, i.e. a combination of cogeneration (gas motor plus generator) and heat pumps, operating in the city of Heidenheim (Spreer, 1982).

Taking the example of space heating, we can compare the thermodynamic yield of the marketable heating systems (Herrmann, 1978). Either coal or gas is used as primary energy (100 units). As the energy flow diagrams in Fig. VII.2 show, electric storage heating has by far the least output of useful energy (= 32). Coke or coal central- heating achieves about twice as much useful energy (= 63). An electric heat pump takes about as much heat out of the surroundings as is lost through waste heat in a power plant and it thus produces about as much useful heat (= 98) as the energy input for a power plant. With the natural gas heat pump, more useful heat (= 133) can be produced than it requires for fuel. This is possible in a thermodynamic process, because a small amount of energy with high temperature (the driving energy of the motor) can be converted into a large amount of energy with low temperature (useful heat). The situation is similar in the case of a combined-heat-and-power-plant with district heating. The yields are even higher than with heat pumps (157 in the high temperature region and 252 in the low temperature region), because, in this case, the entire waste heat of the power plant can be used. These examples show clearly that from the thermodynamic point of view, electrical heating makes no sense at all. The heating of a dwelling with electrical storage heating requires 8

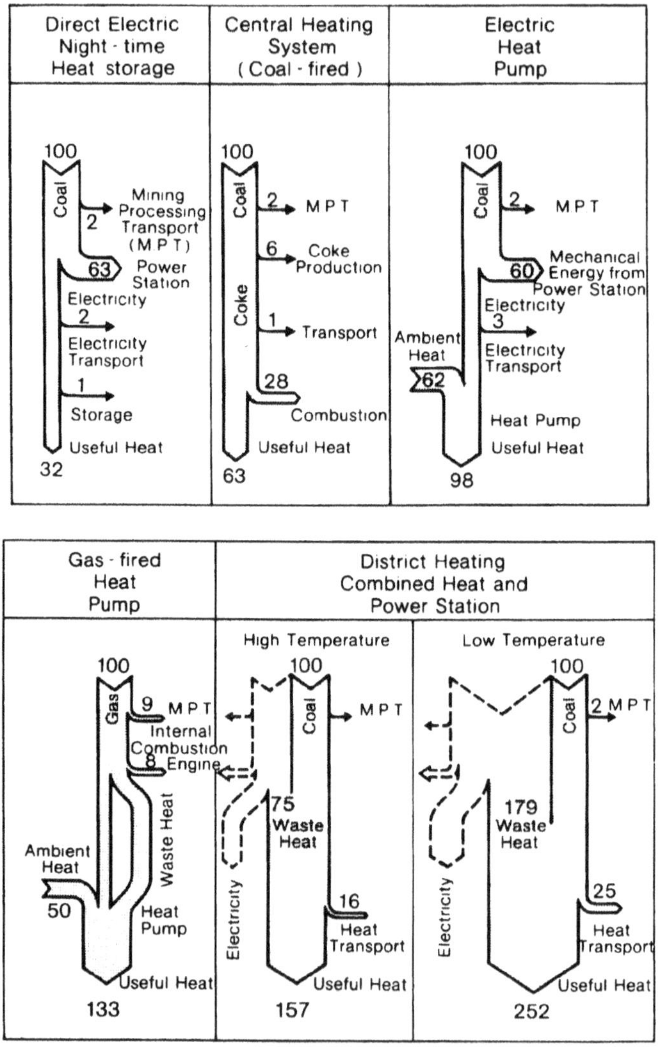

Fig. VII.2 : Comparison of different energy systems for residential heating.
Source: Herrmann (1978).

times as much energy as a low temperature system with district heating from a combined-heat-and-power-plant.

VII.4.3 Examples of product innovation

The more direct solutions for energy conservation and environmental protection, discussed in section 4.2, can be complemented by a series of indirect measures. These include a purposeful selection of materials, reduction of wear and tear and thus an increase of durability, fuel economy, recycling, new types of plastic and materials, as well as energy-saving measuring and control systems. The following brief description can only touch on some of the fascinating possibilities in the various fields of application.

VII.4.3.1 Insulation

In the Federal Republic of Germany about 44% of the final energy consumption is for the residential and commercial sectors and 80% of this is in the form of low temperature heat (50–70°C) for space heating (Bossel and Bossel, 1976). Insulation is therefore a very effective and versatile means of reducing the energy consumption. Ceilings and walls can be insulated with glass fibre and polyurethane foam and the windows with the aid of double- and triple-glazing (Herrmann, 1979).

If the thickness of the insulation in a sloping roof is increased from 8 cm to 12 cm, the demand for heating oil is reduced from 5.9 litre/yr to 4.1 litre/yr per m^2 roof area (Erfurth, 1981). If one builds exterior walls with two shells and a filling of 8 cm thick insulation instead of a single shell, the annual heating oil consumption is reduced from 14.8 litre/m^2 to 4.2 litre/m^2 exterior wall. Using thermography, points of heat loss can be found and eliminated. The continuing fuel price increases are bound to lead to better house designs and the future energy-efficient home will probably differ from current designs in four major areas: (1) It will have an air-vapour barrier instead of a vapour barrier; (2) it will have far higher levels of insulation; (3) passive solar use will increase; and (4) it will be equipped with an energy-efficient air management system (Eyre and Jennings, 1982).

The possible health dangers due to insulation materials (e.g. through formaldehyde emissions from plastic foams or chipboard) appear to be technically controllable, but will be intensively studied further (Energiediskussion, 1981). In 17 houses incorporating such energy conserving measures as weatherstripping and caulking, radon-222 and radon-daughter concentrations have been found to range from 0.6 to 22 nCi/m^3, 5 to 170 times the outdoor background levels (Hollowell et al., 1980). Tests have shown that the potentially serious health risks to occupants can be prevented by increasing the air exchange through a cost-effective mechanical ventilation system with an air-to-air heat exchanger (Roseme et al., 1980). These problems could be avoided in the first place, if only building material devoid of such hazardous substances were used.

VII.4.3.2 Energy-saving light bulbs

A conventional light bulb converts only 10% of the incoming energy into light, the remainder is lost as heat (Naturwissenschaftl. Rundschau, 1981). A newly developed metal-halogen vapour bulb has a 75% smaller electricity consumption and 4-5 times longer durability than a conventional lamp. In the approximately 24 million households in the F.R. Germany, 2 billion KWh or the annual electricity production of a medium-sized power plant could be saved.

VII.4.3.3 Fuel economy of light vehicles

In the F.R. Germany every fifth tonne of crude oil that is imported is lost through the car exhaust (Seiffert, 1981). An improvement of the fuel utilization thus means a large energy saving potential. For the purposes of resource preservation the Porsche "long-term car" was developed, for which the durability is trebled, the reuse of the building materials is made easier and the fuel consumption is reduced (Herrmann, 1978). Through a series of measures, such as, for example, reduction of the weight (bodywork, e.g. out of plastic or aluminium), use of turbocharged diesel engines, five gears, continually variable transmission and stratified-charge low-rpm engines, recovery of the exhaust heat, reduction of aerodynamic drag, use of radial tyres, improved motor, lubricants and drive-train components, regenerative braking, use of a smaller motor made of ceramics instead of steel, microelectronic control of fuel consumption (see also section 4.3.5) and, finally, an upper speed limit, the fuel consumption could be reduced from the present value of 8-10 litres/100 km to about 2 litres/100 km (Herrmann, 1979; Gray and von Hippel, 1981). Since 1978 the US gasoline consumption has been declining. An in-depth study by Greene and Kulp (1982) has revealed that the reduction in demand can be attributed to gasoline price increases (over 50%), improvements in fuel economy (about 24%), a decrease in real disposable income per household (about 20%) and the penetration of the diesel engine (about 3%).

The favourable energy and environmental effects of an innovative fuel utilization can also be complemented by a coordinated traffic, urban and regional planning policy (Kommission für Umwelt and Ökologie, 1981). In addition, the environmental and climatic risk can be reduced by the transition to a **hydrogen economy** in which hydrogen replaces fossil fuels. In particular this would reduce the air pollution at the level where breathing occurs, since the exhausts would only consist of water vapour (Bockris and Justi, 1980; Griesbaum and Hönicke, 1980).

VII.4.3.4 Feedstock uses of carbon

The exhaustible fossil carbon is indeed too valuable to be simply burned. In 1979, in the F.R. Germany, only 7% of the oil consumption was for petrochemicals and only 2% was for coal chemistry (Vester, 1980). With the carbon element, a building brick for all life, about 4 million different chemical compounds have been made so far, while from all of the other 92 elements on the Earth only about 100,000 compounds are known. Carbon chemistry has a firm place in almost all sectors of civilised life, such as in medicine (medicines, vitamins, hormones, cosmetics, artificial organs), the textile industry (synthetic materials, such as nylon, perlon and dralon), building in the widest sense (foams, construction and insulation materials, foils, paints, asphalt, cleaning, thinning, dissolving, lubricating and glueing materials and many others), and is used everywhere where good sound and heat insulation, high resistance to corrosion, tearing and attrition, extreme lightness and material- and energy-saving building components are required.

With the help of synthetic materials based on carbon, lighter cars and better insulated houses can be built - significant factors in energy conservation. The production of plastic pipes consumes only half as much energy as the production of pipes from metal (Flavin, 1980). For health and energy reasons it is important that the synthetic materials are not simply burned (because of poisonous gas fumes) or deposited in a waste tip (because of groundwater pollution). They must be **recycled**, i.e. used as the starting material for new products or decomposed in the natural environment. It is now possible to decompose the synthetic materials using sunlight or through inbuilt bacteria (Vester, 1980).

VII.4.3.5 Microelectronics

Studies by three institutes of the Fraunhofer Society show that, as a result of systematic deployment of microelectronics by the middle of the 1990s in the F.R. Germany, about 8-9% of the present primary energy consumption could be saved and the balance of payments could be cut by 12-13 billion DM (based on 1980 energy prices) (Frisch, 1981). For instance, the energy demand for space heating could be reduced by 1995 by 95-130 billion KWh/yr through improved measurement and control technology. Through microelectronic control of fuel consumption after 1995 about 50 billion kWh/yr or about 4 million tonnes oil could be saved. Finally, if the roughly 17 million electric motors were to run depending on microelectronic control, about 12 billion kWh/yr less of valuable electric energy would have to be produced.

It is estimated that the increased use of microelectronics can save energy on the order of 13-22% by reducing conversion and distribution losses, 10-13% in traffic, 10-12% in households, and 15-25% in industry (Heinrich, 1982). The Federal German Government envisages, in addition to the effect of energy saving and the resulting reduction of the balance of payments deficit, a particularly favourable effect on the international competitiveness and the creation of thousands of jobs in Germany (Energiediskussion, 1980b).

VII.4.4 Removal of institutional barriers

A reorientation means, above all, a removal of the institutional barriers. A few examples show clearly the hindrances to the carrying out of a sensible energy policy. One main reason for many problems in Germany is the **Energy Economy Law** of 1935 that is still valid today. As a result, the eight large utility corporations of the F.R. Germany are granted area, production and management monopolies that remain prohibited to other economic branches for good reasons (Ruske and Teufel, 1980). In this way the utilities can pass on almost all of their investment costs and thus the entire business risk to the consumers through the connection costs and fixed charges. According to the present tariff structure, large consumers or energy wasters are rewarded by a special tariff, while small consumers and those who save energy are punished. With this system the utilities can have no interest in a sensible energy utilisation; since, as the consumption or waste increases, the profit margins also increase.

A further hindrance to a rational energy policy is that the utilities are not prepared to take electricity or heat from small enterprises or private persons. For instance, small hydropower plants in the Black Forest receive only 3.2 Pfennig/kWh during the day and 2.6 Pfennig/kWh at night in summer for the electricity and in winter 4.9 Pfennig/kWh in the day and 3.2 Pfennig/kWh at night (Ruske and Teufel, 1980). For electricity from nuclear power plants, the same utilities pay about 10 Pfennig/kWh and sell it for about 18 Pfennig/kWh to the consumer. Of course, it is true that there are different prices for electricity, depending on whether the system is running on base load, medium load or peak load, but the criticism is nevertheless justified. A change of philosophy might have begun, since the utilities have said in an agreement with the Union of Industrial Power Companies in 1979, that they are willing to take electricity from industrial production, as has long been the case in the gas industry (Der Spiegel, 1979).

As a result of the **Import Tarriff Law** for the protection of the Ruhr coal industry, the principle of equal competitive opportunity is broken. Combined-heat-and-power-plants must burn two tonnes of expensive Ruhr coal for every additional tonne of cheap imported coal (Der Spiegel, 1981). Thus the combined-heat-and-power- plants, and not the nuclear power plants, have to bear the burden of subsidizing the coal, which necessarily distorts the competition.

The Enquete Kommission (1980) presented a detailed catalogue of measures for the removal of barriers which hinder the energy-saving policy of the F.R. Germany. For instance, it recommends that **insulation regulations** of the **Energy Conservation Law** should be correspondingly changed for the improvement of the insulation of buildings. For more support of energy saving investments, the continuation of the **Modernisation and Energy Conservation Law** after 1982 and an increase of the available funds from the present 500 million DM/yr to at least 2 billion DM/yr is recommended. By amending the **income tax regulations** the competitive disadvantage of energy saving technologies should be corrected. The **Rental Housing Law** must also be changed, since at present neither the tenant nor the landlord has sufficient incentive to save residential space heat.

One-sided subsidies and sales-supporting tariffs not only distort the actual market situation, they also hinder the investments in more sensible energy use technologies. Consequently, Sant (1979) believes that an energy policy that is disadvantageous to precisely those fuels and efficiency technologies, through which an energy problem or climate risk could be avoided, is really perverse. It is therefore absolutely necessary, in the interests of a reasonable energy supply system, that both the correct **price signals** are set and the existing **market distortions** are removed.

This list could be continued much further, but for our purposes it is clear that the removal of institutional barriers is absolutely necessary for a reorientation. The state can only regain its **credibility** "if a credible chance is also given to the alternative path to nuclear energy (and fossil fuel) use" (Meyer-Abich, 1981). Up until now, the preference for the addition of large coal and nuclear power plants is a very large hindrance to the chances for more efficient energy use, especially through cogeneration and district heating (Der Rat der Sachverständigen für Umweltfragen, 1981).

With all of these recommendations and measures, we actually have available a great selection of possibilities which are hitherto unused. They could all provide an important contribution to the avoidance of a CO_2/climate threat.

Mankind...is faced now with an unprecedented responsibility and is placed in the new role of a captain guiding all life on the planet - including his own.

Aurelio Peccei,
President of the Club of Rome

VII.5 Ways out of the CO_2/climate threat

The scientific information brought together in this book makes it clear that the activities of mankind have now reached an order of magnitude that could adversely influence the global climate. With the aid of scientific methods we can estimate that an anthropogenic global climatic effect will probably be detectable in the climatic data from the natural climatic variability around the turn of this century at the latest. Moreover, in the course of the next century a serious threat to the global climate is to be expected, largely as a result of the CO_2 influence.

Future climate could develop in this way if the present, or a similar, energy and economic policy is maintained. The CO_2/climate threat is, however, not a fate to which we are helplessly destined. Mankind has possibilities to influence global climate through appropriate actions. These possibilities are not in the area of natural sciences, but rather in the socio-economic-institutional realm.

Both the subtitle of this book and the questions at the beginning of this Chapter suggest the necessity of a counteraction. This requires a reorientation and the setting of new priorities. It is our energy policy which sets the points for the future development of our climate. Let us consider the following possible scenarios:

- fossil fuel use increases again and soon reaches the original historical exponential growth rate of about 4%/yr
- fossil fuel use continues to grow at the reduced rate of about 2%/yr, that is experienced since the energy crisis of 1973/74
- fossil fuel use stabilises at the present level, or
- fossil fuel use is reduced.

Generally, it is agreed that a renewed growth of the fossil fuel consumption to the original exponential growth rate of 4%/yr is to be prevented because of the serious climatic impacts that would be expected in the first half of the 21st century. In spite of the efforts of some industrialised countries to revitalise coal through coal liquefaction and gasification, a sharp increase on a global scale is unlikely. The distribution of fossil fuel resources between a few countries, which favours cartel agreements, means that most of the other nations in the future will hardly be able to find enough fossil energy carriers at a price that they can afford.

The reduction of the growth rate is a first important step in the right direction. However, a characteristic of exponential growth is that the gain of time remains small. With a stabilisation of the fossil fuel consumption, e.g. at the present value, considerably more time is gained. But even with this zero growth rate, the consumption continues to increase by a fixed amount, so that a CO_2 accumulation that must be taken seriously could still occur, even if only after centuries. One must also not forget that the CO_2 remains for a long time in the atmosphere, because it can only be taken up very slowly by the main sink, the ocean. Past experience shows that conventional energy policy is not very effective in reducing the CO_2/climate threat.

This leads to the question: what possibilities exist to reduce the fossil fuel consumption to the extent that the CO_2/climate problem is not only postponed but perhaps even totally avoidable? Many have not given this any consideration because they believed from the very start that economic growth and also energy growth are necessary, and because they could not imagine a maintenance of or even an increase of the standard of living of a growing world population without a simultaneous energy growth.

The realisation that economic growth is not coupled to energy growth, and that a more rational approach to energy use leads to a higher standard of living, together with the fact that official energy demand projections in recent years had to be continually corrected downwards, has instigated a change of philosophy. In addition there are now methodically convincing empirical studies from about a dozen countries that have very useful results for the purposes of a reorientation. These show, if energy supply is managed economically efficient, then energy needs, and hence fossil fuel use can be expected to go down, not up. The same or more energy services can be offered with a lower energy use simply by increasing energy productivity. This has far-reaching consequences for the entire economy and especially for the balance of payments of a country. In contrast to the usual inflation-driven energy policy, with a better energy utilisation, the energy sector can become a net exporter of capital to the rest of the economy, thereby lowering inflation.

This encouraging result shows us the path to a healthy energy and economic policy and at the same time gives us the key to the avoidance of a CO_2/climate problem. Such a low-risk climate and energy policy characteristically uses energy more efficiently, which

- requires less fossil fuels, which, in turn
- saves these irreplaceable energy resources for purposes less wasteful than mere combustion, and which furthermore

- permits the expeditious development of CO_2-free renewable energy sources, all of which
- leads to a reduction of CO_2 emissions into the atmosphere.

This strategy has, in contrast to all other procedures, the decisive advantage that efficient energy use can be introduced into active energy policies without great lead times and has immediate visible success. Moreover, there is enough time for a systematic and well-planned introduction of renewable energy sources. The non-renewable fossil resources can then be preserved for non-energy purposes or provide the necessary capital for investment in a stable future during this historically unique transition period. With such a strategy there is also no need to argue about whether and by when the fossil fuels can be completely replaced by renewable energy sources. For even if the substitution of the fossil fuels is only partly completed during the next 30-50 years, the more efficient utilization of energy and the resulting reduction of fossil energy consumption, would postpone the onset of unacceptable climate risks at least by several decades. Thus time is gained for a systematic transition from a fossil to a CO_2-free renewable energy economy and the foundations laid for a permanent elimination of the CO_2/climate problem.

This low-risk climate energy policy must be supplemented by a land use policy that is characterised by

- a balance between deforestation and reforestation, and
- soil conservation to reduce the carbon emissions.

Strategies that have multiple benefits are to be preferred over one-sided actions. For example, reforestation of a region would not only remove considerable amounts of CO_2 from the atmosphere, it would also improve the water balance, reduce soil erosion, hinder floods, landslides and avalanches, and have favourable effects on the climate. Thus, strategies that have many ecological advantages are to be preferred over those that only bring short-term profit, since these mostly create many new problems as they solve one problem.

The strategies presented here would lead us from the present throughput economy to a sustainable cyclic economy, thereby ensuring a stable social order. The advantages of such a policy are many, ranging from a reduced dependence on external resources, reduced deficits in the balance of payments, reduced technological risks, acceptance and siting problems, as well as job security and the removal of socio-political sources of conflict, to the reduction of environmental destruction and, not least, the avoidance of an irreversible climatic change.

The results presented in this book give rise to cautious optimism, because a policy with the least economic costs also causes the least CO_2/climatic threat. The chances of success are by no means small because we are already forced to take a sensible approach to our resources due to the high energy prices. Economics is our most powerful ally in the quest for a sustainable future with a low climatic risk. No doubt such a rational economic and energy policy will have to fight against considerable opposition by influential lobbies. Therefore every citizen is called upon to help. That we succeed is entirely up to us.

BIBLIOGRAPHY

I. Introduction

Coope, G.R.(1977): Fossil Coleopteron assemblages as sensitive indicators of climatic changes during the Devensian (last) cold stage, Philosoph. Transactions Roy. Soc. London, Series B (280), 313-340.

Flohn, H.(1978): Die Zukunft unseres Klimas: Fakten und Probleme, Promet 2/3, 1-21.

Müller, H.(1974): Pollenanalytische Untersuchungen und Jahresschichtenzählungen an der holsteinzeitlichen Kieselgur von Munster-Breloh, Geol. Jahrb. A, 21, 107-140.

Revelle, R. and H.E. Suess(1957): Carbon dioxide exchange between atmosphere and ocean and the question of an increase of atmospheric CO_2 during past decades, Tellus 9, 18-27.

World Meteorological Organization (WMO, 1979): Proceedings of the World Climate Conference, WMO-No. 537, Geneva.

II. Climate and Climatic Change

Bach, W.(1976): Changes in the composition of the atmosphere and their impact upon climatic variability - an overview, Bonner Meteorol. Abhlgn. H. 26.

Barry, R.G. et al.(1979): Climatic Change, Revs. Geophys. & Space Phys. 17 (7), 1803-1813.

Chervin, R.M., W.M. Washington and S.H. Schneider(1976): Testing the statistical significance of the response of the NCAR general circulation model to North Pacific ocean surface temperature anomalies, J. Atmos. Sci. 33, 413-423.

Dickinson, R.E.(1982): Modeling climate changes due to carbon dioxide increases. In: W.C. Clark (ed.) Carbon Dioxide Review: 1982, 103-142, Oxford Univ. Press, New York.

Etkins, R. and E.S. Epstein(1982): The rise of global mean sea level as an indication of climatic change, Science 215, 287-289.

Flohn, H.(1978): Die Zukunft unseres Klimas: Fakten und Probleme, Promet, 2/3, 1-21.

Flohn, H.(1979): Eiszeit oder Warmzeit? Naturwissenschaften 66, 325-330.

Flohn, H.(1980): Possible climatic consequences of a man-made global warming, RR-80-30, Int. Inst. Applied Systems Analysis, Laxenburg.

Fraedrich, K.(1980): Einfache Klima-Modelle, Promet 10 (1/2), 2-6.

Gates, W.L.(1979): The physical basis of climate. In: WMO World Climate Conf., 112-131, WMO-No. 537, World Meteorol. Org., Geneva.

Grassl, H.(1980): Energiebilanz-Klimamodelle, Promet 10 (1/2),6-12.

Hansen et al.(1983): Efficient three dimensional global models for climate studies: Models I and II, Mo. Wea. Rev., 111(4), 609-662.

Hantel, M.(1980): Zonal gemittelte Klimamodelle, Promet 10 (1/2), 12-19.

Haxel, O.(1976): Beitrag der Physik zur Klimageschichte. Naturwissenschaften, 63, 16-22.

Herterich, K.(1980): Stochastische Klimamodelle, Promet 10 (1/2), 19-23.

Hoffert, M.I., A.J. Callegari and C.-T. Hsieh(1980): The role of deep sea heat storage in the secular response to climate forcing, J. Geophys.Res. 85 (C21), 6667-6679.

Holloway, J.L. Jr. and S. Manabe(1971): Simulation of climate by a global general circulation model, Mon. Wea. Rev. 99, 335-369.

Kellogg,W.W.(1977/1978): Effects of human activities on global climate, WMO Bulletin pt. I,229-240; pt. II, 3-10.

Klimaforschungsprogramm der Bundesrepublik Deutschland(1980): Bericht einer Arbeitskonferenz, Febr. 1980 und Entwurf des Ausschusses Klimaforschung.

Kutzbach, J.E.(1974): Fluctuations of climate monitoring and modeling, WMO Bulletin 23, 47-54.

Kutzbach, J.E.(1975): Diagnostic studies of past climates.In: GARP. The Physical Basis of Climate and Climate Modeling, 117-126, GARP Publ. Series No. 16, WMO, Geneva.

Lemke, P.(1977): Stochastic climate models Pt. 3: Application to zonally averaged energy models, Tellus 29, 385-392.

Lorenz, E.N.(1975): Climatic predictability, In: The Physical Basis of Climate Modeling, GARP No. 16, 132-136, WMO, Geneva.

Manabe, S.(1983): Carbon dioxide and climatic change. In: B. Saltzman (ed.) Advances in the Theory of Climate, Adv. in Geophys. Vol. 25, 39-82, Academic Press, New York.

Mason, B.J.(1977): Zum Verständnis und zur Vorhersage der Klimaschwankungen, Promet 7/4, 1-22.

Mitchell, J.F.B.(1983): The seasonal response of a general circulation model to changes in CO_2 and sea surface temperature, Q.J.R. Met. Soc. 109, 113-152.

Mitchell, Jr. J.M.(1977): Records of the past, lessons for the future. In: Proceedings of the Symp. on Living with Climatic Change. Phase II, 15-25. The MITRE Corp. McLean, USA.

Müller, H.(1979): Climatic changes during the last three interglacials. In: W. Bach et al. (eds.) Man's Impact on Climate, 29-41, Elsevier Publ. Co., Amsterdam.

Pfister, C.(1980): The Little Ice Age: Thermal and wetness indices for Central Europe, J. Interdiscipl. History X (4), 665-696.

Pfister, C., D. Messerli, P. Messerli and H. Zumbühl(1980): Die Bedeutung verschiedener Datentypen für die Rekonstruktion des Klima -und Witterungsverlaufs der letzten Jahrhunderte. Gletscher und Klima, Jahrbuch 1978 der Schweiz. Naturforsch. Ges. Wiss. Teil. Briger Gletschersymposium.

Reiser, H. und V. Renner(1980): Zirkulationsmodell, Promet 10 (1/2), 23-29.

Rotty, R.M.(1975): Energy and climate, IEA (M)-75-3, Institute for Energy Analysis, Oak Ridge, USA.

Schlesinger, M.E.(1983): Simulating CO_2-induced climatic change with mathematical climate models - capabilities, limitations and prospects, III.3 - III.139, US DOE 021, Washington, D.C.

Schneider, S.H. and T. Gal-Chen(1973): Numerical experiments in climate stability, J. Geophys. Res. 78, 6182-6194.

Schneider, S.H. and R.E. Dickinson(1974): Climate modeling, Revs. Geophys. & Space Phys. 12, 447-493.

Schönwiese, C.D.(1979): Klimaschwankungen, Springer Verlag, Berlin.

Schwarzbach, M.(1974): Das Klima der Vorzeit, Enke, Stuttgart.

US NAS (National Academy of Sciences)(1975): Understanding Climatic Change, A Program for Action, Washington D.C.

US NAS(1982): Carbon dioxide and climate: A second assessment, Washington, D.C.

Washington, W.M. and G.A. Meehl(1983): General circulation model experiments on the climatic effects due to a doubling and quadrupling of carbon dioxide concentration, J. Geophys. Res. (in press).

Williams, J. and H. van Loon(1976): The connection between trends of mean temperature and circulation at the surface: pt. III spring and autumn. Mon. Wea. Rev. 104, 1592-1596.

WMO (World Meteorological Organization)(1975): The Physical Basis of Climate and Climate Modelling, GARP Publ. Series No. 16, Geneva.

WMO (1979): Proceedings of the World Climate Conference, WMO-No. 537, Geneva.

WMO (1980): World Climate Program 1980-1983, WMO-No. 540, Geneva.

***

III. Sociopolitical Aspects of the CO_2/Climate Problem

Anderheggen, E.(1980): Haben wir genug Kohle für die Zukunft? Umschau 80 (8),230-231.

Armentano, T.V. and J. Hett (eds.)(1980): The role of temperate zone forests in the world carbon cycle. CONF. 7903105 UC-11, U.S. Dept. of Energy, No. 003, Washington, D.C.

Armentano, T.V.(ed.)(1980): The role of organic soils in the world carbon cycle, CONF. 7905135 UC-11, U.S. Dept. of Energy, No. 004, Washington, D.C.

Bach, W.(1979a): Impact of world fossil fuel use on global climate: Policy implications and recommendations. In: Symp. on Carbon Dioxide Accumulation in the Atmosphere, Synthetic Fuels and Energy Policy, 121-160. Committee on Governmental Affairs, US Senate, USGPO, Washington, D.C.

Bach, W.(1979b): Klimaänderung durch Energiewachstum? Brennst., Wärme, Kraft 31 (2),49-56.

Bach, W. (1980): Der ungleiche Verbrauch von Ressourcen: Eine Bedrohung des Weltfriedens? Umschau 80 (13), 401-402.

Bach, W. and W. H. Matthews(1980): Exploring alternative energy strategies. In: W. Bach et al. (eds.) Renewable Energy Prospects, 711-722, Pergamon Press, Oxford.

Bach, W.(1981): Fossil fuel resources and their impacts on environment and climate, Int. J. Hydrogen Energy 6, 185-201.

Beaujean, J.-M. and J.-P. Charpentier(1978): A review of energy models, RR-78-12, Int.Inst. Appl. Systems Analysis, Laxenburg.

Bischoff, G. und W. Gocht(1979): Energietaschenbuch, Vieweg, Braunschweig.

Bottke, H.(1979): Schatzkammer Ostsibirien, Umschau 79 (5), 150-152.

Brandt, W. et al.(1980): Das Überleben sichern. Bericht der Nord-Süd-Kommission, Kiepenheuer & Witsch, Köln.

Brown, L.R.(1980): Food or fuel: New competition for the world's cropland. Worldwatch paper No. 35, Washington, D.C.

Brown, L.R.(1981): World population growth, soil erosion, and food security, Science 214, 995-1002.

Colombo, U. and O. Bernardini(1979): A low energy growth 2030 scenario and the perspectives for Western Europe, Report for the Commission of the European Communities.

Delcourt, M.R. and W.F. Harris(1980): Carbon budget of the U.S. biota: Analysis of historical change in trend from source to sink, Science 210, 321-323.

Doblin, C.C.(1980): The growth of energy consumption and prices in the USA,FRG, France and the UK 1950-1975, WP-80-126, Int. Inst. for Appl. Systems Analysis, Laxenburg.

Forrester, J.W.(1971): World Dynamics, Wright-Allen Publ. Co., Cambridge, Mass.

Fritz, M.(1980): Nichtkommerzielle Energie, Energiewirtschaftl. Tagesfragen, 30 (8), 583-585.

Gerwin, R.(1980): Die Welt-Energieperspektive, DVA, Stuttgart.

Global 2000(1980): Report to the President, US GPO, Washington, D.C.

Grenon, M. and B. Lapillonne(1976): The WELMM approach to energy strategies and options, RR-76-19, Int. Inst. Appl. Systems Analysis, Laxenburg.

Griffith, E.D. and A.W. Clarke(1979): World coal production, Scientif. American 240 (1), 28-37.

Häfele, W.(1980): A global and long-range picture of energy developments, Science 209, 174-182.

Häfele, W. und H.-H. Rogner(1980): Energie - die globale Perspektive, Angew. Systemanalyse 1 (2), 57-67.

Häfele, W. et al.(1981): Energy in a Finite World, A Global Energy Systems Analysis, 2 vols., Ballinger, Cambridge, USA.

Kemmer, H.-G.(1981): Der Schatz vom Orinoko, Die Zeit, Nr. 12, S. 20.

Keyfitz, N.(1977): Population of the world and its regions 1975-2050, WP-77-7, Int. Inst. Applied Systems Analysis, Laxenburg.

Klauder, W.(1980): Sterben die Deutschen aus? Umschau 80 (21), 649-656.

Laconte, P.(1980): Effects and dangers of extreme urbanization. In: N. Polunin (ed.) Growth without Ecodisaster? 235-246, The MacMillan Press Ltd., London.

Lapillonne, B.(1978): MEDEE2: A Model for long-term energy demand evaluation. RR-78-17, Int. Inst. Appl. Systems Analysis, Laxenburg.

Leontief, W.W.(1977): Structure of the World Economy: Outline of a Simple Input-Output Formulation, Amer. Economic Rev., 823-834.

Leontief, W.W.(1980): The world economy of the year 2000, Scientific American 243 (3), 166-181.

Lovins, A.B., L.H. Lovins, F. Krause and W. Bach(1981): Least-Cost Energy: Solving the CO_2-Problem, Brick House, Andover, USA.

Marchetti, C.(1977a): Sind die bisherigen Energie-Berechnungen falsch? Bild d. Wiss. 7, 68-77.

Marchetti, C.(1977b): Primary energy substitution models: On the interaction between energy and society, Technological Forecasting and Social Change 10, 345-356.

Marchetti, C.(1978): Energy systems - the broader context, RM-78-18, Int. Institute Applied Systems Analysis, Laxenburg.

Marchetti, C. and N. Nakicenovic(1979): The dynamics of energy systems and the logistic substitution model, RR-79-13, Int. Institute for Applied Systems Analysis, Laxenburg.

Mauldin, W.P.(1980): Population trends and prospects, Science, 209, 148-157.

Meadows, D.H. et al.(1972): The Limits to Growth, Potomac Assoc., Washington, D.C.

Mesarovic, M. und E. Pestel(1974): Menschheit am Wendepunkt, DVA, Stuttgart.

Müller, W. und B. Stoy(1978): Entkopplung. Wirtschaftswachstum ohne mehr Energie? DVA, Stuttgart.

Peterson, J.(1980): Grappling with Urban Giants, Ambio 9 (5), 250-252.

Rogner, H.-H. and W. Sassin(1980): High energy demand and supply scenario. In: W. Bach et al.(eds.) Interactions of Energy and Climate, 33-52, Reidel, Dordrecht.

Rotty, R.M. and G. Marland(1980): Constraints on fossil fuel use. In: W. Bach et al. (eds.) Interactions of Energy and Climate, 191-212, Reidel, Dordrecht.

Sassin, W.(1980a): Urbanization and the Energy Problem, Options, IIASA,News Rpt. 3, 1-4.

Sassin, W.(1980b): Energy,Scientific American, 243 (3),107-117.

Spinrad, B.I.(1980): Market substitution models and economic parameters, RR-80-28, Int. Inst. Appl. Systems Analysis, Laxenburg.

U.N.(1975): Recent population trends and future prospects, report of the Secretary General. In: The Population Debate: Dimensions and Perspectives, Prcdgs. World Population Conf., Bucharest 1974, vol. 1, UN, New York.

U.S. Dept. of Agriculture(1977): Cropland erosion, Soil Conservation Service, Washington, D.C.

Voss, A.(1980): Entwicklung der Energieversorgung in der Bundesrepublik Deutschland und anderer Regionen. In: H.A. Oomatia (Hrsg.) Energie und Umwelt, 102-107, Vulkan-Verlag Essen.

WOCOL(World Coal Study)(1980): Coal - Bridge to the Future, Ballinger Publ. Co., Cambridge, Mass.

World Energy Conference(1978): World energy resources 1985-2020: Coal resources. An appraisal of world coal resources and their future availability. IPC Press, Guildford, England.

Ziegler, A. und R. Holighaus(1979): Gas, Kraftstoff und Heizöl aus Kohle, Umschau 79 (12), 367-376.

IV. Influence of Society on Climate (Section 1)

Angell, J.K.(1981): Comparison of variations in atmospheric quantities with sea surface temperature variations in the Equatorial Eastern Pacific, Mon. Wea. Rev. 109, 230-243.

Angell, J.K. and J. Korshover(1977): Estimate of the global change in temperature surface to 100mb, between 1958-1975, Mo. Wea. Rev. 105, 375-385.

Armentano, T.V.(ed., 1980): The role of organic soils in the world carbon cycle, US DOE 004, Washington, D.C.

Armentano, T.V. and J. Hett(eds., 1980): The role of temperate zone forests in the world carbon cycle, US DOE 003, Washington, D.C.

Armentano, T.V. and O.L. Loucks(1982): Prospective significance of temperate zone carbon pool transients, 1980-2010. In: G. Brown (ed.) Global Dynamics of Biospheric Carbon, 73-95, US DOE 019, Washington, D.C.

Arrhenius, S.(1896): On the influence of carbonic acid in the air upon the temperature of the ground, Philosophical Magazine and J. of Science, S.5, 41 (251), 237-276.

Arrhenius, S.(1903): Lehrbuch der kosmischen Physik, Bd. 2, Hirzel Verlag, Leipzig.

Bacastow, R.B.(1976): Modulation of atmospheric carbon dioxide by the Southern Oscillation, Nature 261, 116-118.

Bacastow, R.B.(1979): Dip in the atmospheric CO_2 level during the arid 1960's, J. Geophys. Res. 84, 3108-3114.

Bacastow, R.B. and C.D. Keeling(1973): Atmospheric CO_2 and radiocarbon in the natural carbon cycle. In: G.M. Woodwell and E.V. Pecan (eds.) Carbon in the Biosphere,86-134, Springfield, USA.

Bach, W.(1979): Impact of increasing atmospheric CO_2 concentrations on global climate: Potential consequences and corrective measures, Environm. Int., 2, 215-228.

Bach, W.(1980a): Klimaeffekte anthropogener Energieumwandlung. In: H.A. Oomatia (Hrsg.) Energie u. Umwelt, 84-98, Vulkan-Verlag, Essen.

Bach, W.(1980b): Climatic effects of increasing atmospheric CO_2 levels, Experientia 36 (7), 796-806.

Bach, W.(1981): Fossil fuel resources and their impacts on environment and climate, Int. J. Hydrogen Energy 6, 185-201.

Bach, W. u. G. Breuer(1980): Wie dringend ist das CO_2-Problem? Umschau 80 (17), 520-525.

Bach, W., J. Pankrath and W.W. Kellogg(eds., 1979): Man's Impact on Climate, Elsevier Publ Co., Amsterdam.

Baes Jr., C.F. et al.(1977): Carbon dioxide and climate: The uncontrolled experiment, Amer. Scientist, 65, 310-320.

Baes Jr., C.F.(1981): The response of the oceans to increasing atmospheric CO_2, ORAU/IEA-81-6(M), Institute for Energy Analysis, Oak Ridge Assoc. Universities, Oak Ridge, USA.

Baes, Jr., C.F.(1982): Effects of ocean chemistry and biology on atmospheric carbon dioxide. In: W.C. Clark (ed.) Carbon Dioxide Review: 1982, 189-206, Oxford Univ. Press, New York.

Baes, Jr.,C.F.(1983): The role of the oceans in the carbon cycle. In: W. Bach et al. (eds.) Carbon Dioxide, 31-56, Reidel Publ. Co., Dordrecht.

Berger, W.H.(1982): Increase of carbon dioxide in the atmosphere during deglaciation: The coral reef hypothesis, Naturwissenschaften 69, 87-88.

Bernard, H.W.(1980): The Greenhouse Effect, Ballinger Publ. Co., Cambridge Mass., USA.

Berner W., H. Oeschger and B. Stauffer(1980): Information on the CO_2 cycle from ice core studies, Radiocarbon, 22, 227-235.

Björkström, A.(1983): Models of the ocean as a link in the global carbon cycle. In: W. Bach et al. (eds.) Carbon Dioxide, 57-92, Reidel Publ. Co., Dordrecht.

Bohn, H.L.(1978): On organic soil carbon and CO_2, Tellus 30, 472-475.

Bolin, B.(1977): Changes of land biota and their importance for the carbon cycle, Science 196, 613-615.

Bolin, B., E.T. Degens, S. Kempe, P. Ketner(eds., 1979): The Global Carbon Cycle, SCOPE 13, J. Wiley & Sons, New York.

Bolin, B.(ed., 1981): Carbon Cycle Modelling, SCOPE 16, John Wiley & Sons, New York.

Boville, B.W. and B.R. Döös(1981): Why a world climate programme? Nature and Resources XVII (1), 2-7.

Breuer, G.(1981): Landvegetation: Quelle oder Senke? Naturwiss. Rdsch. 34 (1), 32-33.

Broecker, W.S.(1974): Chemical Oceanography, Harcourt Brace Jovanovitch, New York.

Broecker, W.S.(1975): Climatic change: Are we on the brink of a pronounced global warming? Science, 189, 460-463.

Broecker, W.S. et al.(1979): Fate of fossil fuel carbon dioxide and the global carbon budget, Science 206, 409-418.

Broecker, W.S., T.-H. Peng and R. Engh(1980): Modeling the carbon system, Radiocarbon 22 (3) 565-598, and in: Proceedings of the CO_2 and Climate Res. Program Conf., 43-99, US DOE 011, Washington, D.C.

Broecker, W.S.(1982): Glacial to interglacial changes in ocean chemistry, Progress in Oceanography 11, 151-197.

Brown, J. and J.T. Andrews(1982): Influence of short-term climate fluctuations on permafrost terrain, vol. II pt. 3, US DOE 013, Washington, D.C.

Brown, S., A.E. Lugo and B. Liegel(eds., 1980): The role of tropical forests on the world carbon cycle, US DOE 007, Washington D.C.

Brown, S., A.E. Lugo and G. Gertner(1982): Ecological interpretation of atmospheric CO_2 concentration at Mauna Loa. In: S. Brown (ed.) Global Dynamics of Biospheric Carbon, 1-8, US DOE 019, Washington, D.C.

Bryan, K., F.G. **Komro,** S. **Manabe and M.J. Spelman**(1982): Transient climate response to increasing atmospheric carbon dioxide, Science 215, 56-58.

Bryan, K., S. Manabe and R.C. Pacanowski(1975): A global ocean-atmosphere climate model, Part II. The oceanic circulation, J. Phys. Oceanogr. 5, 30-46.

Budyko, M.I.(1974): Climate and Life, Int. Geophys. Series Vol. 18, Academic Press, New York.

Callendar, G.S.(1938): The artificial production of carbon dioxide and its influence on temperature, Quart. J. Roy. Meteorol. Soc. 64, 223-237.

Cess, R.D. and S.D. Goldenberg(1981): The effect of ocean heat capacity upon global warming due to increasing atmospheric carbon dioxide, J. Geophys. Res. 86 (C1), 498-502.

Chamberlin, T.C.(1897): A group of hypotheses bearing on climatic changes, J. Geol. 7, 653-683.

Chervin, R.M.(1980): Estimates of first - and second - moment climate statistics in GCM simulated climate ensembles, J. Atmos. Sci. 37, 1889-1902.

Clawson, M.(1979): Forests in the long sweep of American history, Science 204, 1168-1174.

Corby, G.A., A. Gilchrist and P.R. Rowntree(1977): United Kingdom Meteorological Office 5-level general circulation model. In: Adler, B. et al.(eds.) Methods in Computational Physics, 67-110, Academic Press, New York/London.

Crane, A.J.(1981): Comments on recent results about the CO_2 greenhouse effect, J. Appl. Meteorol. 20, 165-167.

Dahlman, R.C.(1982): Carbon cycle research plan, US DOE Conf. Berkeley Springs, W.Va., Sep. 19-23, 1982.

Dansgaard, W. et al.(1969): One thousand centuries of climatic record of Camp Century on the Greenland ice sheet, Science 166, 377-381.

Dansgaard, W.(1981): Ice core studies dating the past to find the future, Nature 290, 360-361.

Degens, E.T.(ed.,1982): Transport of carbon and minerals in major world rivers, Mitteilungen des Geol.-Paläontol. Inst. Univ. Hamburg, Heft 52, Hamburg.

Degens, E.T. und S. Kempe(1979): Heizen wir unsere Erde auf? Bild der Wissenschaft, 8, 52-58.

Delmas, R.J., J.-M. Ascensio and M. Legrand(1980): Polar ice evidence that atmospheric CO_2 20.000 yr BP was 50% of present, Nature 284, 155-157.

Desjardins, R.L. et al.(1982): Aircraft monitoring of surface carbon dioxide exchange, Science 216, 733-735.

Detwiler, R.P., C. Hall and P. Bogdonoff(1982): Simulating the impact of tropical land use changes on the exchange of carbon between vegetation and the atmosphere. In: S. Brown (ed.) Global Dynamics of Biospheric Carbon, 141-159, US DOE 019, Washington, D.C.

Dietrich, G., K. Kalle, W. Kraus u. G. Siedler(1975): Allgemeine Meereskunde, Gebr. Borntraeger, Berlin.

Dittberner, G.J.(1978): Climatic change: Volcanoes, man-made pollution, and carbon dioxide, IFEE Transactions on Geoscience Electronics, Vol. GE-16 (1), 50-61.

Epstein, E.S.(1982): Detecting climatic change, J. Appl. Meteorol. 21, 1172-1182.

Etkins, R. and E.S. Epstein(1982): The rise of global mean sea level as an indication of climatic change, Science 215, 287-289.

Flohn, H.(1941): Die Tätigkeit des Menschen als Klimafaktor, Ztschr. f. Erdkunde, 13-22.

Flohn, H.(1980): Possible climatic consequences of a man-made global warming, RR-80-30, Int. Inst. Applied Systems Analysis, Laxenburg.

Flohn, H.(1981): Major climatic events as expected during a prolonged CO_2-induced warming, Report of the Institute of Energy Analysis, Oak Ridge Assoc. Universities, Oak Ridge, USA.

Flohn, H.(1983): Major climatic events expected during a CO_2-induced warming. In: W. Bach et al. (eds.) Carbon Dioxide, 299-314, Reidel Publ. Co., Dordrecht.

Francey, R.J.(1981): Tasmanian tree rings belie suggested anthropogenic $^{13}C/^{12}C$ trends, Nature 290, 232-235.

Franck, H.-G. u. A. Knop(1980): Kohleveredlung an der Schwelle der 80er Jahre. Die Naturwiss. 67 (9), 421-430.

Fraser, P.J., P. Hyson and G.I. Pearman(1981): Some considerations of the global measurements of background atmospheric CO_2 In: World Climate Programme, Analysis and Interpretations of Atmospheric CO_2 Data,179-186, WCP-14, WMO, Geneva.

Freyer, H.D.(1978): Preliminary evolution of past CO_2 increase as derived from C - 13 measurements in tree rings In: J. Williams (ed.) Carbon Dioxide, Climate and Society, 69-87, Pergamon Press, Oxford.

Freyer, H.D.(1979): Variations in the atmospheric CO_2 content. In: Bolin, B. et al.(eds.) The Global Carbon Cycle, 79-99, SCOPE 13, John Wiley & Sons, Chichester/New York.

Gates, W.L.(1975): The January global climate simulated by a two-level general circulation model: a comparison with observation, J. Atmos. Sci. 32, 449-477.

Gates, W.L.(1980): Modeling the surface temperature changes due to increased atmospheric CO_2. In: W. Bach et al. (eds.) Interactions of Energy and Climate, 169-190, Reidel Publ. Co., Dordrecht.

Gates, W.L., K.H. Cook and M.E. Schlesinger(1981): Preliminary analysis of experiments on the climatic effects of increased CO_2 with an atmospheric general circulation model and a climatological ocean, J. Geophys. Res. 86, 6385-6393.

Gilchrist, A.(1983): Increased carbon dioxide concentrations and climate: The equilibrium response. In: W. Bach et al. (eds.) Carbon Dioxide, 219-258, Reidel Publ. Co., Dordrecht.

Gilliland, R.L.(1982): Solar, volcanic and CO_2 forcing of recent climatic changes, Climatic Change 4, 111-131.

Goldman, J.C. and M.R. Dennett(1983): Carbon dioxide exchange between air and seawater: No evidence for rate catalysis, Science 220, 199-201.

Hampicke, U. und W. Bach(1979): Die Rolle terrestrischer Ökosysteme im globalen Kohlenstoff-Kreislauf, Bericht im Auftrag des Umweltbundesamtes 153 S.+ XXII.

Hansen, J. et al.(1981): Climatic impact of increasing atmospheric CO_2,Science 213, 957-966.

Heimann, M., D. Kamber, H. Oeschger u. U. Siegenthaler(1981): Das globale Kohlenstoffsystem: Reservoire u. Flüsse. In: Die Auswirkungen von CO_2-Emissionen auf das Klima, Bd. 1, 105-142. Bericht des Battelle-Instituts, Frankfurt.

Hoffert, M.I., A.J. Callegari and C.-T. Hsieh(1980): The role of deep sea heat storage in the secular response of climatic forcing, J. Geophys. Res. 85 (C 11), 6667-6679.

Hoffert, M.I., A.J. Callegari and C.-T. Hsieh(1981): A box diffusion carbon cycle model with upwelling, polar bottom water formation and a marine biosphere. In: B. Bolin (ed.) Carbon Cycle Modelling, 287-305, SCOPE 16,J.Wiley & Sons, Chichester/New York.

Holland, W.R.(1979): The general circulation of the ocean and its modeling, Dynamics of Atm. and Oceans 3, 111-142.

Idso, S.B.(1980): The climatological significance of a doubling of earth's atmospheric CO_2 concentration, Science 207, 1462-1463.

Jacoby, G.(ed.,1980): Proceedings of the Int. Meeting on Stable Isotopes in Tree-Ring Research, US DOE 012, Washington, D.C.

JASON,(1980): The long term impacts of increasing atmospheric CO_2 levels, Techn. Report JSR-79-04, Stanford Res. International.

Jones, P.D., T.M.L. Wigley and P.M. Kelly(1982): Variations in surface air temperatures: Part 1. Northern Hemisphere, 1881-1980. Mo. Wea. Rev. 110(2), 59-70.

Junge, C.(1978): Die CO_2-Zunahme und ihre mögliche Klimaauswirkung, Promet 2/3, 21-32.

Kaplan, L.D.(1961): Reply to Plass, Tellus 13, 301-302.

Keeling, C.D.(1973): Industrial production of CO_2 from fossil fuels and limestone, Tellus 28, 538-550.

Keeling, C.D. et al.(1976a): Atmospheric carbon dioxide variations at Mauna Loa Observatory, Hawaii, Tellus 28, 538-551.

Keeling, C.D. et al.(1976b): Atmospheric carbon dioxide variations at the South Pole, Tellus 28, 552-564.

Keeling, C.D.(1980): The oceans and biosphere as future sinks for fossil fuel CO_2 .In: W. Bach et al. (eds.) Interactions of Energy and Climate, 129-147, Reidel Publ. Co., Dordrecht.

Keeling, C.D.(1983): The global carbon cycle: What we know and could know from atmospheric, biospheric and oceanic observations, II.3-II.75, US DOE 021, Washington, D.C.

Kellogg, W.W., and R.D. Bojkov(1982): Report of the JCS/CAS meet ing of experts on detection of possible climate change, WCP-29, WMO, Geneva.

Kelly, P.M. et al.(1982a): Variations in surface air temperatures: Part 2. Arctic Regions, 1881-1980, Mo. Wea. Rev. 110 (2), 71-83.

Kelly, P.M. et al.(1982b): Commentary on detecting carbon dioxide effects on climate. In: W.C. Clark (ed.) Carbon Dioxide Review: 1982, 246-251, Oxford Univ. Press, New York.

Kester, D.R. and R.M. Pytkowicz(1977): Natural and anthropogenic changes in the global CO_2 system. In: W. Stumm (ed.) Global Chemical Cycles and their Alterations by Man, 99-120, Dahlem Konferenzen, Berlin.

Kohlmaier, G.H. et al.(1983): The role of the biosphere. In. W. Bach et al.(eds.) Carbon Dioxide, 93-144, Reidel Publ. Co., Dordrecht.

Kukla, G. and J. Gavin(1981): Summer ice and carbon dioxide, Science 214, 497-503.

Lamb, H.H.(1977): Supplementary volcanic dust veil assessments, Clim. Monitor 6, 57-67.

Laurmann, J.A. and R.M. Rotty(1983): Exponential growth and atmospheric carbon dioxide, J. Geophys. Res. 88(C2) 1295-1299.

Laurmann, J.A. and J.R. Spreiter(1983): The effects of carbon cycle model error in calculating future atmospheric carbon dioxide levels, Climatic Change 5(2), 145-181.

Lawson, M.P. et al.(1982): Analysis of the climatic signal in the south dome, Greenland ice core, Climatic Change 4(4), 375-384.

Leavitt, S.W. and A. Long(1983): An atmospheric $^{13}C/^{12}C$ reconstruction generated through removal of climate effects from tree-ring $^{13}C/^{12}C$ measurements, Tellus 35B, 92-102.

Lieth, H. et al.(1980): Die CO_2 -Frage aus geoökologischer und energiewirtschaftlicher Sicht, Brennst.-Wärme-Kraft 32 (9), 393-400.

Lorius, C. and D. Raynaud(1983): Record of past atmospheric CO_2: Tree ring and ice core studies. In: W. Bach et al. (eds.) Carbon Dioxide, 145-176, Reidel Publ. Co., Dordrecht.

Loucks, D.L.(1980): Recent results from studies of carbon cycling in the biosphere. In: Proceedings of the CO_2 and Climate Res. Program Conference, 3-42, US DOE 011, Washington, D.C.

Lovins, A.B., L. H. Lovins, F. Krause and W. Bach(1981): Least-ost Energy: Solving the CO_2 Problem, Brick House, Andover, USA.

MacCracken, M.C.(1983): Have we detected CO_2 -induced climatic change? Problems and prospects. US DOE, 021, V.3-V.46, Washington, D.C.

MacCracken, M.C., and H. Moses(1982): The first detection of carbon dioxide effects: Workshop summary, US DOE 018, 3-44, Washington, D.C.

MacCracken, M.C.(1983): Is there climatic evidence now for carbon dioxide effects? UCRL-88613, Lawrence Livermore National Laboratory.

Machta, L.(1972): Mauna Loa and global trends in air quality, Bull. Amer. Meteorol. Soc. 53, 402-421.

Machta, L.(1973): Prediction of CO_2 in the atmosphere. In: G.M. Woodwell and E.V. Pecan (eds.) Carbon and the Biosphere, 21-31, Springfield, USA.

Madden, R.A. and V. Ramanathan(1980): Detecting climate change due to increasing CO_2, Science 209, 763-768.

Manabe, S., K. Bryan and M.J. Spelman(1979): A global ocean-atmosphere climate model with seasonal variation for future studies of climate sensitivity, Dyn. Atmos. Oceans 3, 393-426.

Manabe, S. and R.J. Stouffer(1979): A CO_2 -climatic sensitivity study with a mathematical model of the global climate, Nature 282, 491-493.

Manabe, S. and R.J. Stouffer(1980a): Sensitivity of a global climate model to an increase of CO_2 concentration in the atmosphere, J. Geophys. Res. 85 (C 10), 5529-5554.

Manabe, S. and R.J. Stouffer(1980b): Study of climatic impacts of CO_2 increase with a mathematical model of the global climate. In: US DOE 009, 127-140, Washington, D.C.

Manabe, S. and R.T. Wetherald(1967): Thermal equilibrium of the atmosphere with a given distribution of relative humidity, J. Atmos. Sci. 24, 241-259.

Manabe, S. and R.T. Wetherald(1975): The effects of doubling the CO_2 concentration on the climate of a general circulation model, J. Atmos. Sci. 32, 3-15.

Manabe, S. and R.T. Wetherald(1980): On the distribution of climatic change resulting from an increase in CO_2 content of the atmosphere, J. Atmos. Sci. 37, 99-118.

Manabe, S. and D.G. Hahn(1981): Simulation of atmospheric variability, Mo. Wea. Rev. 109, 2260-2286.

Manabe, S., R.T. Wetherald and R.J. Stouffer(1981): Summer dryness due to an increase of atmospheric CO_2 concentration, Climatic Change 3, 347-386.

Manabe, S.(1983): Carbon dioxide and climatic change. In: B. Saltzman (ed.) Advances in the Theory of Climate, Adv. in Geophys. vol. 25, 39-82, Academic Press, New York.

Meleshko, V.P. and R.T. Wetherald(1980): The effect of a geographical cloud distribution on climate:A numerical experiment with an atmospheric general circulation model, J. Geophys. Res. 86 (C 12), 11995-12014.

Michael, P., M. Hoffert, M. Tobias and J. Tichler(1981): Transient climatic response to changing carbon dioxide concentration, Climatic Change 3, 137-153.

Miller, P.C.(ed., 1981): Carbon balance in northern ecosystems and the potential effect of CO_2-induced climatic change, US DOE 015, Washington, D.C.

Miller, P.C.(1982a): Carbon balance in northern ecosystems and the potential effect of CO_2-induced climatic change. In: S. Brown (ed.) Global Dynamics of Biospheric Carbon, 56-72, US DOE 019, Washington, D.C.

Miller, P.C.(1982b): Research needed to determine the present carbon balance of northern ecosystems, and the potential effect of CO_2-induced climate change, US DOE 013, vol. II, pt. 14, Washington, D.C.

Millero, F.J. et al.(1976): The density of North Atlantic and North Pacific deep waters, Earth & Planet. Sci. Letters 32, 468-472.

Mitchell Jr., J.M.(1961): Recent secular changes of global temperature, Annals New York Academy of Sciences 95, 235-250.

Mitchell Jr., J.M. and M.C. MacCracken(1980): Climate effects workshop. In: Proceedings of the CO_2 and climate research program conference, 195-214, US DOE 011, Washington, D.C.

Mitchell, J.F.B.(1983): The seasonal response of a general circulation model to changes in CO_2 and sea surface temperature, Q.J.R. Met. Soc. 109, 113-152.

Möller, F.(1963): On the influence of changes in the CO_2-concentration in the air on the radiation balance at the earth's surface and on the climate, J. Geophys. Res. 68, 3877-3886.

Moore, B. et al.(1980): A simple model for analysis of the role of terrestrial ecosystems in the global carbon budget. Report Marine Biology Lab. Woods Hole, Mass., USA.

Mulholland, P.J. and J.W. Elwood(1982): The role of lake and reservoir sediments as sinks in the perturbed global carbon cycle, Tellus 34(5), 490-499.

Neftel, A. et al.(1982): Ice core sample measurements give atmospheric CO_2 content during the past 40 000 years, Nature 295, 220-222.

Newell, R.E., A.R. Navato and J. Hsiung(1978): Long-term global sea surface temperature fluctuations and their possible influence on atmospheric CO_2 concentration. Pageoph. 116, 351-371.

Newell, R.E. and T.G. Dopplick(1979): Question concerning the possible influence of anthropogenic CO_2 on atmospheric temperature, J. Appl. Met. 18, 822-825.

Newhall, C.G. and S. Self(1982): The Volcanic Explosivity Index (VEI): An estimate of explosive magnitude for historical volcanism, J. Geophys. Res. 87 (C 2), 1231-1238.

Oeschger, H.(1980): In der Natur gespeicherte Geschichte von Umweltvorgängen. In: H. Oeschger et al. (eds.) Das Klima, 209-236, Springer Verlag, Berlin.

Oeschger, H., U. Siegenthaler and M. Heimann(1980): The carbon cycle and its perturbation by man. In: W. Bach et al. (eds.) Interactions of Energy and Climate, 107-127, Reidel Publ. Co., Dordrecht.

Oeschger, H. u. M. Heimann(1981): Modelle des globalen Kohlenstoff-Systems. In: Die Auswirkungen von CO_2-Emissionen auf das Klima, Bd. 1, 205-236, Bericht des Battelle-Instituts, Frankfurt.

Oeschger, H. and M. Heimann(1983): Uncertainties of predictions of future atmospheric CO_2 concentrations, J. Geophys. Res. 88 (C 2), 1258-1262.

Paltridge, G.W.(1974): Global cloud cover and surface air temperature, J. Atmos. Sci. 31, 1571-1576.

Parkinson, C.L. and W.W. Kellogg(1979): Arctic sea ice decay simulated for a CO_2-induced temperature rise, Climatic Change 2, 149-162.

Pearman, G.I.(1980): Atmospheric CO_2 concentration measurements. A review of methodologies, existing programmes and available data, WMO Report No. 3, Geneva.

Plass, G.N.(1956): The carbon dioxide theory of climate change, Tellus 8 (2) 140-208.

Plass, G.N.(1961): Comments on the "Influence of carbon dioxide variations on the atmospheric heat balance" by L.D. Kaplan, Tellus 13, 296-300.

Pollard, D., M.L. Batteen and Y.-J. Han(1980): Development of a simple oceanic mixed-layer and sea-ice model for use with an atmospheric GCM. Rept. No. 21, Climatic Res. Inst., Oregon State Univ., Corvallis.

Pollard, D.(1982): The performance of an upper ocean model coupled to an atmospheric GCM: preliminary results, Rept. No. 31, Climatic Res. Inst., Oregon State Univ., Corvallis.

Post, W.M. et al.(1982): Summaries of soil carbon storage in world life zones. In: S. Brown (ed.) Global Dynamics of Biospheric Carbon, 131-139, US DOE 019, Washington, D.C.

Raisbeck, G.M. et al.(1981): Cosmogenic ^{10}Be concentrations in Antarctic ice during the past 30 000 years, Nature, 292, 825-826.

Ramanathan, V., M.S. Lian and R.D. Cess(1979): Increased atmospheric CO_2 :Zonal and seasonal estimates of the effect on the radiation energy balance and surface temperature, J. Geophys. Res. 84 (C 8), 4949-4958.

Ramanathan, V.(1981): The role of ocean-atmosphere interactions in the CO_2-climate problem, J. Atmos. Sci. 38, 918-930.

Revelle, R. and H.E. Suess(1957): Carbon dioxide exchange between atmosphere and ocean and the question of an increase of atmospheric CO_2 during past decades, Tellus 9, 18-27.

Riches, M.R., M. MacCracken and F.M. Luther(1982): CO_2 climate research plan, US DOE Conf. Berkeley Springs, W.Va., Sept. 19-23, 1982.

Roether, W.(1980): The effect of the ocean on the global carbon cycle, Experientia 36 (9), 1017-1025.

Rosenberg, N.J.(1982): The increasing CO_2 concentration in the atmosphere and its implication on agricultural productivity II. Effects of rough CO_2 -induced climatic change, Climatic Change 4, 239-254.

Rossow, W.B., A. Henderson-Sellers and S.K. Weinreich(1982): Cloud feedback: A stabilizing effect for the early earth? Science 217, 1245-1247.

Rotty, R.M.(1980): Past and future emission of CO_2 , Experientia 36 (7), 781-783.

Rotty, R.M.(1981): Distribution and changes in industrial carbon dioxide production. In: Analysis and Interpretation of Atmospheric CO_2-Data, WCP-14, World Meteorol. Organ., Geneva.

Sadler, J.C., C.S. Ramage and A.M. Hori(1982): Carbon dioxide variability and atmospheric circulation, J. Appl. Meteor., 21, 793-805.

Schlesinger, M.E. and W.L. Gates(1980): the January and July performance of the OSU two-level atmospheric general circula - tion model, J. Atmos. Sci. 37, 1914-1943.

Schlesinger, M.E. and W.L. Gates(1981): Preliminary analysis of four general circulation model experiments on the role of the ocean in climate, Rept. No. 25, Climatic Res. Inst., Oregon State Univ., Corvallis.

Schlesinger, M.E.(1982): the climatic response to doubled CO_2 simulated by the OSU atmospheric GCM with a coupled swamp ocean (unpubl. manuscript).

Schlesinger, M.E.(1983a): Simulating CO_2-induced climatic change with mathematical climate models: Capabilities, limitations and prospects, III.3 - III.139, US DOE 021, Washington D.C.

Schlesinger, M.E.(1983b): A review of climate models and their simulation of CO_2-induced warming, Int. J. Env. Studies 20, 103-114.

Schlesinger, W.H. and J.M. Melack(1981): Transport of organic carbon in the world's rivers, Tellus, 33, 172-187.

Schneider, S.H.(1975): On the carbon dioxide-climate confusion, J. Atmos. Sci. 32, 2060-2066.

Schneider, S.H. and S.L. Thompson(1981): Atmospheric CO_2 and climate: Importance of the transient response, J. Geophys. Res. 86 (C 4), 3135-3147.

Schönwiese, C.-D.(1983): Climatic variability within the modern instrumentally-based period. In: W. Bach et al. (eds.) Carbon Dioxide, 315-336, Reidel Publ. Co., Dordrecht.

Schuurmans, C.J.E.(1983): On the detection of CO_2-induced climatic change. In: W. Bach et al. (eds.) Carbon Dioxide, 337-352, Reidel Publ. Co., Dordrecht.

Seiler, W. and P.J. Crutzen(1980): Estimates of gross and net fluxes of carbon between the biosphere and the atmosphere from biomass burning, Climatic Change 2 (3), 207-247.

Sellers, W.D.(1974): A reassessment of the effect of CO_2 variations on a simple global climate model, J. Appl. Meteorol. 13, 831-833.

Shaver, G.R. et al.(1982): The role of terrestrial biota and soils in the global carbon budget. In: S. Brown (ed.) Global Dynamics of Biospheric Carbon, 160-165, US DOE 019, Washington D.C.

Siegenthaler, U. and H. Oeschger(1978): Predicting future atmospheric CO_2 levels, Science 199, 388-395.

Smith, S.V.(1981): Marine macrophytes as a global carbon sink, Science 211, 838-839.

Stuiver, M., H.G. Östlund and T.A. McConnaughey(1981): GEOSECS Atlantic and Pacific C-14 distributions. In: B. Bolin (ed.) Carbon Cycle Modelling, 201-209, SCOPE 16, John Wiley & Sons, New York.

Stuiver, M.(1982): The history of the atmosphere as recorded by carbon isotopes. In: E.D. Goldberg (ed.) Atmospheric Chemistry, 159-179, Springer Verlag, Berlin.

Stuiver, M., P.D. Quay and H.G. Östlund(1983): Abyssal water carbon-14 distribution and the age of the world oceans, Science 219, 849-851.

Sundquist, E.T. and G.A. Miller(1980): Oil shales and carbon dioxide, Science 208, 740-741.

Takahashi, T., W.S. Broecker and A.E. Bainbridge(1981): The alkalinity and total carbon dioxide concentration in the world oceans. In: B. Bolin (ed.) Carbon Cycle Modelling, 271-286, SCOPE 16, John Wiley & Sons, New York.

Tans, P.(1981): $^{13}C/^{12}C$ of industrial CO_2. In: B. Bolin (ed.) Carbon Cycle Modeling, 127-129, SCOPE 16, John Wiley & Sons, New York.

Thompson, S.L. and S.H. Schneider(1981): Carbon dioxide and climate: Ice and ocean, Nature 290, 9-10.

Thompson, S.L. and S.H. Schneider(1982a): Carbon dioxide and climate: Has a signal been observed yet? Nature 295, 645-646.

Thompson, S.L. and S.H. Schneider(1982b): Carbon dioxide and climate: The importance of realistic geography in estimating transient temperature response, Science 217, 1031-1033.

Tolman Jr., C.F.(1899): The carbon dioxide of the ocean and its relation to the carbon dioxide of the atmosphere, J. Geol. 7, 585-618.

Tyndall, J.(1861): On the absorption and radiation of heat by gases and vapours, and on the physical connexion of radiation, absorption, and conduction, Philosophical Magazine and J. of Science, S. 4, 22 (146), 169-194 and 22 (147), 273-285.

US DOE (Department of Energy)(1980): Carbon dioxide research progress report, Fiscal year 1979, DOE 005, Washington, D.C.

US DOE(1981): Flux of organic carbon by rivers to the oceans, Carbon Dioxide Effects Research and Assessment Program, DOE 016, Washington, D.C.

US DOE(1982): Proceedings of the workshop on first detection of carbon dioxide effects, DOE 018, Washington, D.C.

US NAS(National Academy of Sciences)(1979): Carbon dioxide and climate: A scientific assessment, Washington, D.C.

US NAS(1982): Carbon dioxide and climate: A second assessment, Washington, D.C.

US Senate(1979): Carbon dioxide accumulation in the atmosphere, synthetic fuels and energy policy, US GPO, Washington, D.C.

Van Keulen, H., H. van der Laar, W. Louwerse and J. Goudriaan (1980): Physiological aspects of increased CO_2 concentration, Experientia 36 (7), 786-792.

Vinnikov, K. Ya. et al.(1980): Modern changes in climate of the Northern Hemisphere, Meteorologiya i Gidrologiya 6, 5 - 17.

Wallén, C.- C. (1980): Monitoring potential agents of climatic change, Ambio 9 (5), 222-228.

Walsh, J.J. et al.(1981): Biological export of shelf carbon is a sink of the global CO_2 cycle, Nature 291, 196-201.

Washington, W.M. and V. Ramanathan(1980): Climatic response due to increased CO_2: Status of model experiments and the possible role of the oceans. In: Proceedings of the CO_2 and climate research program conference, 107-131, US DOE 011, Washington, D.C.

Washington, W.M. and G.A. Meehl(1982): General circulation model experiments on the climatic effects due to doubling and quadrupling of CO_2 concentration. Report NCAR 10306/81-2(E), Nat. Center f. Atm. Res., Boulder, Co.

Washington, W.M. and G.A. Meehl(1983): General circulation model experiments on the climatic effects due to a doubling and quadrupling of carbon dioxide concentration, J. Geophys. Res. (in press).

Watts, R.G.(1982): Further discussion of "Questions concerning the possible influence of anthropogenic CO_2 on atmospheric temperature", J. Appl. Meteorol. 21, 243-247.

WCP(World Climate Program)(1981a): Analysis and interpretation of atmospheric CO_2 data, WCP-14, WMO, Geneva.

WCP(1981b): On the assessment of the role of CO_2 on climate variations and their impact, WMO, Geneva.

Wetherald, R.T. and S. Manabe(1980): Cloud cover and climate sensitivity, J. Atmos. Sci. 37, 1485-1510.

Wetherald, R.T. and S. Manabe(1981): Influence of seasonal variation upon the sensitivity of a model climate, J. Geophys. Res. 86 (C 2), 1194-1204.

Wigley, T.M.L. and P.D. Jones(1981): Detecting CO_2-induced climatic change, Nature 292, 205-208.

Woodwell, G.M.(1978): The carbon dioxide question, Scientific American, 238, 34-43.

Woodwell, G.M.(ed., 1980): Measurement of changes in terrestrial carbon using remote sensing, US DOE 010, Washington, D.C.

Zimen, K.E., P. Offermann and G. Hartmann(1977): Source functions of CO_2 and future CO_2 burden, Z. Naturforsch. 32a, 1544-1554.

Zimmerman, P.R. et al.(1982): Termites: A potentially large source of atmospheric methane, carbon dioxide, and molecular hydrogen, Science 218, 563-565.

IV. Influence of Society on Climate (Sections 2-6)

Almquist, E.(1974a): An analysis of global air pollution, Ambio 3 (5), 161-167.

Almquist, E.(1974b): Remote sensing of gaseous air pollution, Ambio 3 (5), 168-176.

Arking, A. et al.(1983): The effect of the El Chichón eruption on the Northern Hemisphere climate, Abstract, 63rd AMS Annual Meeting, New Orleans.

Bach. W.(1976): Global air pollution and climatic change, Revs. Geophys. & Space Phys. 14 (3), 429-474.

Bach, W.(1979): Short-term climatic alterations caused by human activities: status and outlook, Progress in Physical Geography 3 (1), 55-83.

Bach, W.(1979/1980): Untersuchung der Beeinflussung des Klimas durch anthropogene Faktoren. Bericht im Auftrag des Umweltbundesamtes, Berlin, 1979; und Münstersche Geographische Arbeiten, Nr.6, 7-34, Schöningh, Paderborn, 1980.

Bach, W.(1980a): Waste heat and climatic change. In: Theodore, L. and A.J. Buonicore (eds.) Energy and the Environment: Interactions, vol. I, Pt. B. chapter 7, 151-190, CRC Press, Boca Raton, Florida.

Bach, W.(1980b): Klimaeffekte anthropogener Energieumwandlung. In: H.A. Oomatia (Hrsg.) Energie und Umwelt, 84-98, Vulkan-Verlag Essen.

Bach, W. and A. Daniels(1975): Handbook of Air Quality in the United States, The Oriental Publ. Co., Honolulu, Hi.

Bartholomäi, G. und W. Kinzelbach(1979): Das Abwärmekataster Oberrheingebiet, Brennst. -Wärme-Kraft 31 (12), 472-478.

Baumgartner, A., M. Kirchner und H. Mayer(1978): Die Wirkungen von potentiellen anthropogenen Veränderungen der Erdoberflächenbedeckung auf die weltweite Verteilung der Oberflächenalbedo, des Rauhigkeitsparameters und anderer wichtiger Parameter des Impulsaustausches an der Grenzfläche der Erde und der Atmosphäre, Promet 2/3, 32-48.

Baumgartner, A. and M. Kirchner(1980): Impacts due to deforestation. In: W. Bach et al. (eds.) Interactions of Energy and Climate, 305-316, Reidel Publ. Co., Dordrecht.

BMI(Hrsg., 1980): Bericht über die Auswirkungen von Luftverunreinigungen auf das globale Klima, Umweltbrief 20. Jan. 1980.

Boeck, W.L.(1976): Meteorological consequences of atmospheric Krypton-85, Science 193, 195-198.

Bryson, R.A. and D. A. Baerreis(1967): Possibilities of major climatic modification and their implications: Northwest India, a case study, Bull. Amer. Meteorol. Soc., 48, 136-142.

CEQ(Council on Environmental Quality)(1975): Fluorocarbons and the Environment, Washington, D.C.

Charlock, T.P. and W.D. Sellers(1980): Aerosol effects on climate: Calculations with time-dependent and steady-state radiative-convective models, J. Atmos. Sci. 37, 1327-1341.

Charney, J.G.(1975): Dynamics of deserts and drought in the Sahel, Quart. J. Roy. Meteorol. Soc. 101, 193-202.

Chervin, R.M.(1980): Computer simulation studies of the regional and global climatic impacts of waste heat emission, In: W. Bach et al. (eds.) Interactions of Energy and Climate, 399-415, Reidel Publ. Co., Dordrecht.

Crutzen, P.J. and D.H. Ehhalt(1977): Effects of nitrogen fertilizers and combustion on the stratospheric ozone layer, Ambio 6 (2-3), 112-117.

DeLaune, R.D. et al.(1983): Methane release from Gulf coast wetlands, Tellus 35B, 8-15.

Dittberner, G.J.(1978): Climatic change:Volcanoes, man-made pollution, and carbon dioxide, IEEE Transactions on Geoscience Electronics, GE-16 (1), 50-61.

Duxbury, J.M. et al.(1982): Emissions of nitrous oxide from soils, Nature 298, 462-464.

Ehhalt, D.H.(1976/1977): Die Verteilung der Chlorfluormethane in der Stratosphäre und ihr Einfluss auf die Ozonschicht, Jahresbericht der Kernforschungsanlage Jülich.

Ehhalt, D.H. and U. Schmidt(1978): Sources and sinks of atmospheric methane, Pageoph. 116, 452-464.

Ehhalt, D.H.(1980): The effects of chlorofluoromethanes on climate, In: W. Bach et al. (eds.) Interactions of Energy and Climate, 243-255, Reidel Publ. Co., Dordrecht.

Ehhalt, D.H. et al.(1982): On the temporal increase of tropospheric CH_4, paper KFA Jülich.

Eiden, R.(1979): The influence of trace substances on the atmospheric energy budget. In: W. Bach et al. (eds.) Man's Impact on Climate, 115-127, Elsevier Publ. Co., Amsterdam.

Ellsaesser, H.W. et al.(1976): An additional model test of positive feedback from high desert albedo, Quart. J. Roy. Meteorol. Soc. 102, 655-666.

Ellsaesser, H.W.(1979): Sources and sinks of stratospheric water vapor, Preprint UCRL-83611, Lawrence Livermore Lab., Livermore, USA.

FAA (US Federal Aviation Administration)(1981): Upper Atmospheric Programs Bulletin, No 81-3, US Dept. of Transportation, Washington, D.C.

Flohn, H.(1973): Geographische Aspekte der anthropogenen Klimamodifikation, Hamburger Georgr. Studien, H. 28, 13-36.

Flohn, H.(1975): Anthropogene Eingriffe in die Landschaft und Klimaänderungen, Beiheft zur Geograph. Ztschr., 137-149.

Flohn, H.(1978): Die Zukunft unseres Klimas: Fakten und Probleme, Promet 2/3, 1-21.

Flohn, H.(1979): Can climate history repeat itself? Possible climatic warming and the case of paleoclimatic warm phases. In: Bach W. et al. (eds.) Man's Impact on Climate, 15-28, Elsevier Publ. Co., Amsterdam.

Flohn, H.(1980a): Das Energieproblem und die Zukunft unseres Klimas, Bonner Universitätsblätter, 9-18.

Flohn, H.(1980b): Geophysikalische Grundlagen einer anthropogenen Klimamodifikation, Veröff. Joachim Jungius-Ges. Wiss. Hamburg, 44,195-218.

Flohn, H.(1981a): Klimaänderung als Folge der CO_2-Zunahme? Phys. Bl. 37 (7), 184-190.

Flohn, H.(1981b): Major climatic events associated with a prolonged CO_2-induced warming. ORAU/IEA-81-8 (M) Institute of Energy Analysis, Oak Ridge Assoc. Universities, Oak Ridge, USA.

Flohn, H.(1981c): Kohlendioxyd, Spurengase und Glashauseffekt: ihre Rolle für die Zukunft unseres Klimas, R.W. Akademie der Wiss., Heft Nr. 304, 46 S., Westdeutscher Verlag, Opladen.

Fraser, P.J. et al.(1983): Global distribution and southern hemispheric trends of atmospheric CCl_3F, Nature 302, 692-695.

Freyer, H.-D.(1979): Atmospheric cycles of trace gases containing carbon, In: B. Bolin et al.(eds.) The Global Carbon Cycle, SCOPE 13, 101-128, J. Wiley & Sons, New York.

Fricke, W.(1980): Die Bildung und Verteilung von anthropogenem Ozon in der unteren Troposphäre, Berichte des Inst. f. Meteorol. u. Geophys. der Univ. Frankfurt Nr. 44, Frankfurt.

Gates, W.L.(1980): Modeling the surface temperature changes due to increased atmospheric CO_2.In: W. Bach et al. (eds.) Interactions of Energy and Climate, 169-190, Reidel Publ. Co., Dordrecht.

Gentry, A.H. and J. Lopez-Parodi(1980): Deforestation and increased flooding of the Upper Amazon, Science 210, 1354-1356.

Global 2000(1980): Report to the President, US GPO, Washington, D.C.

Grassl, H.(1979): Possible changes of planetary albedo due to aerosol particles, In: W. Bach et al. (eds.) Man's Impact on Climate, 229-241, Elsevier Publ. Co., Amsterdam.

Grassl, H.(1980): Mögliche atmosphärische Auswirkungen weltweiter Krypton-85-Bildung, unveröff. Manuskript, Max-Planck-Institut f. Meteorologie, Hamburg.

Groves, K.S., S.R. Mattingly and A.F. Tuck(1978): Increased atmospheric carbon dioxide and stratospheric ozone, Nature 273, 711-715.

Groves, K.S. and A.F. Tuck(1979): Simultaneous effects of CO_2 and chlorofluoromethanes on stratospheric ozone, Nature 280, 127-129.

Hahn, J.(1979): Man-made perturbation of the nitrogen cycle and its possible impact on climate, In: W. Bach et al. (eds.) Man's Impact on Climate, 193-213, Elsevier Publ. Co., Amsterdam.

Haigh, J.D. and J.A. Pyle(1979): A two-dimensional calculation including atmospheric carbon dioxide and stratospheric ozone, Nature 279, 222-224.

Hameed, S., R.D. Cess and J.S. Hogan(1980): Response of the global climate to changes in atmospheric chemical composition due to fossil fuel burning, J. Geophys. Res. 85 (C 12), 75 7-7545.

Hameed, S. and R.D. Cess(1983): Impact of a global warming on biospheric sources of methane and its climatic consequences, Tellus 35B, 1-7.

Hansen, J., D. Johnson, A. Lacis, S. Lebedeff, P. Lee, D. Rind and G. Russel(1981): Climatic impact of increasing atmospheric carbon dioxide, Science 213, 957-966.

Hansen, J., A. Lacis and D. Rind(1983): Climate trends due to increasing greenhouse gases, Proceedings Coastal Zone 83 (in press).

Herrmann, H.(1978): Produktinnovation durch ökologische Bedingungen, VDI-Berichte Nr. 319, 43-48.

Hoffert, M.I., A.J. Callegari and C.-T. Hsieh(1980): The role of deap sea heat storage in the secular response to climatic forcing, J. Geophys. Res., 85 (C 11), 6667-6679.

Hogan, A.W. and V.A. Mohnen(1979): On the global distribution of aerosols, Science 205, 1373-1375.

Hummel, J.R. and R.A. Reck(1977): Development of a global surface albedo model, Paper pres. at the U.S. Geophysical Union, San Francisco.

Husar, R.B., J.P. Lodge Jr. and D.J. Moore (eds.)(1978): Sulfur in the atmosphere, Atmosph. Environment, 12, 1-796.

Isaksen, I.S.A.(1980): The impact of nitrogen fertilization. In: W. Bach et al. (eds.) Interactions of Energy and Climate, 257-268, Reidel Publ. Co., Dordrecht.

Isaksen, I.S.A. and F. Stordal(1981): The influence of man on the ozone layer. Readjusting the estimates, Ambio 10 (1), 9-17.

Johansson, T.B. and P. Steen(1978): Solar Sweden, Secretariat for Future Studies, Stockholm.

Kellogg, W.W.(1980): Aerosols and climate, In: W. Bach et al. (eds.) Interactions of Energy and Climate, 281-296, Reidel Publ. Co., Dordrecht.

Kellogg, W.W. (1983): Feedback mechanisms in the climate system affecting future levels of carbon dioxide, J. Geophys. Res. 88 (2), 1263-1269.

Kerr, R.A. (1981): Mt. St. Helens and a climate quandary, Science 211, 371-372, 374.

Kerr, R.A. (1982): El Chichón forebodes climate change, Science 217, 1023.

Kerr, R.A. (1983): El Chichón climate effect estimated, Science 219, 157.

Ko, M.K.W. and N.D. Sze (1982): A 2-D model calculation of atmospheric lifetimes for N_2O, CFC-11 and CFC-12, Nature 297, 317-319.

Kunkel, G. (1981): The use of the useless, GeoJournal 5.2, 171-178.

Lacis, A. et al. (1981): Greenhouse effect of trace gases, 1970-1980, Geophys. Res. Lett. 8 (10), 1035-1038.

Lettau, H., K. Lettau and L.C.B. Molion (1979): Amazonia's hydrologic cycle and the role of atmospheric recycling in assessing deforestation effects, Mon. Wea. Rev. 107 (3), 227-238.

Llewellyn, R.A. and W.M. Washington (1977): Regional and global aspects, In: Energy and Climate, 106-118, Nat. Academy of Sciences, Washington, D.C.

MacCracken, M.C. and F.M. Luther (1983): The effect of the El Chichón volcano on the seasonal climate of a zonal climate model, Abstract, 63rd AMS Annual Meeting, New Orleans.

Manabe, S. and R.T. Wetherald (1975): The effects of doubling the CO_2 concentration on the climate of a general circulation model, J. Atmos. Sciences, 32, 3-15.

Marchetti, C. (1977): On properties and behavior of energy systems In: H.G. Schlegel and J. Barnes (eds.) Microbial Energy Conversion, 619-642, Pergamon Press, Oxford.

Munn, R.E. and L. Machta (1979): Human activities that affect climate, In: World Climate Conference, 170-209, WMO-No. 537, World Met. Organ., Geneva.

Myers, N. (1979): The Sinking Ark, Pergamon Press Oxford.

Myers, N. (1981): The Hamburger connection: How Central America's forests become North America's Hamburgers, Ambio 10(1), 3-8.

Niehaus, F. (1977): Computersimulation langfristiger Umweltbelastung durch Energieerzeugung, ISR 41, Birkhäuser Verlag, Stuttgart.

Oke, T.R. (1980): Climatic impacts of urbanization. In: W. Bach et al. (eds.) Interactions of Energy and Climate, 339-356, Reidel Publ. Co., Dordrecht.

Potter, G.L. et al. (1980): Climate change due to anthropogenic surface albedo modification, In: W. Bach et al. (eds.) Interactions of Energy and Climate, 317-326, Reidel Publ. Co., Dordrecht.

Ramanathan, V. (1980): Climatic effects of anthropogenic trace gases, In: W. Bach et al. (eds.) Interactions of Energy and Climate, 269-280, Reidel Publ. Co., Dordrecht.

Ranjitsinh, M.K.(1979): Forest destruction in Asia and the South Pacific, Ambio 8 (5), 192-201.

Rasmussen, R.A., M.K. Khalil and R.W. Dalluge(1981): Atmospheric trace gases in Antarctica, Science 211, 285-287.

Robock, A. and C. Mass(1982): The Mt. St. Helens volcanic eruption of 18 May 1980: Large short-term surface temperature effects, Science 216, 628-630.

Sagan, C., O.B. Toon and J.B. Pollack(1979): Anthropogenic albedo changes and the earth's climate, Science 206, 1363-1367.

Sawyer, J.S.(1974): Can man's waste heat affect the regional climate? Paper presented at IAMAP Symposium, Melbourne.

Schikarski, W.O.(1980): Kann die Umwelt unsere Wärmeabfälle verkraften? Umschau 80 (4), 105-107.

Sellers, W.D.(1977): Water circulation on the global scale: Natural factors and manipulation by man, Ambio 6, 10-12.

Smith, I.M.(1982): Carbon dioxide-emissions and effects, Report No ICTIS/TR 18, IEA Coal Research, London.

Toon, O.B. and J.B. Pollack(1980): Atmospheric aerosols and climate, American Scientist, 68, 268-278.

Umwelt(1980): Klimatologische Folgen des Krypton-85-Anstiegs in der Atmosphäre? Nr. 76, 24-25.

USNAS (National Academy of Sciences)(1976): Halocarbons: Effects on stratospheric ozone, Washington, D.C.

USNAS(1979): Protection against depletion of stratospheric ozone by chlorofluorocarbons, Washington, D.C.

Vali, G.M. et al.(1976): Biogenic ice nuclei: Pt. II Bacterial sources, J. Atmos. Sciences 33, 1565-1570.

Wang, W.C., Y.C. Yung, A.A. Lacis, T. Mo. and J.E. Hansen(1976): Greenhouse effects due to man-made perturbations of trace gases, Science 194, 685-690.

Wang, W.C. and P.B. Ryan(1983): Overlapping effect of atmospheric H_2O, CO_2 and O_3 on the CO_2 radiative effect, Tellus 35B, 81-91.

Washington, W.M.(1972): Numerical climatic-change experiments: The effects of man's production of thermal energy, J. Appl. Meteorol. 11, 768-772.

Weiss, R.F.(1981): The temporal and spatial distribution of tropospheric nitrous oxide, J. Geophys. Res., 86 (C 8), 7185-7195.

Williams, J., G. Krömer and A. Gilchrist(1977): Further studies of the impact of waste heat release on simulated global climate, Pt.1, RM-77-15, Pt. 2 RM-77-34, Int. Inst. for Applied Systems Analysis, Laxenburg.

Williams, J. and W. Bach(1979): Energy conversions and climate, Report for the German Federal Environmental Agency, Berlin.

Williams, J. and G. Krömer(1979): A systems study of energy and climate, Status Report SR-79-2A and B. Int. Inst. for Applied Systems Analysis, Laxenburg.

WMO(1982): Report of the meeting of experts on potential climatic effects of ozone and other minor trace gases, WMO-report No. 14, Geneva.

Yung. Y.L., J.P. Pinto, R.T. Watson and S.P. Sander(1980): Atmospheric bromine and ozone perturbations in the lower stratosphere, J. Atmos. Sciences 37, 339-353.
Zimmerman, P.R. et al.(1982): Termites: A potentially large source of atmospheric methane, carbon dioxide, and molecular hydrogen, Science 218, 563-565.

V. Impacts of Climate Change on Society

Aharon, P., J. Chappell and W. Compston(1980): Stable isotope and sea-level data from New Guinea supports Antarctic ice-surge theory of ice ages, Nature 283, 649-651.
Andreae, B.(1980): Weltwirtschaftspflanzen im Wettbewerb. Ökonomischer Spielraum und ökologische Grenzen. Eine produktbezogene Nutzpflanzengeographie, de Gruyter, Berlin, New York.
Andrews, J.T. et al.(1972): Holocene late glacial maximum and marine transgressions in the Eastern Canadian Arctic, Nature 239, 147-149.
Bach, W.(1978): The potential consequences of increasing CO_2 levels in the atmosphere. In: J. Williams (ed.) Carbon Dioxide, Climate and Society, 141-167, Pergamon Press, Oxford.
Bach, W.(1979a): Impact of increasing atmospheric CO_2 concentrations on global climate: Potential consequences and corrective measures, Environm. Int. 2, 215-228.
Bach, W.(1979b): Impact of world fossil fuel use on global climate: Policy implications and recommendations. In: Symp. on Carbon Dioxide Accumulation in the Atmosphere, Synthetic Fuels and Energy Policy, 121-160; Committee on Governmental Affairs, U.S. Senate, Washington, D.C.
Bach, W.(1979c): Wenn die Erde zum Treibhaus wird. Die Zeit, Nr. 40, S. 68, 28 Sept.
Bach, W.(1980a): Klimaeffekte anthropogener Energieumwandlung In: H.A. Oomatia (Hrsg.) Energie und Umwelt, 84-98, Vulkan-Verlag, Essen.
Bach, W.(1980b): Waste heat and climatic change. In: L. Theodore et al. (eds.) Energy and the Environment: Interactions, vol. I., 151-190, CRC Press, Boca Raton, Florida.
Bach, W., J. Pankrath and S.H. Schneider(eds., 1981): Food-Climate Interactions, Reidel Publ. Co., Dordrecht.
Baker, D.N., L.H. Allen, Jr. and J.R. Lambert(1982): Effects of increased CO_2 on photosynthesis and agricultural production, US DOE 013, vol. II, pt. 6, Washington, D.C.
Bardach, J.(1974): Die Ausbeutung der Meere, Fischer Taschenbuch, Frankfurt.
Bardach, J. and R.M. Santerre(1981): Climate and aquatic food production. In: W. Bach et al. (eds.) Food-Climate Interactions, 187-233, Reidel Publ. Co., Dordrecht.

Baumgartner, A.(1979): Climatic variability and forestry. In: World Climate Conference, 581-607, WMO-No. 537, Geneva.

Benci, J.F. et al.(1975): Effects of hypothetical climate changes on production and yield of corn. In: Impacts of Climatic Change on the Biosphere, CIAP Monograph 5, pt. 2, 4-3 to 4-36.

Bentley, C.R.(1982a): Response of the West Antarctic Ice Sheet to CO_2-induced climatic warming, vol. II pt. I, DOE 013, Washington, D.C.

Bentley, C.R.(1982b): The West Antarctic Ice Sheet: Diagnosis and prognosis, DOE, 021, IV.3-IV.50, Washington, D.C.

Bolin, B.(1979): Global Ecology and Man. In: World Climate Conf., 27-50, WMO, No. 537, Geneva.

Brown, A.W.A.(1977): Yellow fever dengue and dengue haemorrhagic fever. In: G.M. Howe (ed.) A World Geography of Human Diseases, 271-317, Academic Press, London.

Brown, L.R.(1978): The Twenty Ninth Day, W.W. Norton, New York.

Brown, L.R.(1980): Food or fuel: New competition for the world's cropland, Worldwatch Paper 35, Washington, D.C.

Brown, L.R.(1981): World population growth, soil erosion, and food security, Science 214, 995-1002.

Brown, L.R.(1982): US and Soviet agriculture: The shifting balance of power, Worldwatch Paper 51, Washington, D.C.

Bryson, R.A., H.H. Lamb and D.L. Donley(1974): Drought and the decline of Mycene, Antiquity, 48 (189), 46-50.

Bryson, R.A.(1975): Cultural sensitivity to environmental change V: Some cultural and economic consequences of climatic change, IES Report 60, Institute for Environmental Studies, Univ. of Wisconsin, Madison.

Bryson, R.A. and T.J. Murray(1977): Climate of Hunger, The University of Wisconsin Press, Madison.

Budyko, M.I.(1982): The Earth's Climate: Past and Future. Int. Geophys. Series 29, Academic Press, New York, London.

Clark, D.L.(1982): Commentary on H. Flohn, Climate change and an ice-free Arctic Ocean. In: W.C. Clark (ed.) Carbon Dioxide Review: 1982, 181-184, Oxford Univ. Press, New York.

Coakley, S.M. and S.H. Schneider(1976): Climate-food interactions: How real is the crisis? Prcdgs. Amer. Phytopathological Soc. 3, 22-27.

Cushing, D.H.(1979): Climatic variation and marine fisheries. In: World Climate Conference, 608-627, WMO-No. 537, Geneva.

Dahlman, R.C.(1982): Vegetation response to carbon dioxide, US DOE Conf. Berkeley Springs, W.Va., Sept. 19-23, 1982.

d'Arge, R.C.(1979): Climate and economic activity. In: World Climate Conference, 625-681. WMO-No. 537, Geneva.

d'Arge, R.C., W. Schulze and D. Brookshire(1980): Benefit-cost valuation of long term future effects: The case of CO_2,Ft. Lauderdale Conf., 24-25, April 1980.

Denton, G.H. and T. Hughes(1981): The Last Great Ice Shield, 2 vols., J.Wiley & Sons, New York.

Diax, H.F. and J.T. Andrews(1982): An analysis of the spatial pattern of July temperature departures (1943-1972) over Canada and estimates of the 700 mb midsummer circulation during middle and late Holocene. J. of Climatol. 2, 251-265.

Eckholm, E.P.(1976): Losing Ground, W.W. Norton, New York.

EDIS (Environmental Data and Information Service)(1981): vol. 12 (1), p. 24.

Etkins, R. and E.S. Epstein(1982): The rise of global mean sea level as an indication of climatic change, Science 215, 287-289.

Flohn, H.(1949/50): Klimaschwankungen im Mittelalter und ihre historisch-geographische Bedeutung, Berichte z. Dt. Landeskunde 7, 347-357.

Flohn, H.(1973): Der Wasserhaushalt der Erde: Schwankungen und Eingriffe, Naturwissenschaften 60, 340-348.

Flohn, H.(1974): Das Wasser als Grundlage unserer Ernährung: Wasserhaushalt und Wasserverbrauch, Ernährungs-Umschau 21 (1), 9-13.

Flohn, H.(1978a): Gefährden Klimaanomalien die Welt-Ernährung? Bild d. Wiss. 12, 132-139.

Flohn, H.(1978b): Die Zukunft unseres Klimas: Fakten und Probleme, Promet 2/3, 1-21.

Flohn, H. et al.(1980): Working group report II: Identification and assessment of the various climatic impacts. In: W. Bach et al. (eds.) Interactions of Energy and Climate, XX-XXIV, Reidel Publ. Co., Dordrecht.

Flohn, H.(1980): Modelle der Klimaentwicklung im 21. Jahrhundert. In: H. Oeschger et al. (eds.) Das Klima, 3-17, Springer-Verlag, Berlin.

Flohn, H.(1981a): Climatic variability and coherence in time and space. In: W. Bach et al. (eds.) Food-Climate Interactions, 423-441, Reidel, Dordrecht.

Flohn, H.(1981b): Major climatic events as expected during a prolonged CO_2-warming, Report Institute of Energy Analysis, Oak Ridge Assoc. Universities, Oak Ridge, USA.

Ft. Lauderdale-Conference: On the methodology for economic impact analysis of climate change, 24.-25. April 1980, Resources for the Future Inc. and U.S. National Climate Program Office.

Gates, W.L.(1980): Modeling the surface temperature changes due to increased atmospheric CO_2. In: W. Bach el al. (eds.) Interactions of Energy and Climate, 169-190, Reidel Publ. Co., Dordrecht.

Global 2000(1980): Report to the President, US GPO, Washington, D.C.

Goudriaan, J. and G.L. Ajtay(1979): The possible effects of increased CO_2 on photosynthesis. In: B. Bolin et al. (eds.) The Global Carbon Cycle, 237-249, SCOPE 13, J. Wiley & Sons, New York.

Harrison, P.(1979): The curse of the tropics, New Scientist, 84, 602-604.

Hasselmann, K.(1979): On the problem of multiple time scales in climate modeling. In: W. Bach et al. (eds.) Man's Impact on Climate, 43-55, Elsevier Publ. Co., Amsterdam.

Hasselmann, K.(1981): Construction and verification of stochastic climate models. In: A. Berger (ed.) Climatic Variations and Variability: Facts and Theories, 481-497, Reidel Publ. Co., Dordrecht.

Hauser, H.K.J.(1977): Süsswasser aus Salzwasser, Umschau 9, 267-273.

Hayes, D.(1979): Alternative Energien, Hoffmann u. Campe, Hamburg.

Haynes, D.L.(1982): Effects of climate change on agricultural plant pests, US DOE 013, vol. II pt. 10, Washington, D.C.

Hekstra, G.P.(1981): Toward a conservation strategy to retain world food and biosphere options. In: W. Bach et al. (eds.) Food-Climate Interactions, 325-359, Reidel Publ. Co., Dordrecht.

Hodges, C.N. et al.(1981): Prospects for seawater-based agriculture. In: W. Bach et al. (eds.) Food-Climate Interactions, 81-99, Reidel Publ. Co., Dordrecht.

Hoffmann, G.(1960): Die mittleren jährlichen und absoluten Extremtemperaturen der Erde. Ihr Bezug auf die Erdoberfläche und Bevölkerung, Meteorol. Abhdlg. 8 (3) und Beiheft.

Hollin, J.T. and R.G. Barry(1979): Empirical evidence concerning the response of the earth's ice and snow cover to global temperature increase, Environm. Int. 2, 437-444.

Hollin, J.T.(1980): Climate and sea level in isotope stage 5: an East Antarctic surge at about 95 000 BP? Nature 283, 629-633.

Holm-Hansen, O.(1982): Effects of increasing CO_2 on ocean biota, US DOE 013, vol. 13 pt. 5, Washington, D.C.

Howell, G.S.(1982): Alleviation of environmental stress on renewable resource productivity, US DOE 013, vol. II pt. 9, Washington, D.C.

Hughes, T.(1977): West Antarctic ice streams; Revs. Geophys. Space Phys. 15, 1-46.

Hughes, T., J.T. Fastook and G.H. Denton(1980): Climatic warming and collapse of the West Antarctic ice sheet. In: Workshop on Environmental and Societal Consequences of a Possible CO_2 -induced Climatic Change, 152-182, US DOE 009, Washington, D.C.

Hughes, T.(1982): The stability of the West Antarctic Ice Sheet: What has happened and what will happen, DOE , 021, IV.51-IV.73, Washington, D.C.

Idyll, C.P.(1973): The anchovy crisis, Scientific American, 228 (6), 22-29.

Johnson, J.H.(1976): Effects of climatic changes on marine food production. In: USNAS Climate and Food, App. A, 181-188, Nat. Academy of Sciences, Washington, D.C.

Karl, T.R. and R.G. Quayle(1981): The 1980 summer heat wave and drought in historical perspective, Mon. Wea. Rev. 109 (10), 2055-2073.

Katz, R.W.(1977): Assessing the impact of climatic change on food production, Climatic Change 1, 85-96.

Kellogg, W.W.(1978): Effects of human activities on global climate, World Meteorol. Organiz. Bulletin, pt. 2, 3-10.

Kellogg, W.W. and R. Schware(1981): Climate Change and Society, Westview Press, Boulder, Col.

Kennett, J.P.(1977): Cenozoic evolution of Antarctic glaciation, the Circum-Antarctic Ocean, and their impact on global pale - oceanography, J. Geophys. Res. 82 (27) 3843-3860.

Klimaforschungsprogramm der Bundesrepublik Deutschland(Mai 1980).

Klingauf, F.(1981): Interrelations between pests and climatic factors. In: W. Bach et al. (eds.) Food-Climate Interactions, 285-301, Reidel Publ. Co., Dordrecht.

Landsberg, H.E.(1976): Weather, climate and human settlements, Special Env. Report No. 7, WMO-No. 448, World Meteorol. Organiz.,Geneva.

Learmonth, A.T.A.(1977): Malaria. In: G.M. Howe (ed.) A World Geography of Human Diseases, 61-108, Academic Press, London.

Lieth, H.(1973): Primary production: Terrestrial ecosystems, Human Ecology, 1, 303-332.

Lindh, G.(1979): Water and food production. In: M.R. Biswas and A.K. Biswas (eds.) Food, Climate and Man, 52-72, J. Wiley & Sons, New York.

Lindh, G.(1981): Water resources and food supply, In: W. Bach et al. (eds.) Food-Climate Interactions, 239-260, Reidel Publ. Co., Dordrecht.

Lovins, A.B.(1978): Sanfte Energie, Rowohlt, Reinbek.

Lvovitch, M.(1977): World water resources present and future, Ambio, 6 (1), 13-21.

Manabe, S. and R.J. Stouffer(1980): Sensitivity of a global climate model to an increase of CO_2 concentration in the atmosphere, J. Geophys. Res. 85 (C 10) 5529-5554.

Markley, O. W. and R. Carlson(1980): Three analytic conclusions having significant implications for the AAAS-DOE Workshop on environmental and societal consequences of a possible CO_2 -induced climate shift. In: US DOE 009, 439-441, Washington, D.C.

McKay, G.A. and T. Allsopp(1980): The role of climate in affecting energy demand/supply. In: W. Bach et al. (eds.) Interactions of Energy and Climate, 53-72, Reidel Publ. Co.,Dordrecht.

McQuigg, J.D. et al.(1973): The influence of weather and climate on United States grain yields: Bumper crops or droughts, Nat. Oceanic and Atmosph. Administr. Report, 30p., Washington, D.C.

Mensching, H.(1978): Die Wüste schreitet voran, Umschau 78 (4) 99-106.

Mercer, J.H.(1978): West Antarctic ice sheet and CO_2-greenhouse effect: a threat of disaster, Nature, 271, 321-325.

Myers, N.(1979): The Sinking Arch, Pergamon Press, Oxford.

Newman, J.E.(1982): Impact of rising atmospheric CO_2 levels on agricultural growing seasons and crop water use efficiencies, US DOE 013, Washington, D.C.

Nicholson, S.H. and H. Flohn(1980): African environmental and climatic changes and the general atmospheric circulation in late Pleistocene and Holocene, Climatic Change, 2(4), 313-348.

Nowicki, J.J. et al.(1980): Working group recreation. In: J.M. Powell (ed.) Socioeconomic impacts of climate, 42-51, Environment Canada, Edmonton.

Odum, E.P.(1971): Fundamentals of Ecology, 3rd ed., Saunders, Philadelphia.

Oerlemans, J.(1982): Response of the Antarctic ice sheet to a climatic warming: A model study, J. of Climatology, 2, 1-11.

Olson, W.P.(1970): Rainfall and plague in Vietnam, Int. J. Biometeorol., 14, 357-360.

Otterman, J.(1974): Baring high-albedo soils by overgrazing: A hypothesized desertification mechanism, Science 186, 531-533.

Parkinson, C.L. and W.W. Kellogg(1979): Arctic sea ice decay simulated for a CO_2-induced temperature rise, Climatic Change 2 (2), 149-162.

Pfister, C.(1981): Die Fluktuationen der Weinmosterträge im Schweizerischen Weinland vom 16. bis ins frühe 19. Jahrhundert. Klimatische Ursachen und sozioökonomische Bedeutung, Schweiz. Ztschr. f. Geschichte 31, 445-491.

Phares, R.E.(1980): Impact of enhanced atmospheric CO_2 on managed forests. In: US DOE, Workshop on Environmental and Societal Consequences of a possible CO_2-induced Climate Change, 295-302, CONF-7904143, No. 009, US DOE, Washington, D.C.

Pimentel, D.(1980): Increased CO_2 effects on the environment and in turn on agriculture and forestry. In: US DOE, Workshop on Environmental and Societal Consequences of a possible CO_2-induced Climate Change, 264-274, CONF-7904143, No. 009, US DOE, Washington, D.C.

Pimentel, D.(1981): Food energy and climate change. In: W. Bach et al. (eds.) Food-Climate Interactions, 303-323, Reidel Publ. Co., Dordrecht.

Pittock, A.B. and M.J. Salinger(1982): Towards regional scenarios for a CO_2-warmed earth, Climatic Change 4 (1), 23-40.

Quirk, W.J. and J.E. Moriarty(1980): Prospects for using improved climate information to better manage energy-systems. In: W. Bach et al. (eds.) Interactions of Energy and Climate, 89-99, Reidel Publ. Co., Dordrecht.

Quirk, W.J.(1981 and pers. communication): Climate and energy emergencies, Preprint Lawrence Livermore Lab., UCRL - 85351.

Ramirez, J.M. et al.(1975): Wheat. In: Impacts of Climatic Change on Biosphere, CIAP Monograph 5, pt. 2, 4-37 to 4-90.

Revelle, R.(1980): Energy dilemmas in Asia: The needs for research and development, Science 209, 164-174.

Roberts, W.O. and H. Lansford(1979): The Climate Mandate, W.H. Freeman & Co., San Francisco.

Roberts, W.O. et al.(1980): Working group report III: Objectives of a climatic impact study program. In: W. Bach et al. (eds.) Interactions of Energy and Climate, XXV-XXXIII, Reidel Publ. Co., Dordrecht.

Rocznik, K.(1981): Der säkulare Wandel von Jahres- und Monatsklima in Mitteleuropa im Zeitraum 1761-1980, Meteorol. Rundsch. 34 (6), 181-185.

Rogers, H.H.(1982): Effects of long-term CO_2-concentrations on field-grown crops and trees. In: S. Brown (ed.) Global Dynamics of Biospheric Carbon, 9-45, US DOE 019, Washington, D.C.

Rosenberg, N.J.(1981): The increasing CO concentration in the atmosphere and its implication on agricultural productivity: I. Effects on photosynthesis, transpiration and water use efficiency, atmosphere and its implication on agricultural productivity: Climatic Change 3(3), 265-279.

Rosenberg, N.J.(1982): The increasing CO_2 concentration in the atmosphere and its implication on agricultural productivity II. Effects through CO_2-induced climatic change. Climatic Change 4(3), 239-254.

Schneider, S.H. and L.E. Mesirow(1976): The Genesis Strategy, Plenum Press, New York.

Schneider, S.H. and W. Bach(1980): Food-Climate Interactions, Report for the Federal Environmental Agency, Berlin, 282p+XL.

Schneider, S.H. and R.S. Chen(1980): Carbon dioxide warming and coastline flooding: Physical factors and climatic impact, Ann. Rev. Energy, 5, 107-140.

Schneider, S.H. and W. Bach(1981): Interactions of food and climate: Issues and policy considerations. In: W. Bach et al. (eds.) Food-Climate Interactions, 1-19, Reidel Publ. Co., Dordrecht.

Schnell, P. und P. Weber(Hrsg.)(1980): Agglomeration und Freizeitraum, Münstersche Geographische Arbeiten, H. 7, 238 S., Schöningh, Paderborn.

Sergin, V.Ya.(1980): A method for estimating climatic fields based on the similarity of seasonal and longer climatic variations. In: J. Ausubel and A.K. Biswas (eds.) Climatic Constraints and Human Activities, 181-202, Pergamon Press, Oxford.

Smith, V.K.(1982): Economic impact analysis and climatic change: A conceptual introduction, Climatic Change 4 (1), 5-22.

Sorensen, B.(1979): Renewable Energy, Academic Press, London, New York.

Stansel, J. and R.F. Huke(1975): Rice, In: Impacts of Climatic Change on the Biosphere, CIAP Monograph 5, pt. 2, 4-90 to 4-132.

Strain, B.R.(1982): Ecological aspects of plant responses to carbon dioxide environment. In: S. Brown (ed.) Global Dynamics of Biospheric Carbon, 46-55, US DOE 019, Washington, D.C.

Strain, B.R. and T.V. Armentano(1982): Response of unmanaged ecosystems, US DOE 013, vol. II pt. 12, Washington, D.C.

Sugden, D.E. and C.M. Clapperton(1980): West Antarctic ice sheet fluctuations in the Antarctic Peninsula area, Nature, 286, 378-381.

Takahashi, K. and J. Nemoto(1978): Relationships between climatic change, rice production, and population. In: Takahashi, K. and M. M. Yoshino (eds.) Climatic Change and Food Production, 183-196, Univ. of Tokyo Press, Tokyo.

The Polar Group(1980): Polar atmosphere-ice-ocean processes: A review of polar problems in climate research, Reviews Geophys. & Space Physics 18 (2), 525-543.

Thomas, R.H., T.J.O. Sanderson and K.F. Rose(1979): Effect of climatic warming on the West Antarctic ice sheet, Nature, 277, 355-358.

Thornes, J.E.(1982): Atmospheric management, Progress in Phys. Geogr. 6 (4), 561-578.

Tucker, H.A.(1982): Effects of climate change on animal agriculture, US DOE 013, vol. II pt. 11, Washington, D.C.

USDOC(Dept. of Commerce)(1980): Climate Impact Assessment U.S., Annual Summary, Washington, D.C.

USDOE(Dept. of Energy)(1980): Climate Program Plan, DOE/EV-0062/1 UC-2, 11, 41, Washington, D.C.

USNAS(National Academy of Sciences)(1973): Weather and Climate Modification Problems and Progress, Washington, D.C.

USNAS(1977a): Energy and Climate, Washington, D.C.

USNAS(1977b): Climate, Climatic Change and Water Supply, Washington, D.C.

USNAS(1977c): World Food and Nutrition Study, Washington, D.C.

USNAS(1978): International Perspectives on the Study of Climate and Society, Washington, D.C.

USNAS(1982): Carbon Dioxide and Climate: A Second Assessment, Washington, D.C.

Uthott, D.(1978): Endogene und exogene Hemmnisse in der Nutzung des Ernährungspotentials der Meere. In: Tag.-Ber. Dt. Geogr. Tags in Mainz 1977, Wiesbaden.

Van Keulen, H., H.H. van der Laar, W. Louwerse and J. Goudriaan (1980): Physiological aspects of increased CO_2 concentration, Experientia 36, 786-792.

Walsh, J.J.(1981): A carbon budget for overfishing off Peru. Nature 290, 300-304.

Walter, H.(1973): Vegetationszonen und Klima, UTB 14, Ulmer, Stuttgart.

Webb III, T.(1982): The use of paleoclimatic data in understanding and possibly predicting how CO_2-induced climatic change may affect the natural biosphere, US DOE 013, vol. II pt. 17, Washington, D.C.

Weihe, W.H.(1979): Climate, Health and Disease. In: World Climate Conf., WMO No. 537, 313-368, Geneva.

Whittaker, R.H.(1975): Communities and ecosystems, MacMillan, New York.

Wigley, T.M.L., P.D. Jones and P.M. Kelly(1980): Scenario for a warm, high-CO_2 world, Nature 283, 17-21.

Williams, J.(1980): Anomalies in temperature and rainfall during warm Arctic seasons as a guide to the formulation of climate scenarios, Climatic Change 2(3), 249-266.

WMO(World Meteorological Organization)(1980): Outline plan and basis for the World Climate Programs 1980-1983, WMO-No. 540, Geneva.

Woodley, W.L. et al.(1982): Clarification of confirmation in the FACE-2 experiment, Bull. Amer. Met. Soc. 63 (3), 273-276.

VI. Strategies for Averting a CO_2/Climate Problem

Albanese, A.S. and M. Steinberg(1980): Environmental control technology for atmospheric carbon dioxide, DOE/EV-0079, DOE, Washington, D.C.

Augustsson, T. and V. Ramanathan(1977): A radiative-convective model study of the CO_2 climate problem, J. Atmos. Sci. 34, 448-451.

Ausubel, J., J. Lathrop, I. Stahl and J. Robinson(1980): Carbon and climate gaming, WP-80-152, Int. Institute Appl. Systems Analysis, Laxenburg.

Bach, W. and G. Schwanhäusser(1978): Can the right energy mix prevent climatic change? In: Proceedings of Conf. on Climate and Energy, 140-145, Amer. Meteorol. Soc., Boston.

Bach, W.(1978a): Das CO_2 -Problem: Lösungsmöglichkeiten durch technische Gegensteuerung? Umschau 78 (4), 117-118.

Bach, W.(1978b): Der Einfluss des Menschen auf das globale Klima, Universitas 33 (4), 399-405.

Bach, W.(1979a): Impact of world fossil fuel use on global climate: Policy implications and recommendations. In: Symposium on CO_2 accumulation in the atmosphere, synthetic fuels and energy policy, 121-160, Committee on Governmental Affairs, US Senate, Washington, D.C.

Bach, W.(1979b): Klimaänderung durch Energiewachstum? Brennst.-Wärme-Kraft 31 (2), 49-56.

Bach, W., W. Manshard, W.H. Matthews and H. Brown (eds.)(1980): Renewable Energy Prospects, Pergamon Press, Oxford.

Bach, W. and W.H. Matthews(1980): Exploring alternative energy strategies. In: W. Bach et al. (eds.) Renewable Energy Prospects, 711-722, Pergamon Press, Oxford.

Bach, W. and H.-J. Jung(1981): Massnahmen zur Reduzierung bzw. Kompensation der Auswirkungen eines erhöhten CO_2 -Gehaltes der Atmosphäre. In: Die Auswirkungen von CO_2-Emissionen auf das Klima, Bd. 2, Teilbericht VI, 154-274, Forschungsbericht im Auftrag des Umweltbundesamtes, Battelle-Institut, Frankfurt.

Bach, W.(1983a): Carbon dioxide/climate threat: Fate or forebearance? In: W. Bach et al. (eds.) Carbon Dioxide, 461-509, Reidel Publ. Co., Dordrecht.

Bach, W.(1983b): Energy strategy for averting a CO_2-induced climatic risk, Energie '83, 604-629, Hamburg.
Baes, C.F.Jr., S.E. Beall, D.W. Lee and G. Marland(1980): The collection, disposal, and storage of carbon dioxide. In: W. Bach et al. (eds.) Interactions of Energy and Climate, 495-519, Reidel Publ. Co., Dordrecht.
Bienewitz, K.-H., S. Hartwig u. B. Oberacher(1981): Energieverbrauch und CO_2-Emission. In: Die Auswirkungen von CO_2-Emissionen auf das Klima, Bd. 1, 38-104, Battelle-Bericht, Frankfurt.
BMFT(1980): Bundesregierung beschliesst Kohleveredlungsprogramm, BMFT-Mitteilungen, 2 Z 7255E, 17-19.
Bossel, H. et al.(1976): Energie richtig genutzt, Umwelt u. Planung Bd. 8, C.F. Müller, Karlsruhe.
Bossel, H. and R. Denton(1977): Energy futures for the FRG, Energy Policy 5 (1), 35-50.
Brennstoff-Wärme-Kraft 30 (12), 451(1978): Energiespeichern mit Kohlendioxid.
Breuer, G.(1979): Can forest policy contribute to solving the CO_2 problem? Environm. Int. 2, 449-451.
Brown, L.R.(1980): Food or fuel: New competition for the world's cropland, Worldwatch Paper 35 , Washington, D.C.
Brown, L.R.(1981): World population growth, soil erosion, and food security, Science 214, 995-1002.
Bruckmann, G.(1978): Sonnenkraft statt Atomenergie, Molden, Wien/München.
Burton, I., R.W. Kates and G.F. White(1978): The Environment as Hazard, Oxford Univ. Press, Oxford.
Budyko, M.I.(1982): The Earth's Climate: Past and Future. Int. Geophys. Series 29, Academic Press, New York, London.
Bzserghi, H.A. et al.(1982): Cost-effective fuel savings in heating applications. Paper pres. at Symp. on the Comparative Merits of Energy Sources in Meeting End-Use Heat Demand, 6-10 Sept. 1982, Ohrid, Yugoslavia.
Caputo, R.S.(1980): Solar energy for the longer term, Int. Inst. Appl. Systems Analysis, Laxenburg, (unpubl. manuscript).
Cess, R. and S.D. Goldenberg(1981): The effect of ocean heat capacity upon global warming due to increasing atmospheric CO_2, J. Geophys. Res. 86 , 498-502.
CEQ (Council on Environmental Quality) (1979): The good news about energy, U.S. Government Printing Office, Washington, D.C.
CEQ(1981): Global energy futures and the carbon dioxide problem, U.S. Government Printing Office, Washington, D.C.
Colombo, U. and O. Bernardini(1979): A low energy growth 2030 scenario and the perspectives for Western Europe. Report prepared for the Commission of the European Communities, Brussels.
Colombo, U.(1982): Alternative energy futures: The case for electricity, Science 217, 705-709.

CONAES (Commission on Nuclear and Alternative Energy Systems) (1980): Energy in Transition, 1985-2010, National Academy of Sciences, Washington, D.C.

Cooper, C.F. (1982): Food and fiber in a world of increasing carbon dioxide. In: W.C. Clark (ed.), Carbon Dioxide Review: 1982, 297-320, Oxford Univ. Press. Oxford.

Craig, P. et al. (1978): Distributed technologies in California's energy future, Interim Report HCP/P7405-01/02, US DOE, Washington, D.C.

Dampier, B. (1981): Sweden's Energy Plan: 50 percent less oil within 10 years, Ambio 10 (5), 216-218.

Die Zeit (1981): Umstrittenes Energieprogramm, Nr. 45, 30. Oktober.

Dunkerley, J. and W. Ramsay (1982): Energy and the oil-importing developing countries, Science 216, 590-595.

Dyson, F.J. (1977): Can we control the carbon dioxide in the atmosphere? Energy 2, 287-291.

Dyson, F.J. and G. Marland (1979): Technical fixes for the climatic effects of CO_2. In: W. P. Elliott and L. Machta (eds.) Workshop on the Global Effects of Carbon Dioxide from Fossil Fuels, 111-118, CONF-770385, 001, US DOE, Washington, D.C.

Energiediskussion (1979): Grundrichtung: Weg vom Öl. Der Beschluss des Berliner Parteitages der SPD, Nr. 5/6, 33-41.

Enquête-Kommission des Deutschen Bundestages (1980): Zukünftige Kernenergie-Politik, Teil I u. II, Drucksache 8/4341, Bonn.

Ford Foundation (1974): A time to choose, Energy Policy Project, Ballinger Publ. Co., Cambridge, USA.

Garcia, R.V. (1981): Nature pleads not guilty, Vol. 1, Pergamon Press, Oxford.

Geberth, R. (1980): Bedeutung von Energieprognosen für energiepolitische Entscheidungen. In: A. Voss u. K. Schmitz (Hrsg.) Energiemodelle für die Bundesrepublik Deutschland, 169-174, Verlag TÜV Rheinland, Köln.

Gerwin R. (1980): Die Welt-Energieperspektive, DVA, Stuttgart.

Glaeser, B. (ed.) (1980): Factors affecting land use and food production, Verlag, Breitenbach, Saarbrücken.

Glantz, M.H. (ed.) (1977): Desertification, Westview Press, Boulder, USA.

Häfele, W. (1978): A perspective on energy systems and carbon dioxide. In: J. Williams (ed.) Carbon Dioxide, Climate and Society, 21-34, Pergamon Press, Oxford.

Häfele, W. (1980): Global perspectives and options for long-range energy strategies. In: W. Bach et al. (eds.) Renewable Energy Prospects, 745-760, Pergamon Press, Oxford.

Häfele, W.u. H.H. Rogner (1980): Energie - die globale Perspektive. Angew. Systemanalyse 1 (2) 57-67.

Häfele, W. et al. (1981): Energy in a Finite World, vol. 1 a. 2, Ballinger Publ. Co., Cambridge, USA.

Hallsworth, E.G. (1981): Soil management and the food supply. In: W. Bach et al. (eds.) Food-Climate Interactions, 261-284, Reidel Publ. Co., Dordrecht.

Hampicke, U. und W. Bach(1979/1980): Die Rolle terrestrischer Ökosysteme im globalen Kohlenstoffkreislauf, Bericht im Auftrag des Umweltbundesamtes, Berlin, und Münstersche Geogr. Arbeiten, Nr. 6, 37-104, Schöningh, Paderborn.

Hauff, V.(1981): Perspektiven einer Energiepolitik für morgen. In: B. Gemper (Hrsg.) Energieversorgung. Expertenmeinungen zu einer Schicksalsfrage, 221-224, Verlag Vahlen, München.

Hekstra, G.P.(1981): Towards a conservation strategy to retain world food and biosphere options. In: W. Bach et al.(eds.) Food-Climate Interactions, 325-359, Reidel Publ. Co., Dordrecht.

Herz, H. u. E. Jochem(1980): Unsicherheiten in der Prognose von Energiebedarfsentwicklungen, Fraunhofer-Gesellschaft, Berichte 4, 3-8.

Hodges, C. et al.(1981): Seawater-based agriculture as a food production defense against climate variability. In: W. Bach et al. (eds.) Food-Climate Interactions, 81-99, Reidel Publ. Co., Dordrecht.

Hoffert, M.I., Y.-C. Wey, A.J. Callegari and W.S. Broecker(1979): Atmospheric response to deepsea injections of fossil-fuel CO_2, Climatic Change 2 (1), 53-68.

Holdren, J.P.(1981): Renewables in the U.S. energy future: How much, how fast? Energy 6 (9), 901-916.

Janssen-Hering, A. und H. Janssen(1981): Kohlendioxid als Chemierohstoff, Umschau, 81 (3), 67-70.

Kellogg, W.W. and R. Schware(1981): Climate Change and Society, Westview Press, Boulder, USA.

Kellogg, W.W.(1981): Data banks for climatological purposes. In: A. Berger (ed.) Climatic Variations and Variability: Facts and Theories, 155-163, Reidel Publ. Co., Dordrecht.

Kellogg, W.W. and R. Schware(1982): Society, science and climatic change, Foreign Affairs 60 (5), 1076-1109.

Kendall, W. and S.J. Nadis(eds.)(1980): Energy strategies toward a solar future, Ballinger Publ. Co., Cambridge, USA.

Kovda, V.A.(1979): To combat salinization of fertile soils - an editorial, Climatic Change, 2 (2), 103-108.

Krause, F.(1978): Alternative Energietechnologien. In: V. Hauff (Hrsg.) Energieversorgung u. Lebensqualität, 173-206, Neckar-Verlag, Villingen.

Krause, F., H. Bossel u. K.F. Müller-Reissmann(1980): Energiewende, S. Fischer Verlag, Frankfurt.

Krause, F.(1982): Higher efficiency or more supplies? An FRG case study of low temperature heating strategies. Paper pres. at Symp. on the Comparative Merits of Energy Sources in Meeting End-Use Heat Demand, 6-10 Sept. 1982, Ohrid, Yugoslavia.

Krickenberger, K.R. and S.H. Lubore(1980): Possibility of utilizing carbon dioxide produced during the high-Btu coal gasification process for enhanced oil recovery, MTR-80W181, The MITRE Corp., McLean, USA.

Kuntze, H., J. Niemann, G. Roeschmann u. G. Schwerdtfeger(1981): Bodenkunde, Uni-Taschenbücher 1106, Verlag Ulmer, Stuttgart.

Lappé, F.M. and J. Collins (1977): Food first, Ballentine, New York.

Leach, G. et al.(1979): A low energy strategy for the United Kingdom, Science Reviews Ltd., London.

Lindh, G.(1981): Water resources and food supply, In: W. Bach et al. (eds.) Food-Climate Interactions, 239-260, Reidel Publ. Co., Dordrecht.

Lockeretz, W., G. Shearer and D.H. Kohl(1981): Organic farming in the corn belt, Science 211, 540-547.

Lovins, A.B.(1975): World Energy Strategies, Friends of the Earth Int., 2nd ed., London.

Lovins, A.B.(1978a): Soft energy technologies, Ann. Rev. Energy 3, 477-517.

Lovins, A.B.(1978b): Sanfte Energie, Rowohlt, Reinbek.

Lovins, A.B.(1979): Re-examining the nature of the ECE energy problem, Energy Policy 7 (3), 178-197.

Lovins, A.B.(1980): Economically efficient energy futures. In: W. Bach et al. (eds.) Interactions of Energy and Climate, 1-31, Reidel Publ. Co., Dordrecht.

Lovins, A.B., L.H. Lovins, F. Krause and W. Bach(1981): Least-cost Energy: Solving the CO_2 Problem, Brick House, Andover, USA.

Marchetti, C.(1976): On geoengineering and the CO_2 problem, RM-76-17, Int. Inst. Appl. Systems Analysis, Laxenburg.

Marchetti, C.(1979): Constructive solutions to the CO_2 problem. In: W. Bach et al. (eds.) Man's Impact on Climate, 299-311, Elsevier Publ. Co., Amsterdam.

Matthöfer, H.(Hrsg.)(1976): Zukünftige Energiebedarfsdeckung und die Bedeutung der nichtfossilen und nichtnuklearen Primärenergieträger, Teil I-VI, Umschau-Verlag, Frankfurt.

Meyer-Abich, K.M.(Hrsg.)(1979): Energieeinsparung als neue Energiequelle, C. Hanser Verlag, München.

Mitchell, Jr. J.M.(1975): A reassessment of atmospheric pollution as a cause of long-term changes of global temperature. In: S. F. Singer (ed.) The Changing Global Environment, 149-173, Reidel Publ. Co., Dordrecht.

Mustacchi, C., P. Armenante and V. Cena(1979): Carbon dioxide removed from power plant exhausts, Environm. Int. 2, 453-456.

Niehaus, F.(1977): Computersimulation langfristiger Umweltbelastung durch Energieerzeugung, ISR 41, Birkhäuser Verlag, Stuttgart.

Niehaus, F.(1979): Carbon dioxide as a constraint for global energy scenarios. In: W. Bach et al. (eds.) Man's Impact on Climate, 285-297, Elsevier Publ. Co., Amsterdam.

Niehaus, F.(1981): The impact of energy production on atmospheric CO_2-concentrations. In: A. Berger (ed.) Climatic Variations and Variability: Facts and Theories, 641-660, Reidel Publ. Co., Dordrecht.

Norgard, J.S.(1979): The gentle path of conservation, Demo-Project, Techn. Univ. Denmark, Lyngby, Denmark.

Norgard, J.S.(1982): Potentials for energy conservation in the rural municipality of Nysted, Demo-Project, Techn. Univ. Denmark, Lyngby, Denmark.

Olivier, D., H. Miall, F. Nectoux and M. Opperman(1983): Energy-Efficient Futures: Opening the Solar Option, Earth Resources Research, Ltd., London.

Pestel, E., R. Bauerschmidt, K.P. Möller und W. Oest(1978): Das Deutschland-Modell, Teil I, Bild d. Wiss. 1, 22-33 u. Teil 2, Bild d. Wiss. 2, 96-104.

Roberts, W.O. et al.(1980): Objectives of a climatic impact study program, working group III. In: W. Bach et al (eds.) Interactions of Energy and Climate, XXV-XXXIII, Reidel Publ. Co., Dordrecht.

Robinson, J. and J. Ausubel(1981): A framework for scenario generation for CO_2 gaming, WP-81-34, Int. Institute for Applied Systems Analysis, Laxenburg.

Rotty, R.M.(1976): Global CO_2 production from fossil fuels and cement AD 1950 - AD 2000, IEA(M)-76-4, Inst. for Energy Analysis, Oak Ridge Assoc. Universities, Oak Ridge, USA.

Rotty, R.M.(1979): Uncertainties associated with global effects of the atmospheric CO_2. In: Carbon dioxide accumulation in the atmosphere, synthetic fuels and energy policy, US Senate Committee on Government Affairs,161-194, USGPO, Washington, D.C.

Rotty, R.M. and G. Marland(1980): Constraints on fossil fuel use. In: W. Bach et al. (eds.) Interactions of Energy and Climate, 191-212, Reidel Publ. Co., Dordrecht.

Sant, R.W.(1979): The least-cost energy strategy, Carnegie-Mellon Univ. Press, Arlington.

Schiffer, H.-W.(1981): Der Ölverbrauchsrückgang und seine Ursachen, OEL-Zeitschrift f.d. Mineralölwirtschaft 11, 301-307.

Schilling, H.-D. u. U. Krauss(1981): Kohlenveredlung und Kohlenverwendung, Brennst.- Wärme-Kraft 33 (4), 130-134.

Schneider, S.H. and W. Bach(1981): Interactions of food and climate: Issues and policy considerations. In: W. Bach et al. (eds.) Food-Climate Interactions, 1-19, Reidel Publ. Co., Dordrecht.

SERI(Solar Energy Research Institute)(1981): A new prosperity: Building a sustainable energy future, Brick House, Andover, USA.

Shah, R.P., E.F. Wittmeyer, S.D. Sharp and R.W. Griep(1978): A study of CO_2 recovery and tertiary oil production enhancement in the Los Angeles Basin, Final Report, SAN/1582-1, US DOE, Washington, D.C.

Sorensen, B.(1979): Renewable Energy, Academic Press, London.

Sorensen, B.(1981): Renewable energy planning, Energy, 6, 293-303.

Sorensen, B.(1982a): An American energy future, Energy 7 (9), 783-799.

Sorensen, B.(1982b): Energy choices: Optimal path between efficiency and cost, paper presented at 1st US-China Conf. on Energy & Environment, Nov. 1982, Peking.

Steinberg, M. and A.S. Albanese(1980): Environmental control technology for atmospheric carbon dioxide. In: W. Bach et al. (eds.) Interactions of Energy and Climate, 521-551, Reidel Publ. Co., Dordrecht.

Stobaugh, R. and D. Yergin(eds., 1979): Energy Future, Random House, New York.

US NAS(National Academy of Sciences)(1980): A strategy for the National Climate Program, Washington, D.C.

US Senate(1979): Carbon dioxide accumulation in the atmosphere, synthetic fuels and energy policy, Comm. on Government Affairs, USGPO, Washington, D.C.

Vester, F.(1980): Neuland des Denkens, Deutsche Verlags-Anstalt, Stuttgart.

Von Weizsäcker, C.F.(1979): Vorwort in K.M. Meyer-Abich (Hrsg.) Energieeinsparung als neue Energiequelle, C. Hanser Verlag, München.

Voss, A. u. F. Niehaus(1977): Die Zukunft des Weltenergiesystems, Umschau 77 (19), 625-632.

Voss, A.(1980): Vorgetäuschte Sicherheit durch Energieprognosen, Umschau 80 (8), 235-236.

Voss, A.(1981):Energiewirtschaftliche Gesamtsituation, Brennst.-Wärme-Kraft 33 (4), 125-128.

Voss, A.(1982): Nutzen und Grenzen von Energiemodellen - einige grundsätzliche Überlegungen, Angew. Systemanalyse 3 (3), 111-117.

WAES(Workshop on Alternative Energy Strategies)(1977): Energy: Global Prospects 1985-2000, McGraw-Hill, New York.

WEC(World Energy Conference)(1978): World energy resources, 1985-2020, IPC Science & Technology Press, Richmond, England.

Weingart, J.M.(1980): The potential role of renewable energy systems, Rapporteur of Report II. In: W. Bach et al. (eds.) Renewable Energy Prospects, 996-998, Pergamon Press, Oxford.

WMO(World Meteorological Organization)(1980): World Climate Programme 1980-1983, WMO-No. 540, Geneva.

Zraket, C.A. and M.M. Scholl(1980): Solar energy systems and resources, The MITRE Corp., McLean, Virginia, USA.

VII. Opportunities for the Future

Asian Action(1980): Newsletter of the Asian Cultural Forum on Development, No. 23, November/December.

Bach, W.(1980): Der ungleiche Verbrauch von Ressourcen: Eine Bedrohung des Weltfriedens? Umschau 80 (13), 401-402.

Baum, G.R.(1981): Risikostrategien im Bereich technologischer Entwicklung, Energiediskussion, 5/6/81, Bonn.

Bischoff, G. u. W. Gocht(1979): Energietaschenbuch, Vieweg, Braunschweig.

Bockris, J.O.M. and E.W. Justi(1980): Wasserstoff, die Energie für alle Zeiten, U. Pfriemer Verlag, München.
Bossel, H. u. U. Bossel(1976): Bessere Energienutzung: Vernünftigste Energiealternative. In: K. Oeser u. H. Zillessen (Hrsg.) Kernenergie, Mensch, Umwelt, 36-46, Verlag Wissenschaft u. Politik, Köln.
Brandt, W.(Hrsg.)(1980): Das Überleben sichern, Kiepenheuer & Witsch, Köln.
Deevey, E.S. et al.(1979): Mayan urbanism: Impact on a tropical karst environment, Science 206, 298-306.
Der Rat der Sachverständigen für Umweltfragen(1981): Energie und Umwelt, Sondergutachten, Wiesbaden, und Umweltbrief Nr. 23, Bonn.
Der Spiegel(1980): Nord-Süd-Dialog: Grosser Krach,Nr. 35,S. 21-22.
Der Spiegel(1981): Fernwärme: Heizung für die halbe Republik, Nr. 16, S. 53-67.
Deuster, G.(1980): Heizkraft spart Devisen. Wachsende Aufgaben für die Fernwärmeversorgung, Z. f. Kommunalwirtschaft, S. 21, September.
Dürr, H.-P.(1977): Sicherung des Weltfriedens geniesst Priorität, Frankfurter Rundschau, Nr. 225,S. 14, 28.September.
Dürr, G. und W. Hähnle(Hrsg.)(1980): Wer macht unsere Zukunft? Sind wir Marionetten? Radius-Verlag, Stuttgart.
Energiediskussion(1980a): Plädoyer für die Wirbelschichtfeuerung, Nr. 1/2, S.2, Februar.
Energiediskussion(1980b): Mikroelektronik dient auch der Heizöleinsparung, Nr. 1/2,S. 3, Februar.
Energiediskussion(1981): Materialien zur Wärmedämmung. Nr. 5/6, S. 5, November.
Enquête-Kommission(1980): Zukünftige Kernenergie-Politik, Deutscher Bundestag, 8. Wahlperiode, Drucksache 8/4341, Bonn.
Eppler, E.(1980): Schlusswort. In: G. Dürr u. W. Hähnle(Hrsg.): Wer macht unsere Zukunft? Sind wir Marionetten? 87-94, Radius-Verlag, Stuttgart.
Erfurth, M.(1981): Heizenergiesparen durch Thermographie,Umschau 81 (3), 85-86.
Eyre, D. and D. Jennings(1982): Air-vapor barriers, Energy, Mines and Resources Canada, Ottawa.
Flavin, C.(1980): The future of synthetic materials: The petroleum connection, Worldwatch Paper No. 36, Washington, D.C.
Flohn, H.(1980): Discussion session. In: Polunin, N. (ed.) Growth without Ecodisasters, p. 593, The MacMillan Press Ltd., London.
Frisch, F.(1981): Intelligent und sensibel sparen, Die Zeit, Nr. 51, S. 62, 11. Dezember.
Ginsburg, T.(1979): Weniger Energie - mehr Lebensqualität. In: M. Grupp (Hrsg.) Energiesucht, 26-29, Bonz Verlag, Fellbach.
Goetzberger, A.(1981): Solarzellen-Kraft-Wärme-Kopplung ein mögliches Energiekonzept, Sonnenenergie 4/5, 42-46.
Gray, Jr. C.L. and F. von Hippel(1981): The fuel economy of light vehicles, Scient. American, 244 (5), 36-47.

Greene, D.L. and G. Kulp(1982): An analysis of the 1978-80 decline in gasoline consumption in the U.S.,Energy 7 (4), 367-375.

Griesbaum, K. und D. Hönicke(1980): Kraftstoffe der Zukunft, Chemie in unserer Zeit 14 (3), 90-101.

Häfele, W. et al.(1981): Energy in a Finite World, Vol. 1 and 2, Ballinger Publ. Co., Cambridge, USA.

Hardin, G.(1968): The tragedy of the commons, Science 162, 1243-1248.

Hein, K.(1981): Elektrischer Strom aus der Heizung? Wärme aus dem Kühlschrank? Umwelt Aktuell, Verlag C.F. Müller, Karlsruhe.

Heinrich, H.J.(1982): Energieeinsparung mit technischen Informationssystemen, Kontakt & Studium, Bd. 95, Verlag TÜV Rheinland, Köln.

Herrmann, H.(1978): Zauberformel Ökotechnik? Umschau 78 (16),501-505.

Herrmann, H.(1979): Energiesparende technische Alternativen, Vortrag auf dem Energiesymposium der Ruhr-Univ. Bochum, 26.-28. 10. 1979.

Hirsch, F.(1980): Die sozialen Grenzen des Wachstums, Rowohlt, Reinbek.

Hoffmann, W.(1981): Inseln der Wärme. Ausbau der Fernwärmenetze spart Energie und schafft Arbeit, Die Zeit, Nr. 14,S. 19, 27. März.

Hollowell, C.D. et al.(1980): Radon in energy-efficient residences, Report No. LBL-9560, Lawrence Berkeley Lab.,Berkeley.

Kommission für Umweltfragen und Ökologie beim SPD-Parteivorstand (1981): Ökologiepolitische Orientierungen der SPD, Bonn.

Langhoff, J. u. H.G. Krischke(1981): Die Wirbelschichtanlagen Flingern und König Ludwig, Brennstoff-Wärme-Kraft 33 (11), 441-443.

Lovins, A.B., L.H. Lovins, F. Krause and W. Bach(1981): Least-cost Energy: Solving the CO_2 Problem, Brick House, Andover, USA.

Mensching, H.(1978): Die Wüste schreitet voran, Umschau 78 (4), 99-106.

Meyer-Abich, K.(1981): Rituale um die Sicherheit, Die Zeit Nr. 23,S. 56, 29. Mai.

Mishan, E.J.(1973): Ills, bads and disamenities: The wages of growth. In: M. Olson and H.H. Landsberg (eds.) The No-Growth Society, 63-87, W.W. Norton & Co. Inc., New York.

Naturwissenschaftliche Rundschau(1981): Energiesparende Glühlampen, 34 (9), S. 394.

Peters, W.(1978): Strom- und Wärmeerzeugung aus Kohle - Die Wirbelschichtfeuerung, eine neue Technologie. In: Bildung u. Gesundheit 7, G.M. Pfaff Gedächtnisstiftung, Kaiserslautern.

Pfeiffenberger, U.(1981): Wärmepumpen in Industrie und Gewerbe, Brennst.-Wärme-Kraft 33 (11), 460-462.

Polunin, N.(ed.)(1980): Growth without Ecodisasters? The MacMillan Press Ltd., London.

Roseme, G.D. et al.(1980): Residential ventilation with heat recovery: Improving indoor air quality and saving energy, Report No. LBL-9749, Lawrence Berkeley Lab., Berkeley.

Rotty, R.M. and G. Marland(1980): Constraints on fossil fuel use. In: W. Bach et al.(eds.) Interactions of Energy and Climate, 191-212, Reidel Publ. Co., Dordrecht.

Ruske, B. u. D. Teufel(1980): Das sanfte Energie-Handbuch, rororo aktuell, Rowohlt, Reinbek.

Sant, R.W.(1979): The least-cost energy strategy. Minimizing consumer costs through competition, 50p., The Energy Productivity Center, Mellon Institute, Arlington, USA.

Schneider, S.H. and W. Bach(1980): Food-climate interactions: Issues and policy implications, Repord for the Federal Environmental Agency, Berlin.

Seiffert, U.(1981): Alternative Kraftstoffe, Umschau 81(2), 49-53.

Spreer,F.(1982): Dezentrale Wärmeversorgung. Neue Wege zur besseren Energie-Nutzung, Bild der Wissenschaft 9, 120-134.

Tinbergen, J.(1977): Wir haben nur eine Zukunft. Der RIO-Bericht an den Club of Rome, Westdeutscher Verlag, Opladen.

Venzky, G.(1981): Werden die Armen reicher, wenn die Reichen ärmer werden? Die Zeit,Nr. 20, S. 14, 8. Mai.

Vester, F.(1976): Wachsende Systeme, Biologisch-kybernetische Grundgedanken zum Wachstum von Bevölkerung, Wirtschaft und Wissen. In: G. Schaefer et al.(Hrsg.) Wachsende Systeme, 28-55, Leitthemen 1/76, Westermann, Braunschweig.

Vester, F.(1980): Neuland des Denkens, Deutsche Verlags-Anstalt, Stuttgart.

Wilhelmy, H.(1981): Welt und Umwelt der Mayas, Piper, München/Zürich.

***

FURTHER READING

1. Books on CO_2 and Climate Themes

Andersen, N.R. and A. Malahoff(eds.)(1977): The Fate of Fossil Fuel CO_2 in the Oceans, Plenum Press, New York/London.

Ausubel, J. and A.K. Biswas(eds.)(1980): Climatic Constraints and Human Activities, Pergamon Press, Oxford.

Bach, W., J. Pankrath and W.W. Kellogg(eds.)(1979): Man's Impact on Climate, Elsevier Publ. Co., Amsterdam.

Bach, W., J. Pankrath and J. Williams(eds.)(1980): Interactions of Energy and Climate, Reidel Publ. Co., Dordrecht/Boston.

Bach, W., J. Pankrath and S.H. Schneider(eds.)(1981): Food-Climate Interactions, Reidel Publ. Co., Dordrecht/Boston.

Bach, W., A.J. Crane, A.L. Berger and A. Longhetto(eds.)(1983): Carbon Dioxide: Current Views and Developments in Energy/Climate Research, Reidel Publ. Co., Dordrecht/Boston.

Berger, A.L.(ed.)(1981): Climatic Variations and Variability: Facts and Theories, Reidel Publ. Co., Dordrecht/Boston.

Bolin, B. et al.(eds.)(1979): The Global Carbon Cycle, SCOPE 13, John Wiley & Sons, New York.

Bolin, B.(ed.)(1981): Carbon Cycle Modelling, SCOPE 16, John Wiley & Sons, New York.

Budyko, M.I.(1982): The Earth's Climate: Past and Future. Int. Geophys. Series 29, Academic Press, New York, London.

Chang, J.(ed.)(1977): General Circulation Models of the Atmosphere, Methods in Computational Physics, Advances in Research and Applications, vol. 17, Academic Press, New York.

Chen, R.S., E. Boulding and S.H. Schneider(eds.)(1983): Social Science Research and Climate Change, Reidel Publ. Co., Dordrecht/Boston.

Clark, W.C.(ed.)(1982): Carbon Dioxide Review: 1982, Oxford Univ. Press, Oxford.

Frakes, L.A.(1979): Climates Throughout Geologic Time, Elsevier, Amsterdam.

Freney, J.R. and I.E. Galbally(eds.)(1982): Cycling of Carbon, Nitrogen, Sulfur and Phosporus in Terrestrial and Aquatic Ecosystems, Springer, Berlin/New York.

Gribbin J.(ed.)(1978): Climatic Change, Cambridge Univ. Press, Cambridge.

Idso, S.B.(1982): Carbon Dioxide: Friend or Foe? IBR Press, Tempe, Arizona.

Kellogg, W.W. and M. Mead(eds.)(1977): The Atmosphere: Endangered and Endangering, U.S. GPO, Washington, D.C.

Kellogg, W.W. and R. Schware(1981): Climate Change and Society, Westview Press, Boulder, USA.

Lamb, H.H.(1972): Climate : Present, Past and Future, vols. 1/2, Methuen & Co. Ltd., London.
Lamb, H.H.(1982): Climate, History and the Modern World, Methuen & Co. Ltd., London.
Likens, G.E.(ed.)(1981): Some Perspectives of the Major Biochemical Cycles, SCOPE 17, John Wiley & Sons, New York.
Lockwood, J.G.(1979): Causes of Climate, Edward Arnold, London.
Oeschger, H., B. Messerli u. M. Slivar(Hrsg.)(1980): Das Klima, Springer-Verlag, Berlin/New York.
Pearman, G.I.(ed.)(1980): Carbon Dioxide and Climate, Australian Academy of Science, Canberra.
Pittock, A.B. et al.(eds.)(1978): Climatic Change and Variability, Cambridge Univ. Press, Cambridge, England.
Roberts, W.O. and H. Lansford(1979): The Climate Mandate, W.H. Freemann & Co., San Francisco.
Rotberg, R.I. and T.K. Rabb(eds.)(1981): Climate and History, Princeton Univ. Press, Princeton.
Schönwiese, C.D.(1979): Klimaschwankungen, Springer-Verlag Berlin.
Slater, L.E. and S.K. Levin(eds.)(1981): Climate's Impact on Food Supplies, AAAS Symposium Series, Westview Press, Boulder USA.
Stumm, W.(ed.)(1977): Global Chemical Cycles and their Alterations by Man, Abakon Verlagsgesellschaft, Berlin.
U.S. National Academy of Sciences(1975): Understanding Climatic Change, Washington, D.C.
U.S. National Academy of Sciences(1977): Energy and Climate, Washington, D.C.
Wigley, T.M.L., M.J. Ingram and G. Farmer(eds.)(1981): Climate and History, Cambridge Univ. Press, London.
Williams, J.(ed.)(1978): Carbon Dioxide, Climate and Society, Pergamon Press, Oxford.
World Meteorological Organization(1975): The Physical Basis of Climate and Climate Modelling, GARP Publ. Series No.16, Geneva.
World Meteorological Organization(1979): Proceedings of the World Climate Conference, WMO-No. 537, Geneva.
World Meteorological Organization(1979): Numerical Methods used in Atmospheric Models, vols. 1/2, GARP Publ. Series No. 17, Geneva.
World Meteorological Organization(1979): Report of the JOC Study Conference on Climate Models: Performance, Intercomparison and Sensitivity Studies, vols. 1/2, GARP Publ. Series No. 22, Geneva.

2. Books on Energy Themes

Bach, W., W. Manshard, W.H. Matthews and H. Brown(eds.)(1980): Renewable Energy Prospects, Pergamon Press, Oxford.
Bischoff. G. und W. Gocht(1979): Energietaschenbuch, Vieweg, Braunschweig.

Bischoff, G. und W. Gocht(1979): Das Energiehandbuch, Vieweg, Braunschweig.

Bockris, J. O'M. und E.W. Justi(1980): Wasserstoff. Energie für alle Zeiten, U. Pfriemer Verlag, München.

Bossel, H. et al.(1976): Energie richtig genutzt, C.F. Müller, Karlsruhe.

Breuer, G.(1980): Energie ohne Angst, Kösel, München.

Chapman, P.(1975): Fuel's Paradise, Penguin Books, Middlesex, England.

Fricke,J. und W.L. Borst(1981): Energie. Ein Lehrbuch der physikalischen Grundlagen, Oldenbourg Verlag, München/Wien.

Gemper, B.B.(Hrsg.)(1981): Energieversorgung, Verlag Vahlen, München.

Gerwin, R.(1980): Die Welt-Energieperspektive, DVA, Stuttgart.

Gorz, A.(1980): Ökologie und Freiheit, Rowohlt, Reinbek.

Grathwohl, M.(1982): World Energy Supply, W. de Gruyter, Berlin/ New York.

Grupp, M.(Hrsg.)(1979): Energiesucht, Bonz-Verlag, Fellbach.

Häfele, W. et al.(1981): Energy in a Finite World, Vol. I and II , Ballinger Publ. Co., Cambridge, USA.

Hatzfeldt, H. et al.(eds.)(1982): Kohle. Konzepte einer umweltfreundlichen Nutzung, Fischer, Frankfurt.

Hauff, V.(Hrsg.)(1978): Energieversorgung und Lebensqualität, Neckar-Verlag, Villingen.

Hayes, D.(1977): Rays of Hope, W.W. Norton, New York; deutsche Fassung(1979): Alternative Energien, Hoffmann und Campe, Hamburg.

Hörster, H.(ed.)(1980): Wege zum energiesparenden Wohnhaus, Philips Fachbücher, Hamburg.

Karweina, G.(1981): Der Megawatt Clan, Stern Buch, Hamburg.

Kendall, H.W. and S.J. Nadis(eds.)(1980): Energy Strategies: Toward a Solar Future, Ballinger Publ. Co., Cambridge, USA.

Krause, F., H. Bossel und K.-F. Müller-Reissmann(1980): Energie-Wende, S. Fischer, Frankfurt.

Krause, F.(1981): Daten und Fakten zur Energiewende, Öko-Institut, Freiburg.

Landsberg, H.H. et al.(1979): Energy: The Next Twenty Years, Ballinger Publ. Co., Cambridge, USA .

Lapedes, D.N.(ed.)(1976): Encyclopedia of Energy, McGraw-Hill Book Co., New York/Düsseldorf.

Leach, G. et al.(1979): A Low Energy Strategy for the United Kingdom, The Int. Institute for Environment and Development, London.

Lienemann, W. et al.(Hrsg.)(1978): Alternative Möglichkeiten für die Energiepolitik, Westdeutscher Verlag, Opladen.

Lovins, A.B.(1975): World Energy Strategies, Friends of the Earth Inc., New York/London.

Lovins, A.B.(1977): Soft Energy Paths, Penguin Books, New York; deutsche Fassung(1978): Sanfte Energie, Rowohlt, Reinbek.

Lovins, A.B., L.H. Lovins, F. Krause and W. Bach(1981): Least-Cost Energy: Solving the CO_2 Problem, Brick House Publ. Co., Andover, USA; deutsche Fassung(1983): Rationelle Energienutzung ohne CO_2-Risiko, Alternative Konzepte 42, Verlag C.F. Müller, Karlsruhe.

Luther, G., M. Horn und H.-J. Luhmann(1979): Stromtarife - Anreiz zur Energieverschwendung? Alternative Konzepte 31, Verlag C.F. Müller, Karlsruhe.

McRae, A. and J.L. Dudas(eds.)(1977): The Energy Source Book, Aspen Systems Corp., germantown, USA.

Meyer-Abich, K.-M.(Hrsg.)(1979): Energieeinsparung als neue Energiequelle, C. Hanser Verlag, München/Wien.

Meyer-Larsen, W.(Hrsg.)(1979): Das Ende der Ölzeit, Heyne-Verlag, München.

Montbrial, T. de(1979): Energy: The Count-Down, Pergamon Press, Oxford.

Nash, H.(ed.)(1979): The Energy Controversy, Amory Lovins and His Critics, Friends of the Earth, San Francisco.

OECD(1979): Facing The Future, Interfutures, Paris.

Olivier, D., H. Miall, F. Nectoux and M. Opperman(1983): Energy-Efficient Futures; Opening the Solar Option, Earth Resources Research, Ltd., London.

Preuss, K.-H.(1981): Wege zur Bescheidenheit, Umschau, Frankfurt.

Rothchild, J.(1981): Stop burning Your Money, Random House, New York.

Ruske, B. und D. Teufel(1980): Das sanfte Energie-Handbuch, Rowohlt, Reinbek.

Schäfer, M.(Hrsg.)(1980): Struktur und Analyse des Energieverbrauchs in der Bundesrepublik Deutschland, Resch, München.

Schurr, S.H. et al.(1979): Energy in America's Future, The Johns Hopkins Univ. Press, Baltimore/London.

Seifritz, W.(1980): Sanfte Energietechnologie - Hoffnung oder Utopie? Verlag K. Thiemig, München.

SERI(1981): A New Prosperity: Building a Sustainable Future, Brick House Publ. Co., Andover, USA.

Slesser, M. and C. Lewis(1979): Biological Energy Resources, E. & F.N. Spon Ltd., London.

Sorensen, B.(1979): Renewable Energy, Academic Press, London/New York.

Stobaugh, R. and D. Yergin(eds.)(1979): Energy Future, Random House, New York.

Tillman, D.A.(1978): Wood as an Energy Source, Academic Press, New York.

Voss, A. und K. Schmitz(Hrsg.)(1980): Energiemodelle für die Bundesrepublik Deutschland, Verlag TÜV Rheinland, Köln.

WAES(Workshop on Alternative Energy Strategies)(1977): Energy, Global Prospects 1985-2000, McGraw-Hill, New York.

Winkler, J.-P.(1980): Sonnenenergie in Theorie und Praxis, C.F. Müller, Karlsruhe.

WOCOL(World Coal Study)(1980): Coal - Bridge to the Future, Ballinger Publ. Co., Cambridge, USA.

WOCOL (World Coal Study) (1980): Future Coal Prospects: Country and Regional Assessments, Ballinger Publ. Co., Cambridge, USA.

3. Books of General Interest

Bach, W. and A. Daniels (1975): Handbook of Air Quality in the United States, The Oriental Publ. Co., Honolulu, Hawaii.

Binswanger, H.C., W. Geissberger und T. Ginsburg (Hrsg.) (1969/1980): Wege aus der Wohlstandsfalle, Fischer, Frankfurt.

Brandt, W. et al. (1980): Das Überleben sichern. Kiepenheuer & Witsch, Köln.

Brown, H. (1978): The Human Future Revisited, W.W. Norton, New York/London.

Brown, L.R. (1974): By Bread Alone, Praeger Publ., New York/Washington.

Brown, L.R. (1978): The Twenty-Ninth Day, W.W. Norton & Co. Inc., New York/London.

Brown, L.R. (1981): Building a Sustainable Society, W.W. Norton, New York/London.

Club of Rome (1976): Reshaping the International Order, Elsevier, Amsterdam; deutsche Fassung (1977): Wir haben nur eine Zukunft, Westdeutscher Verlag, Opladen.

De Witt, S. und H. Hatzfeldt (Hrsg.) (1979): Zeit Zum Umdenken, Rowohlt, Reinbek.

Dürr, G. und W. Hähnle (Hrsg.) (1980): Wer macht unsere Zukunft? Sind wir nur Marionetten? Radius-Verlag, Stuttgart.

Eppler, E. (1981): Wege aus der Gefahr, Rowohlt, Reinbek.

Fetscher, I. (1980): Überlebensbedingungen der Menschheit, Piper, München.

Fromm, E. (1968): The Revolution of Hope, Harper & Row, New York; deutsche Fassung (1974/1980): Die Revolution der Hoffnung, Rowohlt, Reinbek.

Garcia, R.V. (1981): Drought and Man the 1972 Case History, vol. 1: Nature pleads not guilty, Pergamon Press, Oxford/New York/Frankfurt.

Gribbin, J. (1979): Future Worlds, Abacus, Sphere Books, London.

Grupp, M. (Hrsg.) (1980): Wissenschaft auf Abwegen? Verlag Bonz, Fellbach.

Haaf, G. (1981): Rettet die Natur, Bertelsmann, Gütersloh.

Henderson, H. (1978): Creating Alternative Futures, Berkeley Publ. Co., Berkeley, USA.

Hirsch, F. (1976): Social Limits to Growth, Harvard Univ. Press, Cambridge, USA; deutsche Fassung (1980): Die sozialen Grenzen des Wachstums, Rowohlt, Reinbek.

Illich, I. (1973): Tools for Conviviality, Harper & Row, New York; deutsche Fassung (1975/1978): Selbstbegrenzung, Rowohlt, Reinbek.

Jänicke, M.(1979): Wie das Industriesystem von seinen Misständen profitiert, Westdeutscher Verlag, Opladen.
Lovelock, J.E.(1979): Gaia. A New Look at Life on Earth, Oxford Univ. Press, Oxford.
Lovins, A.B. and L.H. Lovins(1982): Brittle Power, Energy Strategy for National Security, Brick House, Andower, Mass.
Meadows, D.L.(ed.)(1977): Alternatives to Growth-I, Ballinger, Cambridge, USA.
Mesarovic, M. und E. Pestel(1974): Menschheit am Wendepunkt, DVA, Stuttgart.
Meyer, N.I., K.H. Petersen und V. Sörensen(1979): Aufruhr der Mitte, Hoffmann & Campe, Hamburg.
Meyer-Abich, K.M.(Hrsg.)(1979): Frieden mit der Natur, Herder, Frieburg.
Mishan, E.J.(1977): The Economic Growth Debate, G. Allen & Unwin Ltd., London.
Müller, W. und B. Stoy(1978): Entkopplung, DVA, Stuttgart.
Müller-Wenk, R.(1980): Konflikt Ökonomie Ökologie, C.F. Müller, Karlsruhe.
Myers, N.(1979): The Sinking Arch, Pergamon Press, Oxford/New York/Frankfurt.
Norman, C.(1981): The God that Limps, W.W. Norton, New York/London.
Olson, M. and H.H. Landsberg(eds.)(1973): The No-Growth Society, W.W. Norton, New York.
Peccei, A. und M. Siebker(1974): Die Grenzen des Wachstums, Fazit und Folgestudien, Rowohlt, Reinbek.
Pestel, E.(1980): Unsere Chance heisst Vernunft, Westermann, Braunschweig.
Polunin, N.(ed.)(1980): Growth Without Ecodisasters? MacMillan Press, London.
Schaefer, G., G. Trommer und K. Wenk(1976): Wachsende Systeme, Leitthemen 1/76, Westermann, Braunschweig.
Schneider, S.H. with L.E. Mesirow(1976): The Genesis Strategy, Plenum Press, New York/London.
Schumacher, E.F.(1973): Small is Beautiful, Blond & Briggs, London; deutsche Fassung(1977): Die Rückkehr zum menschlichen Mass, Rowohlt, Reinbek.
Schumacher, E.F.(1979): Good Work, Harper & Row, New York; deutsche Fassung(1980): Das Ende unserer Epoche, Rowohlt, Reinbek.
Strohm, H.(1979): Politische Ökologie, Rowohlt, Reinbek.
The Global 2000 Report to the President(1980): Entering the 21st Century, Vols. 1/2, U.S. Government Printing Office, Washington, D.C.; deutsche Fassung(1980): Global 2000, Verlag Zweitausendeins, Frankfurt.
Toffler, A.(1980): Die Zukunftschance, C. Bertelsmann, München.
Traube, K.(1978): Müssen wir umschalten? Rowohlt, Reinbek.
Vester, F.(1972/79): Das Überlebensprogramm, Fischer Verlag, Frankfurt.
Vester, F.(1980): Neuland des Denkens, DVA, Stuttgart.

***

Appendix II.1 The use of various types of data for the reconstruction of climate

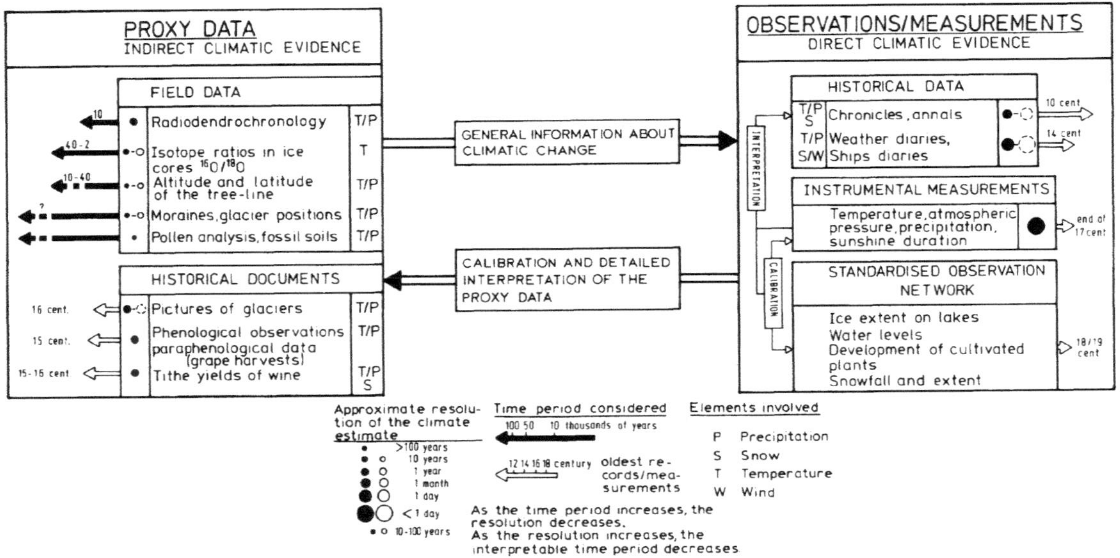

PROBLEMS OF THE VARIOUS TYPES OF DATA		
Field data (indirect climatic evidence)	Historical documents and data (direct and indirect climatic evidence)	Instrumental measurements (direct climatic evidence)
• relatively low temporal resolution • response to several climatic elements (e.g. temperature and precipitation) • imprecise dating (varying according to method)	• observational gaps • only cover the last 5-8 centuries • subjectiveness of observations	• only cover the last 2-3 centuries • low station density (before mid-19th century) • inhomogeneity of the data series
ONLY THROUGH A SYNTHESIS AND INTERCOMPARISON OF THE MANY DATA TYPES WILL IT BE POSSIBLE TO DERIVE A REALISTIC, EXTENSIVE AND DETAILED PICTURE OF PAST CLIMATE. THE KNOWLEDGE OF THE COURSE OF CLIMATE IN THE PAST HELPS US TO RECOGNISE RELATIONS OF USE IN MAKING CLIMATE FORECASTS, AND TO EVALUATE REALISTICALLY THE INFLUENCE OF HUMAN ACTIVITIES ON CLIMATE		

Source: Pfister et al. (1980)

Appendix II.2 Review of the Earth's climatic history

Pre-Cambrian Era (before 600 MYBP*).
Most widespread glaciation ca. 700 MYBP.
Palaeozoic Era (600-225 MYBP).
Gondwana glaciation (300-250 MYBP) with proven inland glaciation in South America, Africa, India, Australia, and Antarctica.
During most of this era, however, the earth was warmer than today and both poles were ice-free 80-90% of the time.
Mesozoic Era (225-65 MYBP).
Much warmer than today with mean annual temperatures of 8-10°C near both poles and 25-30°C in the tropics.
Cenozoic Era (65 MYBP-present).
Tertiary Period (65-2 MYBP)
From the Eocene (60 MYBP) until the Pliocene (2 MYBP), the temperature of Central Europe gradually decreased from 20°C to 10°C above the present temperature. The glaciation of the Antarctic began in the mid-Pliocene (5 MYBP) and the mountain glaciers in the Sierra Nevada, Iceland and Greenland were formed in the late Pliocene (2.5 MYBP).
Quaternary Period (2 MYBP-present).
The Pleistocene, the oldest epoch of the Quaternary, is characterised by the succession of about 20 glacial periods during the last 2 MYBP. We can distinguish:
Primary Cycles with the change from glacial periods to interglacial periods at intervals of about 100,000 years.
Secondary Oscillations with a change from glacial to interglacial periods of about 20,000 years. The maximum of the last glacial in Europe, the Würm, and in North America, the Wisconsin, was about 20,000 YBP. The present interglacial (Holocene) began about 12,000 YBP and reached a climatic optimum (also called Altithermal, Hypsithermal, Atlantic, or postglacial warm period) about 8,000-5,000 YBP.
Tertiary Fluctuations. About 800-1000 AD, the "medieval optimum", when the Vikings cultivated Iceland and Greenland.
About 1500-1700 AD, the "Little Ice Age" with mean annual temperatures in Europe about 1°C lower than today.
About 1880-1940, the "recent warm period" with an increase of the annual mean temperature in the Northern Hemisphere by about 0.6°C.
Since about 1940 a cooling in the Northern Hemisphere of about 0.3°C and in recent years a slight warming; in the Southern Hemisphere (Australia, New Zealand, Antarctica) a continuous warming of about 0.5°C.

---

* MYBP = million years before present.

Based on Bach (1976) and Schönwiese (1979).

Appendix II.3 Basic equations for a selection of climate model types

1-D Horizontally averaged energy balance models

$c\,\frac{\partial T(\phi)}{\partial t}$	$= \frac{1}{4}S_o\,(1-\alpha_p(\phi))$	$-\ \sigma(T(\phi))^4\{1-m\ \tanh[D(T(\phi))^6]\}$	$+\ F_{SO}(\phi)+F_{SA}(\phi)+F_{LA}(\phi)$
Time change of heat content of a zonally averaged vertical column through the land-ocean atmosphere system	Shortwave incoming minus outgoing solar radiation at the top of the atmosphere	Longwave outgoing radiation from the earth's surface and the atmosphere	Northward transport of heat by ocean and atmosphere
c: is the heat capacity (thermal inertia coefficient)	S_o: is the incoming solar radiation at the top of the atmosphere (solar constant)	m: is the atmospheric attenuation coefficient (dependent upon e.g. CO_2 ***concentration***)	F_{SO}: is the sensible heat transport due to ocean currents
T: is the zonally averaged surface temperature	α_p: is the planetary albedo (is a function of the optical properties of the atmosphere and the surface albedo)	D: is an empirical constant	F_{SA}: is the sensible heat transport due to atmospheric motion
ϕ: is the latitude		σ: is the Stefan-Boltzmann constant	F_{LA}: is the latent heat transport in the atmosphere
t: is time			

1-D Stochastic climate models

$c\,\frac{\partial T(\phi)}{\partial t}$	$= \overline{Q}$	$+ Q'$	$+ \overline{A}(F_{SO}(\phi), F_{SA}(\phi), F_{LA}(\phi))$	$+ A'$
See energy balance model	Mean value of the radiation balance [difference between shortwave solar radiation and long-wave radiation (shows e.g. the CO_2 *influence*)]	Deviation from $\overline{Q}$	Means of heat fluxes in the ocean and the atmosphere	Deviation from $\overline{A}$

3-D Atmospheric circulation models

Equation of motion (conservation of momentum)

$\frac{d\underline{V}}{dt} =$	$- 2\underline{\Omega}\times\underline{V}$	$- \frac{1}{\rho}\,\underline{\nabla}p$	$+ \underline{g}$	$+ \underline{F}$
Change in velocity (acceleration)	Coriolis force	Pressure gradient force	Gravitation force	Friction force
$\underline{V}$: is the velocity vector (horizontal components: u, v; vertical component: w)	$\underline{\Omega}$: is the angular velocity of the earth	p: is the pressure		
t: is time		ρ: is the density		

Equation of continuity (conservation of mass)

$\frac{d\rho}{dt} =$	$- \rho\underline{\nabla}\cdot\underline{V}$
Change in the density of a volume of air	Mass divergence of air and water and its aggregate states

Equation of continuity (conservation of water vapour)

$$S = \frac{dq}{dt}$$

Change in water vapour concentration

S: are the sources or sinks of water vapour (evaporation, condensation)

q: is the water vapour mixing ratio

Equation of energy (first law of thermodynamics)

$c_v \frac{dT}{dt} =$	$- p\frac{d\alpha}{dt}$	$+ S$
Change in the internal energy	Work done by the system	Heat added to a specific air parcel [e.g. by condensation, evaporation, radiation (*CO_2 influence*)].
c_v: is the specific heat of air at constant volume	$\alpha = 1/\rho$	

Equation of state

$$p = \rho RT$$

R: is the gas constant for dry air

Hydrostatic approximation

$$\frac{\partial p}{\partial z} = -\rho g$$

z: is the geometrical height

After: Holloway and Manabe (1971); Schneider and Gal-Chen (1973); Lemke (1977).

Appendix III.1 The influence of civilization upon European and Asian Forests

Date	Location	Event
9000–5000 BC	Southwest Asia	Period of early agricultural and pastoral modification
5000 BC–640 AD	Southwest Asia	Widespread and intensive agricultural modification, use for fuelwood and lumber
2600 BC	Southwest Asia	First evidence of an extensive wood export (mostly Cypress) by Phoenicians to Egypt and Mesopotamia
1127–255 BC	China	The Chow Dynasty introduces a forestry policy to end 1,500 years of forest destruction
255 BC	China	Abandonment of the forestry policy of the Chow Dynasty and reduction of the forests to 9% of the total land area
200–400 AD	Italy	Invasion of the Roman Empire by barbarians causes reforestation of many cultivated areas
400–500 AD	England	Conversion of large parts of England by the Saxons from forest to crop and pasture lands
900–1200 AD	Germany	Fourfold increase in population and large forest clearing
1000–1400	Alps, Appenino	Large forest losses from excessive cutting
1000–1800	Northwest Europe	Conversion of extensive areas of formerly deciduous forest to moorland by cutting, burning, grazing, and climatic deterioration
1789–1793	France	Over 3.2×10^6 ha of forest cleared

1878	East-central Europe	Centralized forestry management of the Austro-Hungarian Empire
1878-1913	Germany	Increase of forest land by more than 350,000 ha
World War I	Europe	Widespread forest devastation and excessive harvesting throughout Europe. Some effects still apparent in the 1950s
World War II	Europe	Heavy losses throughout Europe. Cutting was 2 to 3 times beyond sustained yield. Forest losses in the USSR alone were 20 x 10^6 ha
1946-1950	Europe	Reconstruction takes 141 x $10^6 m^3$ of German timber. Eastern European countries develop five-year plans for reforestation
1950-	Europe/Asia	Large-scale reforestation programmes in China, Korea and parts of Europe

Based on Armentano and Hett (1980).

Appendix IV.1 Short historical sketch of CO_2/climate research

Year	Author	Notes [ΔT is the change of the global mean surface air temperature for a CO_2 doubling ($2 \times CO_2$)].
1827	Fourier	Was probably the first to discuss the CO_2/greenhouse effect and compares it with the warming of air isolated under a glass plate.
1846/61	Tyndall	Was probably the first to point out that a change of the atmospheric CO_2 content could have been the cause of climatic changes in the geological past. States that water vapour has a large influence in addition to that of CO_2.
1896/1903	Arrhenius	Was probably the first to calculate the change of mean global temperature at the earth's surface for a change of the atmospheric CO_2 content. His result: a CO_2 increase by a factor of 2.5–3 would increase the temperature by 8–9°C.
1897/99	Chamberlin	Postulates that the increase or decrease of the CO_2 in the atmosphere depends strongly on the ocean as a CO_2 reservoir. Sees a cause-effect relationship between CO_2 changes and ice ages.
1899	Tolman	Probably the first to research and demonstrate the critical role of the ocean in the global CO_2 distribution.
1938	Callendar	Reaches the conclusion that the anthropogenic CO_2 production of 4.5 billion tonnes/yr cannot be entirely absorbed by the ocean and that the CO_2 content has increased by about 6% in 36 years. On the basis of temperature observations at 140 stations he finds a continual temperature increase of 0.6^{o}C since 1875. Was probably the first to connect fossil fuel combustion and the CO_2 increase in the atmosphere with a climatic effect.

1941	Flohn	Points out that anthropogenic CO_2-production perturbs the carbon cycle and leads to a continual CO_2 increase in the atmosphere. Was probably the first to point to the reduction of terrestrial vegetation as a result of expanding civilization.
1956	Plass	Calculates a temperature change of $\pm 3.6^{o}C$ for a doubling or halving of the atmospheric CO_2 concentration assuming average cloudiness.
1957	Revelle and Suess	State that mankind has set in motion a global geophysical experiment due to the rapid consumption of fossil fuels. Attempt to evaluate the CO_2 exchange between the atmosphere, hydrosphere and biosphere with the aid of a box model.
1957	Keeling and co-workers	Within the framework of the International Geophysical Year a CO_2 measurement programme was started on Mauna Loa (Hawaii) and the South Pole. The atmospheric CO_2 concentration on Mauna Loa increased from 315 ppmv in 1958 to 340 ppmv in 1982.
1961	Mitchell	Was probably the first to try to explain the combined effects of CO_2 and volcanism on the global temperature record since 1850.
1961 1961 1963	Plass Kaplan Möller	Further attempts to calculate the effect of a CO_2 doubling on the surface temperature.
1967	Manabe and Wetherald	Develop a 1-D radiative-convective model. $\Delta\bar{T}_s = 2.4^{o}C$ (for constant relative humidity), $\Delta\bar{T}_s = 1.4^{o}C$ (for constant absolute humidity).
1972	Machta	Develops a box model and uses it to make projections of the atmospheric CO_2 concentration up till 2000 (380 ppmv).
1975	Manabe and Wetherald	Develop a 3-D atmosphere-ocean general circulation model (3-D GCM) with 9 levels, annual average insolation, fixed

		cloudiness, and a non-circulating ocean (ocean treated as a flat swamp), idealized geography and topography. $\Delta\bar{T}_s$ = 2.9°C (about 10°C at high latitude).
1975	Schneider	Compiles CO_2/climate model results.
1975/ 80/83	Oeschger and co-workers	Study the carbon cycle and take isotype measurements from Greenland ice cores.
1976	Bryson and Dittberner	Argue that a CO_2-induced warming could possibly be compensated by an increase in turbidity (by aerosols) that reduces solar radiation.
1977	Baes, Goeller, Olson, Rotty	Review the carbon cycle.
1979	MacCracken and Potter	Develop a 2-D atmosphere-ocean (mixed layer) statistical-dynamical model with 9 levels, annually averaged insolation, parameterized cloudiness, and prescribed meridional ocean heat transport. $\Delta\bar{T}_s$ = 1.5°C (global average) $\Delta\bar{T}_s$ = 2.3°C (Northern Hemisphere) $\Delta\bar{T}_s$ = < 1°C (Southern Hemisphere)
1979	Hansen and co-workers	Develop a 7-layer 3-D GCM with realistic topography and geography and variable cloudiness. $\Delta\bar{T}_s$ = 3.9°C (swamp ocean, annual average insolation) $\Delta\bar{T}_s$ = 3.5°C (mixed ocean layer, seasonal insolation).
1979/ 80	Manabe and Stouffer	Couple their 3-D atmospheric GCM to an ocean mixed layer of 68.5 m. For prescribed cloudiness, seasonal insolation and realistic geography they obtain for $\Delta\bar{T}_s$ = 2°C (about 8°C at high latitude).
1980	Manabe and Wetherald	Run their 1975 3-D GCM model version with variable cloudiness. $\Delta\bar{T}_s$ = 3°C.
1980	Gates and co-workers	Develop a 3-D uncoupled atmosphere-ocean model with 2 levels, variable cloudiness, seasonal insolation, hydrologic

		cycle, but with fixed sea surface temperatures, i.e. set at present values as a function of season. $\Delta\bar{T}_s = 0.2^oC$.
1981	Hansen and co-workers	Use a 1-D radiative-convective model with fixed relative humidity, constant cloud height and vegetation-albedo feedback. $\Delta\bar{T}_s = 3.5^oC$.
1981	Michael, Hoffert, Tobias, Tichler	Use an energy balance atmospheric model coupled to a 1-D upwelling diffusion model of the deep ocean, a carbon cycle model, and a social model which considers the societal response to increasing global temperatures in order to assess the transient climatic response to changing atmospheric CO_2 levels.
1981	Cess and Goldenberg	Use a coupled land-ocean (70 m mixed layer) 1-D global climate model to estimate the time-dependent increase in atmospheric CO_2. Find that the ocean heat capacity will cause a lag in CO_2-induced warming of 1-2 decades. $\Delta T_s = 1.5\text{-}1.8^oC$ (with ocean effect) $\Delta T_s = 3.1^oC$ (when ocean heat capacity is neglected).
1981	Wetherald and Manabe	Modifying the computations by Manabe and Stouffer (1980) by using a limited computational domain and idealized geography, they obtain $\Delta\bar{T}_s = 2.4^oC$.
1981	Manabe, Wetherald, Stouffer	Present a concensus scenario for CO_2-induced hydrologic changes obtained from climate model results with simple and realistic geography, limited and global domain, and low and high computational resolution. Identify a significant reduction in zonal mean soil moisture in summer in two separate zones of middle and high latitudes of the Northern Hemisphere, and in winter at about 25°N.
1982	Bryan, Komro, Manabe and Spelman	Use a 3-D coupled ocean-atmosphere GCM with no seasonal variation in solar insolation and a highly idealized geome-

		try of land and sea in a first attempt to predict the transient response of climate to increasing atmospheric CO_2. Conclude that, barring a drastic acceleration of the rate of CO_2 increase, sensitivity studies of climate equilibrium can be used as an approximate guide for predicting the latitudinal pattern of sea surface temperature trends.
1982	Thompson and Schneider	Test the results of Bryan et al. (1982) by including realistic variations of land fraction and ocean mixing with latitude. Find only a limited applicability of equilibrium simulations as approximate guides to the transient temperature response. Believe that it is premature to conclude that equilibrium simulations can be reliably used to predict regional patterns of climatic changes.
1982/ 83	Schlesinger	Uses the 2-layer 3-D GCM of Gates (1980) and couples it with a swamp ocean. $\Delta\bar{T}_s = 2^oC$ (This is an order of magnitude greater than the value obtained with the non-interactive climatological ocean model).
1982/ 83	Washington and Meehl	Use their 9-level atmosphere-ocean coupled 3-D GCM with realistic geography and topography to calculate $\Delta\bar{T}_s = 1.3^oC$ (swamp model, computed clouds) $\Delta\bar{T}_s = 1.4^oC$ (swamp model, fixed clouds) $\Delta\bar{T}_s = 2.9^oC$ (ocean mixed layer, average of each hemisphere's summer and winter values).
1983	Gilchrist/ Mitchell	Deploy their 5-layer coupled atmosphere-ocean 3-D GCM with realistic topography, seasonal insolation but fixed clouds. $\Delta\bar{T}_s = 0.4^oC$ (for prescribed sea surface temperatures, SST) $\Delta\bar{T}_s = 2.6^oC$ (by adding 2^oC uniformly to the SST at each gridpoint). Experiments with latitudinally varying additions to the SST are under way.

Sources: Tyndall (1861); Arrhenius (1896, 1903); Chamberlin (1897); Tolman (1899); Callendar (1938); Flohn (1941); Plass (1956, 1961); Revelle and Suess (1957); Kaplan (1961); Mitchell (1961); Möller (1963); Manabe and Wetherald (1967); Machta (1972); Manabe and Wetherald (1975); Schneider (1975); Oeschger et al. (1975, 1980, 1983); Baes et al. (1977); Dittberner (1978); MacCracken and Potter (cit. Kellogg and Schware, 1981); Hansen et al. (cit. US NAS, 1979); Manabe and Stouffer (1979, 1980); Manabe and Wetherald (1980); Gates (1980); Washington and Ramanathan (1980); Hansen et al. (1981); Michael et al. (1981); Cess and Goldenberg (1981); Wetherald and Manabe (1981); Manabe et al. (1981); Bryan et al. (1982); Thompson and Schneider (1982b); Schlesinger (1982, 1983a); Washington and Meehl (1982, 1983); Gilchrist (1983); Mitchell (1983).

Appendix IV.2 Calculation of the global atmospheric carbon content

The observed CO_2 concentration in 1979 was 336.58 ppmv at Mauna Loa (Hawaii) and 334.83 ppmv in S.E. Australia (Tasmania)(Pearman, 1980; Fraser et al., 1981). Samples have shown that the Mauna Loa measurements are about 0.7 ppmv lower than the average values of the northern hemisphere troposphere, and that the measurements from S.E. Australia are about 0.2 ppmv lower than the average values of the southern hemisphere troposphere. This gives a global CO_2 content of the troposphere of 336.2 ppmv. Samples have further shown that the average global CO_2 content of the troposphere is about 1.7 ppmv higher than the CO_2 content of the entire atmosphere (troposphere plus stratosphere), so that the average value for the latter is 334.5 ppmv.

Given that the total weight of the atmosphere is 5.10×10^{15}t and the mean global CO_2 content is 334.5 ppmv, the total weight G of carbon in the atmosphere can be calculated as follows:

$$G = 5.10 \times 10^{15}\text{t air} \times 334.5 \times 10^{-6} \text{ ppmv } CO_2/\text{ppmm air} \times 1.53 \text{ ppmm/ppmv} \times 0.2725 \text{ C}/CO_2$$

$G = 710 \times 10^9$tC (see also Fig. IV.1 in text)

where, for a molecular weight of C = 12 and CO_2 = 44

$CO_2 = 0.2725$ C and C $= 3.67\ CO_2$

and for the weight of one litre of air L_{air} = 1.2928 g/litre

and L_{CO_2} = 1.9768 g/litre (at 0°C)

1 ppmv = ppmm L_{CO_2}/L_{air} = 1.53 ppmm.

As a result, with a mean global CO_2 content of 334.5 ppmv, 1 ppmv = 2.12×10^9t C or 7.79×10^9t CO_2.

Appendix IV.3 Calculation of the surface air temperature as a function of the atmospheric CO_2 level, other trace gases, and model-derived sensitivity parameters.

Studies have shown that the various climate model results can be compared with the aid of model-dependent parameters, and that there is a simple logarithmic relation between the mean surface air temperature T_s and the CO_2 content of the atmosphere (Ramanathan, 1980; Gates, 1980; Hoffert et al., 1980; Flohn, 1981a, b).

We begin with the radiation balance equation at the upper boundary of the atmosphere.

$$Q = (1 - \alpha_p)(SC/4) = -E \quad (E \sim 240 \text{ W/m}^2 \pm 1\%) \qquad (1)$$

where Q is the net extraterrestrial solar radiation, α_p is the planetary albedo (~0.29±0.01), SC is the solar constant (~1,360 W/m^2), and E is the terrestrial radiation to space.

Assuming a near radiative equilibrium of Q = -E and denoting any deviation from an undisturbed reference level by Δ, then we can write

$$\Delta E(T_s, CO_2) = B \, \Delta T_s - nC \ln A \qquad (2)$$

where A is the normalised CO_2 content ($A = 1 + \Delta CO_2/CO_2{}^*$ is the deviation from an "undisturbed" reference value*, usually 300 ppmv), the parameters B, C and n depend on the respective model assumptions and can be estimated. The parameter B, which is frequently given as $\lambda = CB^{-1}$ is the thermal sensitivity, i.e. the ratio between the change of surface temperature and the change of the radiation balance:

$$B = \Delta E/\Delta T_s \sim 1.8 \ (\pm 0.4) \text{ W/m}^2 \cdot \text{K}.$$

The parameter C depends on B and the temperature change resulting from the CO_2 increase:

$$C = \Delta E/\Delta CO_2 \sim 6.8 \ (\pm 1.2) \text{ W/m}^2 \ (\text{for a } CO_2 \text{ doubling } A = 2).$$

If the other infrared absorbing gases are taken into consideration, then we must modify C by n which corrects the greenhouse effect due to the other trace gases.

$$n = \frac{\Delta E \ (\text{all IR-absorbing gases})}{\Delta E \ (CO_2 \text{ alone})}$$

It is estimated that n has presently a value of ~1.3 which is expected to increase to ~1.7-1.8 in the next 50-60 years. Such a time-dependent increase of n appears to be justified because of the relatively long atmospheric residence times of chlorofluoromethanes (60–80 years for F-11 and 135–180 years for F-12) and nitrous oxide (150–175 years), and because the chlorofluoromethanes have a relatively higher growth rate (9–10%/yr) than CO_2 (0.4%/yr) (see Table IV.3). The influence of water vapour is already considered in C. The relationship between a virtual (all trace gases) and the real CO_2 content (only CO_2) is then:

$$\frac{CO_2(\text{virtual})}{CO_2(\text{real})} = \left(1 + \frac{\Delta CO_2}{CO_2{}^*}\right)^n \qquad (3)$$

Assuming that $\Delta E = 0$, a simple estimate of the near-surface warming ΔT can be derived from equation (2) with the combined parameter $D = nCB^{-1}$:

$$\Delta T_s = D \ln \left(1 + \frac{\Delta CO_2}{CO_2{}^*}\right) \qquad (4)$$

For the present value of $n \sim 1.3$ we find $D \sim 5$, but in the future we can expect $n \sim 1.7$ and thus $D \sim 6$. Equation (4) can be displayed on a semi-logarithmic plot (see Fig. IV.27 in text). This diagram allows to determine the atmospheric CO_2 levels which correspond to critical temperature thresholds.

Appendix V.1 Effects of weather and climate on energy demand and supply

Energy technology or activity	Time scale			
	Days	Months	Years	Decades
Transportation of resources.	River flooding, river and lake ice inhibit barge traffic; ocean weather conditions affect oil and LNG tankers.	Weather and climate extremes affect fuel transportation especially in Arctic regions.	Climate variations determine the options for fuel transport.	
Power plants.	Storms and frost destroy power lines; heat waves and extreme cold overload the system because of unusual energy demands.	Extreme seasonal climate fluctuations may stress distribution capabilities.	Interannual changes of storm tracks may affect the need for distribution capabilities of power over large regions.	
Cooling towers.	River icing limits cooling water supply.	Droughts with higher water temperatures limit cooling potential.	Chronic lack of cooling water may require legislative or judicial action.	Climatic change could make changes in the cooling method necessary.
Fossil fuel power generation.	Extreme weather affects ability to recover resources (e.g. oil and gas fields in offshore areas by hurricanes).	Climate variations influence both the demand for coal, oil and gas and also storage requirements.	Climate variations can influence the choice of the various fuel options.	
Nuclear power generation (including waste management).		Variations in climatic conditions influence energy demand.		Changes of ground-water level may flood radioactive waste repositories.

Appendix V.1 Effects of weather and climate on energy demand and supply (Continued.)

Energy technology or activity	Time scale			
	Days	Months	Years	Decades
Geothermal energy.	Daily weather fluctuations influence noxious gas emissions and groundwater quality.			
Solar energy.	Day to day weather fluctuations influence the amount of solar energy available for conversion.	Changes in storm tracks may alter the number of cloud-free days.	Reduction of atmospheric transmissivity through air pollution reduces the efficiency of solar collectors.	Climatic changes could make changes in the design of solar systems necessary.
Wind energy.	Day to day weather fluctuations influence the wind resource. Storms could damage wind energy systems.		Interannual shifts of cyclone tracks could alter wind resources.	Climatic changes could influence the wind energy resource potential.
Biomass.	Day to day weather fluctuations influence both plant productivity and harvest. Storms, droughts and floods can reduce harvests.	Precipitation and water availability influence biomass growth.	Precipitation quality (e.g. acid rain) can influence the reproductive capability of the biomass.	Climate variations can influence forest management.
Hydropower.	Daily weather fluctuations can cause extremes in river flow and snow melt.	Seasonal precipitation controls water supply.	Multi-year droughts affect available water supplies.	Climatic changes can affect water supply and hydropower.

Appendix V.1 Effects of weather and climate on energy demand and supply (Continued.)

Energy technology or activity	Time Scale			
	Days	Months	Years	Decades
Ocean thermal energy conversion (OTEC).	Daily weather fluctuations, especially severe weather, can reduce efficiency of OTEC by strong ocean mixing.	Seasonal variations in ocean currents change the temperatures, which increases or decreases the energy resource.	Changes of the ocean surface temperature can make a relocation of the OTEC system necessary.	Trends in climate may affect ocean currents and thermal gradients.
Energy conservation.	Weather extremes affect the extent of conservation possible.	Cold winters and hot summers influence energy demand and prices and the need for conservation for fuels in short supply.	Climatic variations from year to year influence the effectiveness of various conservation measures.	
Overall energy demand.	Cold and warm periods induce peak demands.	Prevailing climatic conditions determine the demand (e.g. dry periods increase the demands of irrigation systems).		Long-term climate trends could change regional or even continental energy demand patterns.

After: USDOE (1980).

Explanation of scientific terms

Absolute humidity Is the ratio of the mass of water vapour present to the volume occupied by the mixture (see also relative humidity).

Acid rain Rain containing significant amounts of sulphate and nitrate as a result of SO_2 and NO_x-bearing coal, oil and gas.

Aerosols Colloidal solid (dust) and liquid (fog) suspended matter in the air with a particle size of one thousandth to one millionth of a millimetre.

Albedo Ratio of the reflected to the incoming solar radiation, or reflectivity.

Almost intransitivity The property of a complex and nonlinear system to remain in one mode of behaviour for a time and then, without any external forcing imposed on it, to abruptly shift to another mode for a time.

Ambient air quality standards This refers to the amount of pollutants, radiation, noise and waste heat found in a particular area (for example in a community). In general, the ambient air quality standards of a particular area consist of the emissions from a variety of sources.

Anaerobic Organisms or tissues that live in the absence of oxygen.

Anomaly The deviation of temperature, precipitation and other climatic elements in a given region over a specified period from the normal value for the same region.

Anthropogenic Influenced or caused by mankind.

Assimilation Conversion of the materials taken in by a living being from the air and foodstuffs into its own substances, especially the carbon assimilation in plants. In the latter, the carbon dioxide from the air with the addition of water form the organic substances sugar, carbohydrate, etc., as the assimilate (photosynthesis) with the by-product oxygen.

Atmosphere The envelope of air surrounding the Earth and bound to it by virtue of the Earth's gravitational attraction.

Atmospheric circulation The flow of the atmosphere, especially its large-scale wind systems.

Autotrophs Living things that take in their carbon in the form of CO_2 and produce higher organic compounds from it, above all green plants. Opposite: heterotrophs.

Base load Minimum power plant capacity, that day and night, summer and winter is always used.

Biomass Materials that arise from biological production: plants, flesh of animals (and humans), especially, for example, wood, straw, faeces, parts of household waste, etc. Contains stored solar energy as a result of photosynthesis.

Biome Natural ecosystem.

Biosphere The entire natural living space of the earth with the animals and plants, all living things in the sea, and all micro-organisms. Necessary prerequisites for life on Earth are water, air, and a source of energy (sunlight).

Biota The animal and plant life in a region or of the world.

Biotope Living space characterised by existing plants and animal groups; also the living space of individual types.

Boundary conditions The specified inputs in a mathematical model that are not generated by the model itself.

Buffer factor Is defined as the ratio of the percentage increase of carbon in the atmosphere to the percentage increase of carbon in the surface water, when the total amount of carbon in these two reservoirs increases by an incremental amount, and chemical equilibrium is maintained.

Carbon dioxide, CO_2 Colourless, tasteless non-combustible gas. Is transparent to shortwave, solar radiation but absorbs longwave or infrared radiation. Thereby contributes to the greenhouse effect. Is the most important influence in the expected global warming.

Climate Is the synthesis of weather over the whole of a period, essentially long enough to establish its statistical ensemble properties (means, variances, probabilities of extreme events, etc.) and is largely independent of any instantaneous state.

Climate change Refers to the difference between long-term mean values of a climate variable or climate statistic, in which the mean is taken over a given time interval, usually a number of decades.

Climate models These describe the physical and chemical interactions within the climate system using mathematical equations. There is a hierarchy of models, ranging from zero-dimensional to three-dimensional climate models. The main types are radiative-convective, energy-balance, stochastic and circulation models.

Climate system Consists of five components: the atmosphere (air shell); the hydrosphere (ocean, lakes, rivers and groundwater); the cryosphere (continental and sea ice, ice cover on lakes and rivers, mountain glaciation and snow cover); the lithosphere (landmasses, rocks and soils); and the biosphere (plant and animal kingdom including humans).

Climatic variability Includes the extremes and deviations of monthly, seasonal and annual values compared with the value expected on the basis of climatic means (time average). The deviations are usually referred to as anomalies.

Cogeneration The combined production of electricity and heat, for example, in combined heat- and power-plants, neighbourhood diesel cogeneration systems, or photovoltaic cells that simultaneously function as solar energy collectors.

Concentration Is the fraction of the total of a substance made up by one component, e.g. ppmv of CO_2, that is parts of CO_2 per million parts of air.

Condensation nuclei Liquid or solid particles upon which condensation of water vapour takes place in the atmosphere.

Conductivity A measure of a system's conductance of heat, sound, or electricity.

Convection Mass motions within a fluid resulting in transport and mixing of the properties of that fluid.

Cooling degree-day A one degree departure of the daily mean temperature above a base temperature (usually $24^{o}C$ or $75^{o}F$); used as an indication of fuel consumption for refrigeration or air conditioning (see heating degree-day).

Cryosphere The total snow and ice on the land or floating in the oceans.

Denitrifying bacteria Bacteria which, with a lack of free oxygen, split nitrate and can use the oxygen contained therein for respiration, i.e. for the oxidation of organic carbon compounds. The nitrate (NO_3) is reduced stepwise to laughing gas (N_2O) or nitrogen (N_2).

Dissolved inorganic carbon (DIC) The sum of all soluble chemical species of carbon in ocean water which are part of the carbonic acid system. Includes CO_2, H_2CO_3, HCO_3^-, CO_3^{2-}, etc.

Dissolved organic carbon (DOC) The sum of all organic compounds dissolved or dispersed as colloids in ocean water.

Ecology Science of the interrelations of organisms (plants, animals, humans) among one another and with the environment.

Ecosystem A system of environment that is independent in terms of its material budget and self-supporting in its material circulation.

Efficiency The efficiency shows what percentage of the energy that is introduced is used. The unused remaining energy is referred to as lost energy.

Emissions In the case of energy conversion processes, one speaks of emissions when air or water pollutants, radiation, noise or waste heat are given to the environment from a specified source (see also ambient air quality standards).

Energy carriers This refers to all sources from which energy can be gained directly or after conversion.

Energy service Service that the consumer obtains by using the energy, e.g. heating of a room, transport of x number of people over x number of kilometres per unit of time, or the production of a car, etc.

Erosion Removal of the Earth's surface, especially fertile soil, by water, ice and wind.

Feedback processes A sequence of interactions (shown as loops) in which the final interaction influences the original one (e.g. the ice-albedo-temperature feedback, or the CO_2-ocean circulation-upwelling feedback)(see also negative and positive feedbacks).

Final energy The energy delivered to the final consumer (household, industrial plant, vehicle motor, etc.) after the conversion of the primary energy in refineries, coke plants, power plants, etc. and distribution via the electricity network, petrol stations, etc. to the electric plug, petrol tank, heating oil tank, etc.

Final energy carriers For example, petrol, diesel oil, heating oil, coke, coal, gas, solar heat, electricity, fuel alcohol, district heat, etc.

Fluidised-bed technology This consists of a combustion chamber, in which air is blown in from below keeping a layer of sand particles in suspension. Coal (or garbage etc.) is cut into small pieces and introduced into the fluidised layer from the side and burned. The combustion temperature can be reduced from 1,600°C in conventional plants to 800°C in the fluidised bed. No cinders are formed and sulphur dioxide emissions can be absorbed up to 95% in the fluidised bed, the nitrogen oxide emissions are lower, the efficiency is higher and the plants are more compact than conventional boiler plants and therefore cheaper.

Fossil fuels Fuels stored in the Earth's crust, that come from plants or animals of the geological past and have stored the solar energy in the form of chemical energy (coal, oil, natural gas).

Freons Compounds analogous to methane or other light hydrocarbons in which all of the hydrogen atoms have been replaced by chlorine or fluorine atoms. Chemically, they behave like inert gases. They have no sink in the troposphere. They destroy the ozone in the stratosphere and thereby influence the radiation budget and the climate.

GCMs General circulation models of the atmosphere, the ocean or the biosphere.

Glacials Are cold periods or ice ages with extended ice cover.

Greenhouse effect Refers to the warming of the lower atmosphere by trace gases (e.g. CO_2, H_2O, N_2O, CH_4, etc.) that are transparent to radiation but absorb longwave or infrared radiation from the Earth's surface that would otherwise escape into space. The term has become commonly used although the warming of the air in a greenhouse is primarily a result of the reduction of convective heat losses and only to a smaller extent through reduction of heat radiation.

Gross Domestic Product Sum of the incomes from profits and property, indirect taxes and depreciation.

Gross National Product Value of goods and services, that are produced within one year, but not used again in the indigenous production process in the same time period, deducting the charges for imports.

Heating degree-day A one degree departure of the daily mean temperature below a base temperature (usually 18°C or 65°F); used as an indication of fuel consumption for heating (see cooling degree-day).

Heterotrophs Living things that can take their carbon only in the form of higher organic compounds from the environment - animals, fungi, most bacteria. Opposite: autotrophs.

Humus The material found in the upper plant-bearing layer of the soil due to decay and rotting of plant or animal materials. Humus improves the uptake of water by the soil, dissolves minerals important for the plants, and makes the soil permeable to oxygen, which microbes require for the oxidation of nitrogen-containing waste products to nitrates etc.

Hydrocarbons Compounds of carbon, hydrogen and oxygen.

Hydrological cycle The hydrosphere gives water vapour to the atmosphere, e.g. via evaporation, and the atmosphere returns this water vapour to the hydrosphere via precipitation in liquid form (rain) or solid form (snow, hail).

Ice packs Are large areas of floating ice formed over many years.

Inert gases Are gases that are chemically inert and thus do not enter into reactions with other materials, e.g. helium, neon, argon, krypton, xenon and radioactive radon that comes from the decay of radium.

Interglacials Are warm periods occurring between two glacial periods.

Latent heat transport Is the transport of heat by the atmosphere in the form of water vapour; the latent heat of vaporization is released when the water vapour condenses and falls as rain or snow.

Medium load Power plant capacity that is required for the major part of the year.

Monoculture Single cultivation of a particular crop or cultivated plant over a longer period on the same area. Monocultures favour both pest attacks and erosion through far-reaching changes of the relevant ecosystems.

Negative feedback Includes mechanisms that cause a reduction or damping of the response of a system. Opposite: positive feedback.

Neighbourhood cogeneration systems Usually a group of gas- or oil-driven diesel motors with which electricity and heat are produced for a neighbourhood (several hundred homes). The diesel motors drive generators mechanically, while the heat produced by the motor is distributed via a heat exchanger to a local warm water network.

Nitrifying bacteria Groups of bacteria that derive their energy through "combustion" of ammonia (NH_3) to nitrite (NO_2), or from nitrite to nitrate (NO_3).

Non-energy consumption The use of oil, gas and coal as raw materials for the production of plastics, lubricants, etc.

Ozone Oxygen molecule with three atoms O_3 (normal oxygen: O_2). Very reactive and toxic for living beings. In the stratosphere at about a height of 25 km it is continually destroyed and built-up again. It absorbs the dangerous shortwave part of the ultraviolet solar radiation (cancer risk, damage to vision). Influences the radiation balance and thus the climate through chemical reactions.

Palaeoclimatology The science of the study of the climates of the past.

Parameter Factor included in an experiment or calculation; element influencing a system.

Parameterisation In climate models a statistical or empirical relationship specified between variables.

Peak load The highest load arising only for a few hours or a few days in the year. The power capacity of the electricity supply must be sufficient to cover the peak load.

Perturbation A displacement from equilibrium.

Photolysis Splitting of chemical compounds through the radiative energy of ultraviolet or visible light.

Photosynthesis Production of higher organic compounds from simple basic materials with sunlight as the source of energy, especially carbohydrate production of green plants.

Pluvials Epochs with increased precipitation activity over large regions of the Earth.

Positive feedback Includes mechanisms that cause an amplification of the response of the system. Opposite: negative feedback.

Primary energy Energy content of energy carriers before they are converted.

Primary energy carrier For example, coal, oil, natural gas, biomass, geothermal heat, running water, solar radiation, wind, etc.

Process heat Heat energy used for industrial and commercial processes (melting, drying, hardening, cooking, etc.). Low temperature process heat is generally defined as heat below 600°C.

Proxy data Climatic data inferred from non-climatic records in the absence of instrumental observations.

Recycling Re-use of waste products.

Relative humidity Is the ratio (in %) of water vapour in the air to water vapour in the air at saturation; it decreases if a parcel of air is heated without addition of water vapour (see also absolute humidity).

Renewable energy carriers E.g., biomass, running water, solar radiation, waves, wind.

Reserves Are those supplies that have been unambiguously identified geologically and geographically and could be technically and economically recoverable under present conditions or those to be expected in the near future.

Reservoir Is the total amount of a substance in the atmosphere, in the biosphere, in the oceans, in sedimentary rocks, etc.

Resources Are all reserves and all the supplies assumed and expected on the basis of geological comparisons, for which there is only partial evidence; the recovery of resources would be extremely uneconomical but perhaps possible with the aid of new technology.

Residence time Is the total mass of a substance in a reservoir divided by the rate of inflow or outflow.

Sink Is the storage of the flux of a material into a reservoir; e.g. growing plants are a sink for atmospheric CO_2, because they extract it and fix it in their tissues.

Spectral distribution Intensity of radiation as a function of wavelength.

Steady state The state of a system in which its characteristics do not change with time.

Stratosphere Is the atmospheric layer above the troposphere, in which solar energy is absorbed through photolytic processes and ozone formation leading to a rise in temperature. Contains small amounts of water vapour and has therefore only little cloud formation. The lower boundary over middle latitudes is about 10 km, over the tropics about 18 km above sea level. The upper boundary is at about 50 km.

Suess effect Is the relative decrease in $^{14}C/C$ in the atmosphere caused by the burning of fossil fuels.

Synergism Is the interaction of two agents which together increase each other's effectiveness.

Synthetic fuel production From lignite and bituminous coal synthetic gas and liquid fuels can be produced requiring much additional energy.

Systems analysis Is the investigation of the complex interactions of the elements of a system, its structure and, in some cases, dynamics, as well as its function and organisation within the total system.

Technically recoverable resources Are those parts of the assumed and proven supplies that appear to be technically recoverable with high costs and expected new technology. Only this part of the resources is significant for the long-term global energy perspective.

Teleconnections Are interrelations among events occurring great distances apart.

Troposphere Is the lowest layer of the atmosphere in which nearly all cloud formation and weather processes occur. Reaches to 10 km in middle latitudes, and to 18 km above sea level in the tropics.

Useful energy Is the final energy which, after introduction into the boiler, the electric motor, the vehicle, etc., arrives as useful energy at the radiator in the room, the drive shaft of the motor, etc.

Waste heat Comprises all of the sensible and latent heat leaving a system including the conversion losses, but with the exception of useful energy. The heat that is not recovered is also added to the environment and consequently lost.

Weather Is associated with the complete state of the atmosphere at a particular instant in time and with the evolution of this state through the generation, growth and decay of individual disturbances.

WMO World Meteorological Organization with headquarters in Geneva, Switzerland.

AUTHOR INDEX

Aharon, P. 186, 289
Ajtay, G. L. 174, 291
Albanese, A. S. 201, 202, 204, 205, 297, 303
Allen, Jr., L. H. 174, 289
Allsopp, T. 163, 164, 165, 293
Almquist, E. 129, 283
Anderheggen, E. 56, 269
Andersen, N. R. 307
Andreae, B. 169, 289
Andrews, J. T. 85, 157, 186, 273, 289, 291
Angell, J. K. 76, 119, 271
Arking, A. 138, 283
Armenante, P. 301
Armentano, T. V. 68, 85, 174, 269, 271, 295, 319
Arrhenius, S. 271, 320, 324
Ascensio, J.-M. 274
Augustsson, T. 233, 297
Ausubel, J. 242, 297, 302, 307

Bacastow, R. B. 76, 97, 100, 272
Bach, W. 51, 52, 57, 58, 64, 72, 73, 82, 83, 84, 87, 92, 108, 112, 117, 126, 129, 135, 136, 138, 143, 144, 145, 147, 154, 167, 168, 170, 175, 177, 201, 205, 206, 207, 213, 224, 225, 226, 239, 243, 244, 246, 249, 266, 269, 270, 272, 275, 277, 283, 288, 289, 295, 297, 298, 300, 301, 302, 303, 305, 306, 307, 308, 310, 311, 314
Baerreis, D. 139, 140, 284
Baes, C. F. 89, 91, 201, 202, 204, 205, 272, 298, 322, 325
Bainbridge, A. E. 89, 281
Baker, D. N. 174, 289
Bardach, J. 181, 183, 289
Barry, R. G. 32, 185, 266, 292
Bartholomäi, G. A. 145, 283
Batteen, M. L. 114, 279
Baum, G. R. 245, 303
Baumgartner, A. 139, 142, 159, 283, 284, 290
Baurschmidt, R. 302
Beall, S. E. 298
Beaujean, J.-M. 65, 269
Benci, F. J. 171, 172, 290
Bentley, C. R. 186, 290
Berger, A. L. 307
Berger, W. H. 96, 272
Bernard, C. 30
Bernard, H. W. 117, 272
Bernardini, O. 51, 52, 100, 212, 228, 229, 236, 237, 269, 298
Berner, W. 95, 96, 272
Bienewitz, K. H. 212, 298
Binswanger, H. C. 311
Bischoff, G. 57, 68, 254, 269, 303, 308, 309
Biswas, A. K. 307
Björkström, A. 98, 272
Bockris, J. O. M. 258, 304, 309
Boeck, W. L. 134, 284
Bogdonoff, P. 85, 274
Bohn, H. L. 85, 272
Bojkov, R. D. 121, 276
Bolin, B. 73, 92, 93, 98, 158, 160, 272, 290, 307
Borst, W. L. 309
Bossel, H. 213, 257, 298, 300, 304, 309
Bossel, U. 257, 304
Bottke, H. 56, 269
Boulding, E. 307
Boville, B. M. 124, 273
Brandt, W. 63, 249, 269, 304, 311
Breuer, G. 83, 206, 272, 273, 298, 309
Broecker, W.S. 89, 90, 92, 95, 97, 117, 273, 281, 300
Brookshire, D. 290
Brown, A.W.A. 195, 290
Brown, H. 297, 308, 311

Brown, J. 85, 273
Brown, L.R. 46, 168, 169, 181, 207, 239, 269, 290, 298, 311
Brown, S. 68, 75, 273
Bruckmann, G. 226, 298
Bryan, K. 114, 125, 273, 277, 323, 325
Bryson, R.A. 139, 140, 184, 284, 290, 322
Budyko, M.I. 109, 160, 185, 190, 205, 273, 290, 298, 307
Burton, I. 240, 298
Bzserghi, H.A. 213, 298

Callegari, A.J. 267, 275, 286, 300
Callendar, G.S. 273, 320, 324
Caputo, R.S. 224, 225, 226, 298
Carlson, R. 155, 293
Cena, V. 301
Cess, R.D. 133, 236, 273, 280, 286, 298, 323, 325
Chamberlin, T.C. 273, 320, 324
Chang, J. 307
Chapman, P. 309
Chappell, J. 289
Charlock, T.P. 138, 284
Charney, J.G. 141, 284
Charpentier, J.-P. 65, 269
Chen, R.S. 188, 189, 190, 295, 307
Chervin, R.M. 38, 105, 147, 266, 273, 284
Clapperton, C.M. 186, 296
Clark, D.L. 193, 290
Clark, W.C. 307
Clarke, A.W. 56, 270
Clawson, M. 85, 273
Coakley, S.M. 168, 169, 290
Collins, J. 239, 301
Colombo, U. 51, 52, 100, 212, 228, 229, 236, 237, 269, 298
Compston, W. 289
Cook, K.H. 113, 275
Coope, G.R. 27, 266
Cooper, C.F. 239, 299
Corby, G.A. 113, 273
Craig, P. 223, 299
Crane, A.J. 123, 274, 307
Crutzen, P.J. 86, 92, 129, 281, 284
Cushing, D.H. 182, 290

Dahlman, R.C. 93, 95, 173, 174, 274, 290
Dalluge, R.W. 288
Dampier, B. 219, 299
Daniels, A.P. 135, 283, 311
Dansgaard, W. 95, 97, 116, 274
D'Arge, R.C. 158, 166, 167, 290
Deevey, E.S. 245, 304
Degens, E.T. 86, 92, 272, 274
DeLaune, R.D. 130, 284
Delcourt, M.R. 68, 269
Delmas, R.J. 95, 274
Dennett, M.R. 88, 275
Denton, G.H. 186, 290,292
Denton, R. 213, 298
Desjardins, R.L. 84, 274
Detwiler, R.P. 85, 274
Deuster, G. 254, 304
De Witt, S. 311
Diaz, H.F. 157, 291
Dickinson, R.E. 39, 114, 120, 266, 268
Dietrich, G. 90, 274
Dittberner, G.J. 135, 136, 274, 284, 322, 325
Doblin, C.C. 65, 269
Donley, D.L. 290
Döös, B.R. 124, 273
Dopplick, T.G. 123, 124, 279
Dudas, J.L. 310
Dürr, G. 248, 304, 311
Dürr, H.-P. 247, 304
Dunkerley, J. 220, 299
Duxbury, J. M. 127, 284
Dyson, F.J. 206, 207, 299

Eckholm, E.P. 182, 291
Ehhalt, D.H. 129, 130, 284
Eiden, R. 139, 284
Ellsaesser, H.W. 126, 129, 141, 284, 285
Elwood, J.W. 86, 278
Engh, R. 273

Eppler, E. 248, 304, 311
Epstein, E.S. 39, 117, 119, 120, 187, 266, 274, 291
Erfurt, M. 304
Etkins, R. 39, 117, 187, 266, 274, 291
Eyre, D. 257, 304

Farmer, G. 308
Fastook, J.T. 292
Fetscher, I. 311
Flavin, C. 259, 304
Flohn, H. 27, 32, 33, 34, 38, 97, 103, 110, 140, 141, 142, 150, 151, 157, 159, 175, 177, 178, 180, 184, 185, 186, 187, 190, 191, 192, 193, 248, 266, 267, 274, 285, 291, 294, 304, 321, 324, 327
Forrester, J.W. 59, 269
Fraedrich, K. 39, 267
Frakes, L.A. 307
Francey, R.J. 94, 274
Franck, H.-G. 81, 274
Fraser, P.J. 129, 275, 285, 326
Freney, J.R. 367
Freyer, H.D. 84, 94, 129, 130, 275, 285
Fricke, J. 285, 309
Fricke, W. 131
Frisch, F. 259, 304
Frisch, J.R. 212
Fritz, M. 61, 269
Fromm, E. 311

Galbally, I.E. 307
Gal-Chen, T. 268, 317
Garcia, R.V. 240, 299, 311
Gates, W.L. 36, 111, 113, 114, 123, 157, 267, 275, 280, 285, 291, 322, 325, 327
Gavin, J. 120, 276
Geberth, R. 210, 299
Geissberger, W. 311
Gemper, B.B. 309
Gentry, A.H. 142, 285
Gertner, G. 75, 273
Gerwin, R. 57, 58, 59, 269, 299, 309
Gilchrist, A. 111, 273, 275, 288, 324, 325
Gilliland, R.L. 121, 275
Ginsburg, T. 248, 304, 311
Glaeser, B. 239, 299
Glantz, M.H. 239, 299
Gocht, W. 57, 68, 254, 269, 303, 308, 309
Goeller, H.E. 322
Goethe, J.W. von 27
Goetzberger, A. 252, 304
Goldenberg, S.D. 236, 273, 298, 323, 325
Goldman, J.C. 88, 275
Gornitz, V. 187, 188
Gorz, A. 309
Goudriaan, J. 174, 282, 291, 296
Grassl, H. 39, 129, 131, 134, 138, 267, 285, 286
Grathwohl, M. 309
Gray, C.L. 258, 304
Greene, D.L. 258, 305
Grenon, M. 65, 270
Gribbin, J. 307, 311
Griep, R.W. 302
Griesbaum, K. 258, 305
Griffith, E.D. 56, 270
Groves, K.S. 129, 132, 286
Grupp, M. 309, 311

Haaf, G. 311
Häfele, W. 59, 60, 62, 65, 66, 204, 212, 223, 224, 227, 228, 249, 270, 299, 305, 309
Hahn, D.G. 105, 278
Hahn, J. 127, 286
Hähnle, W. 248, 304, 311
Haigh, J.D. 132, 286
Hall, C. 85, 274
Hallsworth, E.G. 239, 299
Hameed, S. 131, 133, 286
Hampicke, U. 72, 73, 84, 87, 92, 206, 207, 275, 300
Han, Y.-J. 114, 279
Hansen, J.E. 39, 113, 114, 118, 119, 121, 122, 148, 149, 150, 267, 275, 286, 288, 322, 323, 325
Hantel, M. 39, 267

Hardin, G. 246, 305
Harris, W.F. 68, 269
Harrison, P. 196, 291
Hartmann, G. 283
Hartwig, S. 298
Hasselmann, K. 157, 292
Hatzfeldt, H. 309, 311
Hauff, V. 220, 300, 309
Hauser, H.K.J. 178, 292
Haxel, O. 32, 267
Hayes, D. 165, 292, 309
Haynes, D.L. 170, 173, 292
Heimann, M. 88, 89, 90, 97, 98, 275, 279
Hein, K. 252, 253, 254, 305
Heinrich, H.-J. 305
Hekstra, G.P. 173, 239, 292, 300
Henderson, H. 311
Henderson-Sellers, A. 125, 280
Herrmann, H. 148, 255, 256, 257, 258, 286, 305
Herterich, K. 39, 267
Herz, H. 209, 300
Hett, J. 68, 85, 269, 271, 319
Hirsch, F. 248, 305, 311
Hodges, C. 183, 239, 292, 300
Hoffert, M.I. 32, 97, 203, 267, 275, 278, 286, 300, 323, 327
Hoffmann, G. 194, 292
Hoffmann, W. 254, 305
Hogan, A.W. 135, 286
Hogan, J.S. 286
Holdren, J.P. 226, 300
Holighaus, R. 55
Holland, W.R. 114, 275
Hollin, J.T. 185, 186, 292
Holloway, Jr., J.L. 267, 317
Hollowell, C.D. 257, 305
Holm-Hansen, O. 183, 292
Hönicke, D. 258, 305
Hori, A.M. 74, 280
Horn, M. 310
Hörster, H. 309
Howell, G.S..173, 292
Hsieh, C.-T. 267, 275, 286
Hsiung, J. 279
Hughes, T. 186, 193, 290, 292
Huke, R.F. 171, 172, 295
Hummel, J.R. 144, 286
Husar, R.B. 131, 286
Hyson, P. 275

Idso, S.B. 123, 124, 275, 307
Idyll, C.P. 183, 292
Illich, I. 311
Ingram, M.J. 308
Isaksen, I.S.A. 129, 131, 132, 286

Jacoby, G. 94, 276
Jänicke, M. 312
Janssen, H. 205, 300
Janssen-Hering, A. 205, 300
Jennings, D. 257, 304
Jochem, E. 209, 300
Johansson, T.B. 143, 286
Johnson, D. 286
Johnson, J.H. 182, 292
Jones, P.D. 116, 118, 276, 283, 296
Jung, H.J. 201, 205, 298
Junge, C. 276
Justi, E.W. 258, 304, 309

Kalle, K. 274
Kamber, D. 275
Kaplan, L.D. 276, 321, 325
Karl, T.R. 195, 292
Karweina, G. 309
Kates, R.W. 298
Katz, R.W. 172, 293
Keeling, C.D. 74, 75, 76, 80, 97, 100, 101, 102, 103, 272, 276, 321
Kellogg, W.W. 38, 109, 121, 133, 139, 157, 161, 163, 169, 185, 195, 239, 241, 267, 272, 276, 279, 286, 287, 293, 294, 300, 307
Kelly, P.M. 116, 118, 121, 276, 296
Kemmer, H.-G. 57, 270
Kempe, S. 92, 272, 274
Kendall, W. 223, 300, 309
Kennedy, J.F. 249

Kennett, J.P. 191, 293
Kerr, R.A. 137, 138, 287
Kester, D.R. 73, 276
Ketner, P. 272
Keyfitz, N. 46, 270
Khalil, M.A.K. 288
Kinzelbach, W. 145, 283
Kirchner, M. 142, 283, 284
Klauder, W. 49, 270
Klingauf, F. 173, 293
Knop, A. 81, 274
Ko, M.K.W. 129, 287
Kohl, D.H. 301
Kohlmaier, G. 75, 76, 276
Komro, F.G. 273, 323
Korshover, J. 119, 271
Kovda, V.A. 240, 300
Kraus, W. 274
Krause, F. 213, 215, 221, 270, 277, 300, 301, 305, 309, 310
Krauss, U. 226, 302
Krickenberger, K.R. 204, 300
Krischke, H.G. 254, 305
Krõmer, G. 144, 288
Kukla, G. 120, 276
Kulp, G. 258, 305
Kunkel, G. 147, 287
Kuntze, H. 239, 300
Kutzbach, J.E. 32, 38, 267

Lacis, A. 150, 286, 287, 288
Laconte, P. 50, 270
Lamb, H.H. 135, 276, 290, 308
Lambert, J.R. 174, 289
Landsberg, H.E. 197, 293
Landsberg, H.H. 309, 312
Langhoff, J. 254, 305
Lansford, H. 180, 295, 308
Lapedes, D.N. 309
Lapillonne, B. 65, 270
Lappé, F.M. 239, 301
Lathrop, J. 297
Laurmann, J.A. 98, 100, 276
Lawson, M.P. 45, 276
Leach, G. 213, 218, 301, 309
Learmonth, A.T.A. 195, 293
Leavitt, S.W. 94, 276
Lebedeff, S. 286
Lee, D.W. 298
Lee, P. 286
Legrand, M. 274
Lemke, P. 267, 317
Leontief, W.W. 52, 59, 270
Lettau, H. 142, 287
Lettau, K. 287
Levin, S.K. 308
Lewis, C. 310
Lian, M.S. 280
Liegel, B. 273
Lienemann, W. 309
Lieth, H. 73, 86, 92, 166, 277, 293
Likens, G.E. 308
Lindh, G. 176, 177, 181, 239, 293, 301
Llewellyn, R.A. 146, 287
Lockeretz, W. 239, 301
Lockwood, J.G. 308
Lodge, Jr., J.P. 286
Long, A. 94, 276
Longhetto, A. 307
Lopez-Parodi, J. 142, 285
Lorenz, E.N. 43, 267
Lorius, C. 96, 277
Loucks, D.L. 85, 87, 88, 271, 277
Louwerse, W. 282, 296
Lovelock, J.E. 312
Lovins, A.B. 59, 100, 165, 208, 211, 212, 213, 226, 228, 230, 231, 236, 237, 251, 270, 277, 293, 301, 305, 309, 310, 312
Lovins, L.H. 270, 277, 301, 305, 310, 312
Lubore, S.H. 204, 300
Lugo, A.E. 75, 273
Luhmann, H.-J. 310
Luther, F.M. 89, 120, 125, 138, 280, 287
Luther, G. 310
Lvovitch, M. 177, 293

MacCracken, M.C. 75, 89, 117, 119, 120, 125, 138, 277, 278, 280, 287, 322, 325
Machta, L. 99, 140, 277, 321, 325

Madden, R.A. 118, 277
Malahoff, A. 307
Manabe, S. 39, 105, 107, 108, 109, 110, 111, 112, 113, 114, 115, 118, 146, 157, 267, 273, 277, 278, 282, 283, 287, 293, 317, 321, 322, 323, 325
Manshard, W. 297, 308
Marchetti, C. 63, 64, 146, 202, 203, 206, 212, 270, 287, 301
Markley, O.W. 155, 293
Marland, G. 59, 206, 207, 212, 249, 271, 298, 299, 302, 306
Mason, B.J. 37, 267
Mass, C. 137, 288
Matthews, W.H. 51, 64, 225, 269, 297, 308
Matthöfer, H. 224, 301
Mattingly, S.R. 286
Mauldin, W.P. 46, 47, 48, 270
Mayer, H. 283
McConnaughey, T.A. 89, 281
McKay, G.A. 163, 164, 165, 293
McQuigg, J.D. 170, 171, 293
McRae, A. 310
Mead, M. 307
Meadows, D.H. 59, 270, 312
Meehl, G.A. 39, 108, 113, 268, 282, 324, 325
Meier-Leibnitz, H. 247
Melack, J.M. 86, 280
Meleshko, V.P. 105, 278
Mensching, H. 159, 246, 293, 305
Mercer, J.H. 186, 187, 293
Mesarovic, M. 59, 270, 312
Mesirow, L.E. 168, 295, 312
Messerli, B. 268, 308
Messerli, P. 268
Meyer, N.I. 312
Meyer-Abich, K.M. 207, 213, 245, 261, 301, 305, 310, 312
Meyer-Larsen, W. 310
Miall, H. 302, 310
Michael, P. 97, 278, 323, 325
Miller, G.A. 82, 281
Miller, P.C. 85, 278
Millero, F.J. 89, 278
Mishan, E.J. 247, 305, 312
Mitchell, J.F.B. 39, 267, 278, 324, 325
Mitchell, Jr., J.M. 31, 39, 119, 267, 278, 301, 321, 325
Mo, T. 288
Mohnen, V.A. 135, 286
Molion, L.C.B. 287
Möller, F. 278, 321, 325
Möller, K.P. 302
Montbrial, T. 310
Moore, B. 82, 83, 84, 278
Moore, D.J. 286
Moriarty, J.E. 164, 294
Moses, H. 75, 117, 277
Mulholland, P.J. 86, 278
Müller, H. 27, 32, 266, 268
Müller, W. 54, 270, 312
Müller-Reissmann, K.F. 300, 309
Müller-Wenk, R. 312
Munn, R.E. 140, 287
Murray, T.J. 184, 290
Mustacchi, C. 203, 301
Myers, N. 142, 159, 287, 293, 312

Nadis, S.J. 223, 300, 309
Nakicenovic, N. 63, 64, 270
Nash, H. 310
Navato, A.R. 279
Nectoux, F. 302, 310
Neftel, A. 95, 278
Nemoto, J. 182, 296
Newell, R.E. 76, 91, 123, 124, 279
Newhall, C.G. 135, 279
Newman, J.E. 170, 294
Nicholson, S.H. 159, 294
Niehaus, F. 142, 233, 234, 287, 301, 303
Niemann, J. 300
Norgard, J.S. 213, 301, 302
Norman, C. 312
Nowicki, J.J. 197, 294

Oberacher, B. 298
Odum, E.P. 158, 294
Oerlemans, J. 188, 294

Oeschger, H. 95, 97, 98, 272, 275, 279, 281, 308, 322, 325
Oest, W. 302
Offermann, P. 283
Oke, T.R. 143, 287
Olivier, D. 213, 218, 226, 302, 310
Olson, J.S. 322
Olson, M. 312
Olson, W.P. 195, 294
Opperman, M. 302, 310
Östlund, H.G. 89, 281
Otterman, J. 159, 294

Pacanowski, R.C. 114, 273
Paltridge, G.W. 123, 279
Pankrath, J. 272, 289, 307
Parkinson, C.L. 109, 185, 279, 294
Pearman, G.I. 75, 76, 275, 279, 308, 326
Peccei, A. 262, 312
Peng, T.-H. 273
Pestel, E. 59, 210, 270, 302, 312
Peters, W. 254, 305
Petersen, K.H. 312
Peterson, J. 51, 270
Pfeiffenberger, U. 255, 305
Pfister, C. 32, 175, 268, 294, 313
Phares, R.E. 294
Pimentel, D. 173, 174, 177, 294
Pinto, J.P. 289
Pittock, A.B. 157, 294, 308
Plass, G.N. 279, 321, 324
Pollack, J.B. 136, 137, 288
Pollard, D. 114, 279
Polunin, N. 248, 305, 312
Post, W.M. 86, 279
Potter, G.L. 129, 143, 287, 322, 325
Preuss, K.-H. 310
Pyle, J.A. 132, 286
Pytkowicz, R.M. 73, 276

Quaas, M. 212
Quay, P.D. 89
Quayle, R.G. 195, 292
Quirk, W.J. 164, 166, 294

Rabb, T.K. 308
Raisbeck, G.M. 97, 279
Ramage, C.S. 74, 280
Ramanathan, V. 106, 118, 123, 126, 129, 131, 132, 133, 233, 277, 280, 282, 287, 297, 325, 327
Ramirez, J.M. 171, 172, 294
Ramsay, W. 220, 299
Ranjitsinh, M.K. 142, 288
Rasmussen, R.A. 127, 129, 130, 131, 288
Rasool, R.I. 138
Rauschelbach, H. 190
Raynaud, D. 96, 277
Reck, R.A. 138, 144, 286
Reiser, H. 39, 268
Renner, V. 39, 268
Revelle, R. 27, 70, 75, 88, 159, 266, 280, 294, 321, 325
Riches, M.R. 89, 120, 125, 280
Rind, D. 150, 286
Roberts, W.O. 154, 157, 180, 239, 295, 302, 308
Robinson, J. 242, 297, 302
Robock, A. 137, 288
Rocznik, K. 157, 295
Roeschmann, G. 300
Roether, W. 88, 280
Rogner, H.-H. 59, 60, 223, 224, 270, 271, 299
Rogers, H.H. 174, 295
Rose, K.F. 296
Roseme, G.D. 257, 306
Rosenberg, N.J. 112, 174, 280, 295
Rossow, W.B. 125, 280
Rotberg, R.I. 308
Rothchild, J. 310
Rotty, R.M. 36, 59, 74, 79, 80, 81, 92, 99, 100, 212, 227, 232, 249, 268, 271, 276, 280, 302, 306, 322
Rowntree, P.R. 273
Ruske, B. 252, 253, 254, 255, 260, 306, 310
Russel, G. 286
Ryan, P.B. 133, 288

Sadler, J.C. 74, 76, 280
Sagan, C. 143, 288
Saint Exupéry, A. 251
Salinger, M.J. 157, 294
Sander, S.P. 289
Sanderson, T.J.O. 296
Sant, R.W. 226, 261, 302, 306
Santerre, R.M. 183, 289
Sassin, W. 49, 51, 59, 271
Sawyer, J.S. 146, 288
Schaefer, G. 312
Schäfer, M. 310
Schiffer, H.-W. 208, 302
Schikarksi, W.O. 148, 288
Schilling, H.-D. 226, 302
Schlesinger, M.E. 39, 106, 111, 113, 114, 115, 118, 120, 122, 268, 275, 280, 324, 325
Schlesinger, W.H. 86, 280
Schmidt, U. 129, 130, 284
Schmitz, K. 310
Schneider, S.H. 38, 39, 95, 96, 117, 120, 125, 138, 168, 169, 170, 188, 189, 190, 239, 246, 266, 268, 280, 281, 289, 290, 295, 302, 306, 307, 312, 317, 322, 324, 325
Schnell, P. 196, 295
Scholl, M.M. 226, 303
Schönwiese, C.D. 34, 37, 116, 268, 281, 308, 314
Schulze, W. 290
Schumacher, E.F. 312
Schurr, S.H. 310
Schurrmans, C.J.E. 120, 281
Schwanhäusser, G. 205, 224, 297
Schware, R. 157, 161, 163, 195, 239, 293, 300, 307
Schwarzbach, M. 268
Schwerdtfeger, G. 300
Seiffert, U. 258, 306
Seifritz, W. 310
Seiler, W. 86, 92, 281
Self, S. 135, 279
Sellers, W.D. 123, 138, 144, 281, 284, 288
Sergin, V. Ya. 161, 162, 295
Shah, R.P. 204, 302
Sharp, S.D. 302
Shaver, G.R. 93, 281
Shearer, G. 301
Siebker, M. 312
Siedler, G. 274
Siegenthaler, U. 97, 275, 279, 281
Slater, L.E. 308
Slesser, M. 310
Slivar, M. 308
Smith, I.M. 150, 288
Smith, S.V. 91, 281
Smith, V.K. 158, 295
Sörensen, V. 312
Sorensen, B. 165, 213, 222, 223, 226, 295, 302, 310
Spelman, M.J. 114, 273, 277, 323
Speth, G. 70
Spinrad, B.I. 65, 271
Spreer, F. 253, 255, 306
Spreiter, J.R. 98, 276
Stahl, I. 297
Stansel, J. 171, 172, 295
Starr, C. 208
Stauffer, B. 272
Steen, P. 143, 286
Steinberg, M. 201, 202, 204, 205, 297, 303
Stobaugh, R. 213, 219, 303, 310
Stordal, F. 129, 132, 286
Stouffer, R.J. 107, 108, 110, 112, 113, 114, 118, 157, 277, 278, 293, 322, 323, 325
Stoy, B. 54, 270, 312
Strain, B.R. 174, 295
Strohm, H. 312
Stuiver, M. 89, 94, 281
Stumm, W. 308
Suess, H.E. 27, 70, 94, 266, 280, 321, 325
Sugden, D.E. 186, 296
Sundquist, E.T. 82, 281
Sze, N.D. 129, 287

Takahashi, K. 182, 296
Takahashi, T. 89, 281
Tans, P. 94, 281
Teufel, D. 252, 253, 254, 255, 260, 306, 310

Thomas, R.H. 186, 296
Thompson, S.L. 95, 96, 118, 120, 125, 281, 324, 325
Thornes, J.E. 164, 296
Tichler, J. 278, 323
Tillman, D.A. 310
Tinbergen, J. 247, 306
Tobias, M. 278, 323
Toffler, A. 312
Tolman, Jr., C.F. 282, 320, 324
Toon, O.B. 136, 137, 288
Traube, K. 312
Trommer, G. 312
Tuck, A.T. 129, 132, 286
Tucker, H.A. 169, 175, 296
Tyndall, J. 282, 320, 324

Uthott, D. 181, 296

Vali, G.M. 141, 288
Van der Laar, H. 282, 296
Van Keulen, H. 86, 174, 282, 296
Van Loon, H. 38, 268
Venzky, G. 249, 306
Vester, F. 242, 247, 259, 303, 306, 312
Vinnikov, K.Ya. 109, 282
Von Hippel, F. 258, 304
Von Weizsäcker, C.F. 209, 303
Voss, A. 57, 209, 233, 271, 303, 310

Wallén, C.-C. 76, 282
Walsh, J.J. 91, 183, 282, 296
Walter, H. 161, 296
Wang, W.C. 126, 129, 133, 288
Washington, W.M. 38, 39, 106, 108, 113, 146, 266, 268, 282, 287, 288, 324, 325
Watson, R.T. 289
Watts, R.G. 122, 282
Webb III, I. 157, 296
Weber, P. 196, 295
Weihe, W.H. 194, 196, 296
Weingart, J.M. 224, 225, 303
Weinreich, S.K. 125, 280
Weiss, R.F. 127, 129, 288
Wenk, K. 312
Wetherald, R.T. 105, 109, 111, 112, 113, 118, 146, 277, 278, 282, 283, 287, 321, 322, 323, 325
Wey, Y.-C. 300
White, G.F. 298
Whittaker, R.H. 296
Wigley, T.M.L. 116, 118, 157, 276, 283, 296, 308
Wilhelmy, H. 245, 306
Williams, J. 38, 143, 144, 146, 157, 268, 288, 297, 307, 308
Wiliams, R. 204
Winkler, J.-P. 310
Wittmeyer, E.F. 302
Woodley, W.L. 178, 297
Woodwell, G.M. 73, 84, 92, 283

Yergin, D. 213, 219, 303, 310
Yung, Y.L. 130, 288, 289

Ziegler, A. 55, 271
Zimen, K.E. 86, 97, 100, 283
Zimmerman, P.R. 93, 129, 130, 283, 289
Zraket, C.A. 226, 303
Zumbühl, H. 268

SUBJECT INDEX

abandonment, and climate 184, 185
acceptance problems 6, 26, 61
acid rain 8, 87, 134
acidification of soils 134
activities, human 27, 70ff
aerosols 3, 27, 37, 134, 148
- changes 37
- climatic effects of 12, 135ff
- concentrations 135ff
- definition 332
- effective mechanisms 138
- emissions 135ff
- extinction characteristics 138
- injections in the stratosphere 135, 136
- residence time 135
- situation 135
age structure of population 49
agricultural systems 84
agricultural technology 239
airborne fraction 9, 74
- apparent 74
- effective 98
air-vapour barrier 257
albedo change 37
- of the ocean surface 225
- through construction of large-scale solar farms 205
albedo, definition 332
albedo modification 18, 205
algal blooms 91
allothermal processes 21, 82, 233
almost-intransitivity 44
Altithermal 32, 38, 157, 314
see also Hypsithermal
ambient air quality standards, definition 332
ammonia 11
- influence on climate 129, 131
- production 204
anaerobic, definition 332
analysis
- of hypothetical scenarios 156ff
- of ice cores 95ff
of tree rings 93ff
anchovy catch and climate 16, 182
animal pests 15, 173
annual isotherms 163
Antarctic bottom water 90
Antarctic krill 181
Antarctic polar front 135
anthropogenic, definition 332
anthropogenic climate factors 7, 70ff
appropriate energy form 201
aquaculture 16, 183
aragonite 89
Arctic ocean, ice-free 190
Arctic sea ice 16, 109
- melting 16, 185ff
- thickness 110
ash content, reduction by fluidised-bed 254
assimilation
- definition 332
autothermal processes 8, 82
autotrophs, definition 333
availability of alternative energy 65
avoidance of an irreversible climatic change 263

background air pollution monitoring network (BAPMoN) 76, 77
balance of trade 254
bank policies 65
barriers, removal of, 25, 260ff
base load, definition 333
bicarbonate ions 88
bioenergy 61, 224
biological methods, for CO_2 reduction 18, 206ff
biomass 82, 165
- definition 333
- increase 8
biome
- definition 333

biome (continued)
- influence of humans 158
- response to climate changes 159ff
bioplantations 12, 143, 225
biosphere 3, 33
- definition 333
- role in carbon cycle 8, 82ff
biota growth factor 86
biotope, definition 333
black water rivers in Amazon area 86
box-diffusion models 97
box models 97
breeder reactors 61
bubonic plague 195
buffer factor 88, 200, 204 see also Revelle factor
burning of forests 86
Byrd station 95

CAGE 89
calcite 89
Camp Century 95
capital investments in unconventional oil mining 5, 57
capital requirements 64ff
carbon
- bank 18, 206
- chemistry 18, 25
- content of the atmosphere 326
- fluxes 82
- loss in organic soils 68
- material 259
- sinks 6, 66
- sources 6, 66
carbon budget, total 91ff
carbon content
- of the Arctic 85
- of the Taiga 85
carbon cycle
- atmospheric-biological-oceanic 72ff
- causes of the perturbation 72ff
- charcoal formation 86
- evidence of the perturbation 8ff, 74ff
- for 1978 91
- geological 72
- in polar latitudes 85
- in temperate latitude forests 85
- in tropical ecosystems 84
- models 9, 97ff
- natural 8, 72ff
- organic soils 85
- perturbation of the natural 8ff, 72ff
- reduction of sequestering by air pollution 87
- river sediments 86
- stimulation of photosynthesis 86
carbon dioxide, see CO_2
carbon monoxide
- concentration 130
- influence on climate 128
- production 130
- sinks 131
carbonate chemistry 88
carbonate ions 88
carbonate solution 8, 74
carbonate weathering 72
carrying capacity of the Earth 46, 246
cartel agreements 263
catastrophic climatic changes 13, 151
cause-effect flow diagram for CO_2/climate 155
cause-effect linkage for CO_2/climate 120
CCCO, Committee for Climatic Change and the Oceans 243
cement production and CO_2 emission 8, 79, 100
Cenozoic 314
central England, abrupt warming 27
change of behaviour 248
chlorine content, stratospheric 131
chlorofluoromethanes 3, 11, 127
- background concentration 128
- global production 127
- influence on climate 127, 131
- sinks 128

circulation models 42ff
- hydrological response 111ff
- numerical procedure 106ff
- structure 104ff
- thermal response 108ff
- verification 107ff
civilisation
- decline 16, 245
- influence on forests 318
- Mayan 23, 245
climate 2ff, 30ff
- anomalies 28, 30, 198
- classification 41
- components 2
- damage in agricultural sector 169
- data 3, 30ff
- definition 33, 333
- deterioration 185
- elements 30
- evidence 2, 30 , 313
- history 2, 27, 30ff, 314
- impact studies 13, 154ff
- indicators, indirect 156
- influence 1, 7, 27
- marginal areas 71
- noise 28, 38, 118
- optimum 32, 314
- past 30ff, 116ff, 314
- prediction 3, 39ff, 104ff
- predictions of the first kind 43
- predictions of the second kind 44
- processes 3, 33
- programme of the WMO 240ff
- protection 30
- research 3, 39, 44ff
- researcher 39
- research programme, Federal Republic of Germany 44, 154
- risk 26, 263
- scenario analysis 32, 156ff
- signal 28, 38, 118
- simulation 3, 39ff, 104ff
- state 44
- system 2ff, 33ff, 334
- theories 37, 43
- variability 28, 38, 118
climate
- a common good 244
- a hazard 163
- a source of energy 163
climate, in Africa 159
climate and weather patterns, reconstruction 313
climate anomalies
- cause 156
- regional distribution 112
climate models 3, 39ff, 104ff, 315
- basic equations 315ff
- calculations 39ff, 104ff, 156, 315ff
- circulation models 42ff, 104ff, 315
- coupled atmosphere/ocean models 43, 113ff
- energy balance models 40, 104, 315
- hierarchy 104
- radiative-convective models 104, 315
- simple models 39
- statistical-dynamical models 104
- stochastic models 41, 316
- time-dependent models 121ff
- time-independent models 104ff
- zonally averaged models 41
climate threat 262
climate zones
- asymmetry 190
- shift 13, 16, 169
climatic change 1, 2ff, 27, 30ff, 37, 38
- abrupt 27
- catastrophic 13, 151
- causes 37ff
- definition 333
- early detection 37ff, 117ff, 152
- effects on society 154ff
- irreversible 7
climatic effects due to
- aerosols 12, 134
- anthropogenic trace gases 11, 126ff
- CO_2 10, 70ff, 152
- land use changes 12, 139ff
- waste heat 12, 144

climatic effects on
- comfort 193ff
- disease 17, 194ff
- ecosystems 12, 158ff
- energy costs 166ff
- energy supply and demand 14, 162ff
- fisheries 16, 181ff
- food supply 14, 168ff
- global 7, 154ff
- health 17, 193ff
- pest management 173
- population and settlement 16, 183ff
- recreation 17, 196ff
- society 13ff, 154ff
- water supply 15, 176ff
- welfare 17, 193ff
- world food situation 168ff
climatic factors
- anthropogenic 7ff, 70ff
- contribution to greenhouse effect 148ff
climatic variation
- causes 30
- definition 292
- natural 27, 156
"closed-cycle" method 225
cloud physics 134
cloud seeding 16, 178
CO_2, an interaction scheme 200
CO_2, definition 70, 333
CO_2 and temperature change
- through efficiency scenario 235, 236, 238
- through fossil fuel strategy 233ff
- through IIASA scenario 235, 236, 238
- through synthetic fuel strategy 232
CO_2/climate effects 70ff
- in equilibrium state 10, 104ff
- temporal course 11, 116ff
CO_2/climate experiment 23
CO_2/climate future 101
CO_2/climate interactions 155
CO_2/climate problem 7, 25, 28, 60, 70ff, 244
- avoidance through precautionary measures 18, 198ff
- high priority 13, 152
- keys for avoidance 262
- reduction 6, 208, 239
- sociopolitical aspects 4ff, 46ff
CO_2/climate research 70
- historical review 320ff
CO_2/climate risk 244
CO_2/climate threat
- reduction 18, 198ff
- urgency 26, 152
- ways out 25, 262ff
CO_2 absorptive capacity
- of seawater 88ff
- reduction 8, 88
CO_2 assimilation rate, change 174
CO_2 concentration change
- conversion of forest to farmland 142
- Greenland ice 9, 95
- Mauna Loa (Hawaii) 74ff
- model projections 203
- South Pole 74
- various energy scenarios 234ff
CO_2 content of atmosphere
- doubling 100, 108ff
- future development 97ff
- increase 9, 74ff
- preindustrial 9, 74, 96
- real 12, 150
- virtual 12, 150
CO_2 control system 201
CO_2 development
- calculation 9ff, 234ff
- in the future 97ff
- in the past 93ff
- measurement 9ff, 74ff, 237
CO_2 disposal 199
CO_2 doubling
- moisture change 111ff
- temperature change 108ff, 115, 151
CO_2 effect, early detection 37ff, 117ff, 152
CO_2 emission factors 79
CO_2 emissions
- future 99ff

CO_2 emissions (continued)
- growth rate 79ff
- latitudinal distribution 80
- net reduction 201
CO_2 equilibrium concentration in the atmosphere 204
CO_2 exchange, between ocean and atmosphere 88
CO_2fertilization effect 86, 174
CO_2 fixing on clay-containing layers 204
CO_2-free fuels 100
CO_2-free renewable energy economy 264
CO_2 greenhouse effect 3, 35ff, 70, 111, 131, 205, 336
CO_2 growth trend 99ff, 195
CO_2 hydration 205
CO_2 increase
- atmosphere 74ff, 200, 203
- climatic effects 10ff, 103ff
- Mauna Loa 9, 74ff
- sensitivity of the climatic equilibrium 103ff
- South Pole 9, 74
- temporal course 75, 117, 125
- time-dependent 11, 121
- time-independent 104
- troposphere 32
CO_2-induced warming
- identification 117
- in the tundra 161
CO_2 injection into the ocean 202
CO_2, limiting value for long-term exposure 195
CO_2, long-term trends 76, 93ff
CO_2 monitoring network 76ff
CO_2 output 6, 64
- of British energy scenarios 218
- of Energy Path of Enquête-Kommission 216
- of scenarios of the Oko-Institut 213
- of Swedish energy policies 218
- of various fuels 78
CO_2 production
- for various logistic growth rates 99ff, 203
- from fossil fuels 78ff
CO_2 quadrupling, pack ice changes 110ff
CO_2 recycling 204
CO_2 reduction
- decision aids 240
- flanking measures 239
CO_2 removal 18, 199ff
- costs 201
CO_2, seawater solution 202
CO_2 signal 117ff, 152
CO_2 storage 202ff
- in abandoned gas and oil wells 204
- in soil humus 18, 207
- in terrestrial plants 18, 206
- in the ocean 18, 202ff
- in water plants 18, 207
CO_2 transfer
- from the ocean surface water into the deep sea 8, 88
- in the ocean 89ff
- projections 88
- through biological processes in the ocean 8, 91
CO_2 utilization 18, 204
coal 5, 54ff
- cartel 56
- chemistry 259
- deposits 54
- export 56
- fields 58
- processing 57
- production 56
- reserves 5, 54ff
- resources 5, 54ff
- revitalisation 226
coal central heating 255
coal gasification 21, 81, 226
coal liquefaction 21, 81, 226
coastal protection 240
cod, and climate 16, 182
cogeneration 24, 252, 261
- definition 334
cohort-component method 48
coke central heating 255
coking gas 81
combined greenhouse effect 12, 150ff
combined-heat-and-power-generation 24, 148, 252

combined-heat-and-power-plant with district heating 254
comfort and climate 30, 193ff
comfort zone 193
Commission of the European Communities 229
common good, climate as 246
compensation payments for CO_2/climate effect 80
computer costs for climate model calculations 105
conflict potential 23, 248
consequences of CO_2/climate effects, estimation of 13, 154ff
constraints 247
continental drift 38
continuity equation 104, 317
control methods
- biological/climatological for pest management 173
- biological for CO_2 reduction 206ff
- technical for CO_2 reduction 199ff
conurbations 143
conversion of forest into agricultural and grassland 142
cooling, abrupt 27
cooling demand 163
cooling trend 2, 32
corn yield, see 'maize'
cost-benefit analysis 158
cost estimates for climate effects
- for sea level rise 188ff
- in energy sector 166ff
- in food sector 168, 175
- in illness sector 195
- in leisure sector 196ff
Council for Environmental Quality 47
countermeasures, to reduce the CO_2/climate threat 36, 101, 153, 198ff, 262ff
coupling of energy and economic growth 251
coupling technique for climate models 106

credibility gap 23, 241, 245, 261
critical threshold values for the CO_2 and temperature increase 12, 150ff, 238
crop damage 15, 173
cryosphere 3, 33, 34, 334
cultivation conditions 153
cultivation of new grain types 173
cybernetic system 242
cyclic economy 26, 264

death of plants 87
decision aids 22, 240ff
decision makers 4, 154
decision making 154, 243
decision-making process 22
decline of cultures, climate as contributing cause 184
deep sea circulation system 90
Deep Sea Drilling Programme 191
deep water formation 97
defence expenditure 65
deficit in balance of payments 26, 208, 260, 263, 264
deforestation 1, 2, 7, 26, 27, 35, 66, 140ff, 318
- Amazon 142
- of tropical forests 8, 84
denitrification 127
denitrifying bacteria, definition 334
depopulation through climatic deterioration 184
desalination of seawater 178
desertification 12, 140ff
desulphurisation in fluidised-bed technology 254
developing countries 4, 71, 220
development aid 249
Devonian Shale, natural gas reserves 5, 58
discount rate 158
disease and climate 17, 194ff
disparities
- between industrialised and developing countries 249
- in energy consumption 61ff

district heating 25, 254
- islands 254
- network 25, 254
- programme 254
- supply 261
diversion of rivers 16, 144, 180
DOC, dissolved organic carbon in ocean 91
downwelling processes 97
drainage of swamps 144
drainage schemes 12, 139
drinking water supply 15, 176
drought periods 15, 30, 171, 180, 246
dry extraction methods, analysis of ice core samples 95
dry ice, CO_2 removal 202

early warning system, to protect against climatic effects 30
early detection, of climatic changes 37ff, 117ff, 152
earth's surface characteristics 35, 37
East Antarctic, influence on the ice sheet 185
eccentricity 38
ecology 44
- definition 335
econometric models 158
economic development 5, 51ff
economic plans of OPEC countries 65
economic policy, new emphasis 29, 251
economics of energy carriers 61
ecosystem
- damage through waste water 91
- climatic effects on 13, 158ff
- definition 335
- dependence on temperature and precipitation 159
- terrestrial 7, 66
eddy fluxes 41
Eem-Sangamon Interglacial 157
effects-chain for CO_2 increase 155
efficiency, definition 335
efficiency considerations 230
efficiency increase 230
efficiency scenario 19, 21, 230
- effect on CO_2- and temperature change 234ff
efficiency technologies 261
efficient energy use, most important energy source 19, 26, 213ff, 220, 230, 262ff
El Chichón, volcanic eruption in Mexico 137ff
El Niño 138, 182
electric heat pumps 255
electric motors 259
electric storage heating 255
electrical loading in clouds 134
electrostatic precipitators 254
emissions
- definition 335
- of trace gases 37, 126ff
employment 25, 26, 255, 264
- opportunities 4, 51
- programme 25, 255
energy and economic policy 262
energy budget 35
energy carriers
- alternative 64
- definition 335
- shares 6, 63ff
energy centres 144
energy conservation 19, 25, 208ff, 251ff
energy consumption
- disparities 6, 61ff
- per capita 62ff
- total 62ff
energy consumption level 230
energy conversion 143
energy costs
- and climate 166ff
- in household for various temperature changes 167
energy crises 19, 99
energy demand
- future 58
- increase 5, 227
energy demand and climate 163ff
energy demand projections 62, 226ff, 263
energy development 6ff, 58ff

energy development costs 64
energy distribution 220
energy economic law 25, 260
energy economics models 65
energy equation 104, 317
energy fluxes 3, 255
energy fluxes, natural and anthropogenic 144
energy intensity 230
energy manager 23
energy market shares 63ff
energy mix 58ff
energy paths of the Enquête Kommission of the German Parliament 19, 216ff
energy plans 20, 222
- California
- France
- Germany
- India
- Quebec
- Sardinia
- Saskatchewan
- Scandinavia
- Sweden
- U.S.A.
energy plantations 68, 82
energy policy 6, 19, 262ff
Energy Policy Project of the Ford Foundation 213
energy price
- development 64ff
- increase 65
- level 65
energy production
- climatic effects 13, 162ff
- economic path 263
- influence of weather and climate 329
energy production projects 12, 143, 144
energy productivity 26, 46, 208, 263
- in developing countries 20, 220
- in industrialised countries 20, 220
Energy programme, continuation of
- 1st to 3rd German National 210
- 2nd U.S. National 227
energy projections 209ff
energy saving 148, 208ff, 230
energy scenarios
- alternative 262
- analyses 58
- characteristics 226ff
- effects on CO_2 and temperature changes 232ff
- efficiency 228, 230
- IIASA 227ff
- zero growth 228, 229
energy sector, price increases 7, 64ff
energy services 26, 208, 230, 263
- definition 335
- supply of 59
energy sources, renewable and CO_2-free 2, 26, 223ff
energy strategies
- based on synthetic fuels 232
- characteristics 21ff, 226ff
- effects on CO_2 and temperature change 232ff
- with exclusive use of fossil energy sources 233ff
energy studies
- CONAES 20, 219
- Enquête-Kommission 19, 216ff
- Great Britain 19, 218
- Harvard School of Business 20, 219
- Öko-Institut 19, 213ff
- Sweden 19, 218
energy supply and demand and climate 162ff
energy supply security 30
energy systems for space heating 256
energy technologies 230
energy use, more rational/efficient 2, 6, 19, 26, 59, 60, 208ff, 251ff, 264
energy use technology 19, 214, 261
energy utilisation, improved 65
energy utilities 260
energy waste 24, 208
ENSO 89
environment
- changes in Africa 159
- destruction 26, 264
- indicators 120

environment (continued)
- monitoring network (GEMS) 76
- protection 5, 6, 71
- research 44
- risks 26, 264
equation of motion 104, 316
equation of state 104, 317
equatorial upwelling water 192
equilibrium response experiments 104ff
erosion, definition 290
European Community 44
eutrophication 91
exhaustion time of energy resources 65
extraction
- oil shales 8, 57
- tar sands 8, 57
extraction capacity of fossil fuels 5, 54ff
extraction technology 5, 54

famines
- climatically-induced 168
- in Sahel countries of Africa 175
FAO, Food and Agriculture Organisation 243
Federal Environmental Agency 230
feedback
- biogeophysical 141
- mechanisms 38
- negative, definition 337
- neglect 123
- positive, definition 338
- processes 123
fertility rates 4, 48
final energy, definition 335
final energy consumption 214
final energy carriers, definition 335
fingerprint, of CO_2-induced climatic change 121
firewood plantations 225
firn area, change 120
fish catch, climatic effects 16, 181ff
fish farming, climatically-independent commercial 16, 183
fish larvae and climate 16, 183
flanking measures to CO_2 reduction 22, 239ff
flaring, and CO_2 emissions 8, 79, 100
floods 15, 26, 30, 66, 180
flow diagram for CO_2/climate impact analysis 155
fluidised-bed combustion 24, 82, 253ff
- definition 336
food production and climate 7, 69, 168ff
food reserves 239
food supply security 14, 26, 30
- climatic effects 14, 167ff
forest
- cover, global 67
- destruction 7, 26, 27, 35, 66, 84, 140ff, 318
formic acid as fuel 204
fossil fuel consumption
- future 99ff
- growth rates 99ff
- reduction 64, 262
fossil fuel strategy, effect on CO_2 and temperature change 233ff
fossil fuels
- availability 5, 54ff
- definition 336
- global resources 54ff
- role of conventional 8, 78ff
- role of unconventional 8, 80ff
fossil pollen 30
fossil soils 30
Fraunhofer Society 258
French Revolution 175
freons 127ff, 336
freshwater fish and climate 181
fuel consumption of personal vehicles 25, 258
fuelwood 159
- demand 68
full employment 248
future
- expectations 65
- opportunities 22ff, 244ff
- prospects: extrapolating old trends 247ff
- prospects: shifting emphases 251ff

future potential threat, assessment of CO_2/climate effects 152
future energy projections 209ff
future risks 13

gaming, analysis of CO_2/climate problems 242
gaps in knowledge, in the CO_2 question 28, 71
gas resources
- as gas hydrate 5, 58
- in Devonian Shales 5, 58
- in geopressurised zones 5, 58
- unconventional 5, 58
generator gas 81
genetic diversity 173
Geochemical Ocean Sections Study (GEOSECS) 89
geophysical experiment, through CO_2 increase 7, 27, 70
Geophysical Fluid Dynamics Laboratory, Princeton University, U.S.A. 104
geothermal energy 224, 330
glacials 27, 32, 314
- definition 336
glacial power 224
global food situation and climate 10, 112, 168ff
global income 52
global potential for conflict 250
Glomar Challenger 191
goals for avoiding a CO_2/climate-problem 198ff
grains
- price index 175
- reserves 168
- surpluses 168, 175
"Green Revolution" 239
greenhouse cultures, increase of biomass through CO_2 86
greenhouse effect 1, 3, 7, 11, 27, 35, 70, 111, 131
- definition 336
Greenland ice cores, 180 yr cycle 117
Gross National Product 5, 51
- definition 336
- estimates and projections 52ff
groundwater
- pollution 259
- use 177
growth 247ff
- controversy 247
- direction 248
- momentum 49
- qualitative 23, 248
- quantitative 23, 248
- rates of fossil fuel use 99
- term 248
growth factors for global primary energy 60
Guano
- birds 183
- fertilizer 183

harvest yields, dependence on climate 15, 169ff
health, and climatic effects 17, 30, 193ff
heat
- budget 35
- diffusion, into the thermocline 121
- insulation 25, 148, 257
- island effect 143
- loss, influence on climate 37, 144ff
- loss points 257
- storage capacity of the ocean 32
- transport in ocean currents 3, 35
heat production
- anthropogenic sources 145
- selected objects 145
- urban areas 145
heat pumps 25, 148, 255
heat waves 17, 195
heating degree days 163, 336
heating, energy demand for 163
heating system, thermodynamic efficiency 255
heavy metals 134
heavy oil sediments 57
heliostats 143
herbicides 173
herring catch and climate effects 182

heterotrophs, definition 337
high Btu gas 81
historical climate research 30ff, 156ff, 313
Holocene 9, 32, 95, 157
hookworm disease and climate 195
human problems, in context 244
Humboldt current 182
humus, definition 337
hydrocarbons, definition 337
hydrogen economy 25, 204, 258
hydrological cycle 111, 176
- definition 337
- intensification 111
hydrosphere 3, 33
hydrostatic approximation 104, 317
hygiene problems in water supply 15, 176
Hypsithermal 32, 157, 314
see also Altithermal

ice age 2, 32, see also glacials
ice-albedo feedback effect 138
ice cores 2, 9, 11, 30, 95ff
ice cover 30
ice deposits
- Antarctica 95
- Greenland 95
ice surge theory 186
ice volume 31
icebergs, transport from Antarctica 178
Iceland, settlement 16, 184
ICSU, International Council of Scientific Unions 243
IIASA, International Institute for Applied Systems Analysis 6, 59, 223
IIASA energy scenarios 21ff, 227ff
- effect on CO_2 and temperature change 234ff
impacts, of climatic change on society 154ff
Import Tariff Law 260
income loss, in water sport areas through climatic influence 197
income tax regulations 261
industrialisation 4, 12, 49ff
- effect on climate 139, 143
inequality of opportunity 23, 248
inequities in energy use 61ff, 248
inertia effect of ocean 117, 238
inexhaustible resources 14, 20, 165, 221ff
inflation, reduction through efficient energy use 263
influencing factors on climate 128ff
- anthropogenic sources
- atmospheric residence time
- background concentration
- development trend
- order of magnitude
- sinks and removal processes
- temperature changes
informing the public 22, 241
ingenuity 7
injection strategy, for CO_2 storage in ocean 202ff
innovation 65, 257ff
input/output analysis 35
Institute for Energy Analysis, Oak Ridge, U.S.A. 59
Institute of Applied Systems Research and Prognosis, Hanover, Germany 210
institutional barriers for rational energy use 25, 260ff
instrumental observations 2, 30, 313
insulation regulations of the Energy Conservation Law 261
integrated pest management 15, 173
interactions of CO_2/climate effects 155, 200
interactions within the climate system 3, 33, 34
interactions, chemical-climatic 131ff
interglacial 27, 32
- definition 337
international cooperation to reduce CO_2 242
International Decade for Drinking Water Supply and Hygiene 176

International Geophysical Year 76
intertropical convergence 135
- shift 191
intransitivity 43
investments 65
IOC, Interparliamentary Oceanographic Commission 243
ionisation rate 131
irrigation, using groundwater 177
isotope analysis 9, 93ff

jet streams 135
job security 264

Krypton-85
- from reprocessing 131
- from nuclear power plants 131
- from nuclear weapons tests 131
- influence on climate 129

labour productivity 195
lack of trust 23, 245
lake level variations 30
lake varves 30
land ecosystems 82
LANDSAT-data 84, 193
landscape protection 5
landslide 66
land use changes 27
- climatic effects 12, 139ff
land use policy 26, 264
life quality
- deterioration 51
- in towns 4
light bulbs energy saving 257
light water reactors 60, 228
liquid gas tankers 58
lithosphere 3, 33, 34
Little Ice Age 32, 185, 314
living standard 65
location changes due to climatic changes 169
long-term climatic change 161
low Btu gas 81
low risk energy and climate policy 26, 244, 263ff
Lüneburg Heath, abrupt cooling 27

MAB, Man and the Biosphere Programme 243
magneto-hydrodynamic generators 82
maize production in the U.S.A. 15, 170
maize yield changes with temperature and precipitation changes 15, 170ff
malaria and climate 195
market distortions 25, 261
market penetration times 6, 63ff
mass spectroscopy 97
maximum likelihood estimation technique 119
Mayas, collapse of a civilisation 23, 245
measurement methods 313
measures, catalogue of, for removing barriers 25, 260ff
measures for avoiding a CO_2/ climate problem 198ff
medium load, definition 337
megacities 51
megalopoli 51, 143
meridional temperature gradient 108, 112
Mesozoic 2, 32, 314
metal-halogen vapour bulbs 25, 258
meteorological conditions 33
meteorological variables 33
methane 11
- concentration 130
- influence on climate 128
- production 58, 130
- source 130
methanol production 201, 204
microelectronics, use for energy saving 25, 259
Mid-Ocean Dynamics Experiment (MODE) 89
migrations 184
Mill Creek Indian culture 16, 184
mini power plants 253
mining process for fossil fuels 5
mixing processes in ocean 97
MIZEX 89

models
- AGCM with climatological ocean 113
- AGCM with OGCM 114
- AGCM with slab 113
- AGCM with swamp 113
- AGCM with variable depth mixed layer 114
- finite differencing 106
- grid point 42
- spectral 42, 106
- types 113ff
models for climate simulation 104
- calculations 10, 106ff
- development 103
- numerical procedure 106
- results, critical comparison 112ff
- structure 104ff
- verification 107ff
monitoring, of CO_2 76ff
monoculture 169
- definition 337
monoethanolamine 201
monopoly 242
mortality rate 4, 48
Mt. Agung explosion on Bali, effect on temperature 136ff
Mt. St. Helens, eruption in NW U.S.A. 137
mountain building 38
Mycenaean civilisation 16, 184
myth, coupling of energy and economic growth 24, 251

natural gas 5, 57ff
- availability 58
- exploitation 57
- pipelines 57
natural gas heat pump 255
natural gas reserves, conventional 57
natural gas resources, unconventional 58
natural gas sources in geo-pressurised zones 58
natural radiation budget, anthropogenic effects on 35ff
neighbourhood cogeneration systems 24, 253
- definition 337
net-C-transfer 84
net export of grains 168
net import of grains 168
net migration, population 4, 48
net primary production 76
new perspective in the CO_2/climate problem 24, 29, 251, 260ff
nitrate aerosol formation 135
nitrifying bacteria, definition 338
nitrogen fertilizer consumption 127
nitrogen oxide 11, 128
- concentration 128
- influence on climate 128
- production 128
- overview 128
nitrous oxide 3, 11
- concentration 127, 128
- emission 127, 128
- influence on climate 127, 128
noise, in climate data 28, 118
non-energy uses, definition 338
North Atlantic Bottom Water 90
North Pacific Experiment (NORPAX) 89
north-south conflict 71, 249ff
Norwegian Sea 90
nuclear moratorium 227
nuclear weapons tests, radioactive trace gases in ocean 89

obligation to advise 242
obligation to consult 242
observation network for CO_2 76ff
ocean
- damping effect 121
- models 113ff
- role in C cycle 87ff
- sink for atmospheric CO_2 8, 87ff
- surface temperatures 76
- thermal capacity 2, 87ff
- thermal inertia 117, 238
ocean bed cores 30

ocean circulation system, influence on fish stocks 16, 181
ocean energy (OTEC) 165, 224, 331
oecumenopoli 51, 143
oil 5, 57
oil crisis 213, 226
oil from tar sands 57
oil production
- conventional 57
- unconventional 57
oil reserves, economically exploitable 5, 57
oil shales 5, 16, 57, 82
oil supplies 5, 57
options to avoid a CO_2/climate threat 25, 262ff
orbital changes 3, 33
Oregon State University, Corvallis, U.S.A. 113
organic waste 224
overall effect, due to human activities 27, 148ff
overall efficiency of cogeneration 252
overfishing 181
overgrazing 12, 35, 140ff, 159
oxygen deficiency 95
ozone 11, 128
- content of the troposphere 125, 131
- destruction by CFM 130
- definition 338
- endangerment 130
- influence on climate 128

pack ice
- annual pattern for CO_2 quadrupling 110
- model 109
palaeoclimatology 38
- analogue studies 10, 103
- definition 338
- indicators 13, 151, 156
Palaeozoic 72, 314
parameter, definition 338
parameterisation 40, 106, 112
- definition 338
particle deposition in ocean 91
particulate emissions, anthropogenic and natural 135
particulates in fluidised-bed combustion 254
peak load, definition 338
periodicity, weather fluctuations 43
pest management and climate 173
pesticides 173
petrochemistry 259
phenological observations 313
photochemical processes 126
photochemical smog 135
photolysis
- definition 338
- sink for CFMs 128
photosynthesis 70
- definition 338
- production in surface water 207
- stimulation 8, 86, 174
photosynthesis-respiration cycle 74
physical processes, interactions in climate system 33
phytomass, net loss 84
phytoplankton, growth 225
pioneer effect 82
plant
- diseases 173
- exhalations (terpenes) 135
- productivity and CO_2 fertilization 174
- protection problems 173
Pleistocene 32, 314
pluvials, definition 338
POC, particulate organic carbon, in the ocean 91
POLYMODE 89
population and climate 16, 183ff, 195
population growth 4, 46ff
- in developing countries 46
- in industrialised countries 46
- in the future 46
- in the past 46
population projections 47
potato famine in Ireland 175
potential energy 147
power densities, future development 147
Pre-Cambrian 314

precaution
- measures to avoid a CO_2/ climate problem 2, 13, 18, 26, 28, 45, 71, 121, 198ff, 234, 244, 262ff
- strategy 18, 199
- system 240
precession 38
precipitation 2, 32
- rate of change 111
predictability of climate 43ff
price
- agreements 65
- development in energy sector 7, 64ff
- increases 250
- increases and climate anomalies 175
- signals 261
primary energy
- carriers 338
- consumption 62
- consumption, non-commercial 61
- demand, for efficient energy use 230
- demand projections
  - FRG 210
  - USA 211
  - world 212
- future demand 59
- supply, for various energy sources and scenarios 228
- use for the FRG 214ff
process heat, definition 339
product innovation 257
progress, credence in 244ff
property loss through sea level rise 188ff
protection of the top soil 239
proxy data 2, 30, 313
pure chemistry, opportunities for producing higher C compounds 205
pure fossil fuel strategy, effect on CO_2 and temperature increase 21, 233
pyrolysis 81

quasi-equilibrium climate 106
Quaternary 314

radiation balance model 109
radiation balance on the earth-atmosphere system 35ff
radiative power of the sun 145
radiodendrochronology 313
radio-echo investigations 193
Radon-222 in insulated houses 257
Rajasthan desert 140
rational energy use 209
- in industrialised and developing countries 220
- selected studies 213ff
reconstruction of climate 30ff, 156ff, 313
recreation and climatic effects 196
recreation industry 196
recycling
- definition 339
- synthetic materials 259
reducing technological risks 264
reduction gas 81
reduction of sociopolitical conflicts 264
reforestation 2, 19, 26, 66, 85, 206, 319
regional planning policy 25, 258
remote sensing programme 193
renewable energy carriers, definition 339
renewable energy sources 14, 20, 165ff, 221ff
- development 221
  - California
  - Federal Republic of Germany
  - France
  - India
  - Quebec
  - Sardinia
  - Saskatchewan
  - Scandinavia
  - Sweden
  - USA
- global potential 20, 223
- increased use 165
- influence of climate 165
- substitution of non-renewable 221
- system effects 224

renewable energy sources (cont.)
- UN Conference 225
reserves
- coal 54
- definition 339
reservoirs 16, 144
residence times
- in atmosphere 33, 128ff
- in ice 33
- in soil 33
- in water 33
resources
- availability 64
- coal 54
- definition 339
- distribution 262
respiration cycle 74
response times in the climate system 3, 33
responses to CO_2 increase 112
rethinking in CO_2/climate problem 26, 147, 251ff, 262ff
Revelle factor 88, see also buffer factor
rice yields 15
- change with temperature and precipitation changes 171ff
river deposits, carbon in 8, 86
river regulation 144
Rockefeller Foundation 47
rotation velocity of the Earth 39
roughness parameter 139
runoff rate 32, 111
rural exodus 4, 49

Sahel area 23
- desertification 141
- famines 23, 246
salt production 144
satellite pictures 37, 84
scattering, forward and backward 134
scenario analysis 156ff, 209ff
- business as usual 214
- climate 156ff
- coal and gas 214
- efficiency 230ff
- fossil fuel 233
- IIASA 227
- palaeoclimatological 38, 156
- socioeconomic 158
- sun and coal 216
- synthetic fuel 226
- zero growth 229
SCOPE, Scientific Committee for Problems of the Environment 234
seafish and climate 181
sea ice, distribution of thickness 110, 115
sea level rise 17, 185ff
- damage estimates 188
- effects 188
- in Germany 190
- in the USA 188
sea surface temperatures 32, 113
sea water desalination 16, 178
secondary oscillations 314
sensitivity experiments 4, 10, 40, 43, 104ff, 115, 141, 147
sensitivity of equilibrium temperature 146
service sector 5
settlement and climate 16, 183ff
settlement, influence on environment and energy development 4, 49ff
shift of precipitation pattern through CO_2 increase 159
signal in climate data 11, 28, 118ff
silting of rivers 66
silviculture 68
sinks for CO_2
- charcoal 86, 92
- ocean 87, 92
- reforestation 87, 92, 319
- river sediments 86, 92
- storage in land and water plants 206ff
siting problems for energy production 6, 264
skin cancer rate, increase due to O_3 reduction 130
snow accumulation in Antarctic 188
society, climatic effects 13ff, 154ff
soil
- conservation 26, 263

soil (continued)
- conservation methods 68
- cultivation methods 159
- destruction 7, 68
- encroachment on 7, 68
- erosion 26, 66, 68
- moisture 2, 32
- organic 85
- oxidation 7, 27
solar constant 144
- changes 149
solar energy collection 60, 165, 224
solar farms 12, 143
solar insolation 3, 35
solar power plant in Spain 143
solar radiation, effect of an increase 149
solar towers 143
'Solar Sweden' study 143
Southern Oscillation 76
space heating 257, 259
spawning grounds, influence of climate on 183
stabilisation of fossil fuel consumption 262
stability considerations for climate 43
stable social order 264
state of aggregation of water 176
statistical-dynamical climate models 40
strategies for avoiding a CO_2/climate problem 18, 198ff, 262ff
stratosphere 34
- definition 339
stunted growth due to acid rain 87
subsistence agriculture 242
Suess effect 94
sulphate aerosol formation 131, 135
sulphur dioxide, influence on climate 129, 131
"supply-side" economics 247
surface albedo 12, 134, 150
- large-scale changes 139
surface characteristics 35, 139ff
surface circulation system in ocean 90
surface temperature
- changes of the annual 191
- geographical distribution 108
surface warming 122
survival problem 70
sustainable energy future 238
sycamores, CO_2 storage in 18, 206
synergism, definition 340
synthetic fuel production 8, 16, 21, 80
- definition 340
- influence of the expansion on the CO_2 content 232
- products 226
- Programme of the Federal Republic of Germany and the U.S.A. 226
- processes 64
- scenario 226
synthetic gas 81, 204
synthetic natural gas 76
systems analysis 13, 154ff, 226ff
- definition 340

tar sands 5, 57
tariff structure 25, 260
technically recoverable resources, definition 340
technological approaches for CO_2 reduction 18, 199ff
technological knowledge, use of 252ff
technological risks 26, 264
teleconnections 147, 166
temperature changes
- calculated 121
- detectability of CO_2 signal 119
- effects on welfare 193
- for CO_2 doubling 108ff
- for various climatic factors 128, 129, 149
- for various energy scenarios 234ff
- natural 11, 118
- Northern Hemisphere 117

temperature changes (continued)
- observed 118, 121
- out-of-phase 120
temperature cycle, 180 yr 11, 116
temperature thresholds, critical 12, 150ff, 327
terrestrial net radiation 147
Tertiary 314
thermal inertia of the ocean 117, 238
thermocline 73, 90, 114
thermography 257
thermohaline processes 90, 202
throughput economy 26, 264
tidal energy 224
timber 159
time gain, by reduced CO_2 growth rate 26, 100, 264
TOPEX 89
total investment for future energy supply 64
tourism and climatic effects 196
town gas 81
trace element cycle 33, 35
trace gases 126ff
- climate-chemistry interactions 131ff
- concentration 126ff
- estimate of climatic effects 133ff
- overview 128ff
- production 126ff
tragedy of the commons 246
transient response studies 116ff
Transient Tracer in the Ocean (TTO) Programme 89
transition time, for fuels 252
transitivity 43
transport of icebergs 16, 178
tree rings 2, 9, 27, 30, 93ff
tropical rainforests, destruction 12, 67, 142
troposphere 33
- definition 340
two-pronged strategy for avoidance of a CO_2/climate problem 20, 208, 230, 251, 263ff
typhus and spotted fever epidemics and climate 194

uncertainty in the CO_2question 28, 71
unconventional fuels 80ff
UN Conferences
- Desertification 140
- Renewable Energy Sources 225
- World Climate Conference 28, 44
UN consumption statistics, for C production 79
UNEP, United Nations Environment Programme 44, 76, 243
UNESCO, United Nations Education Science and Culture Organisation 243
Union of Industrial Power Companies 260
urban agglomerations, rough projections 51
urban landscape 35
urban planning 25, 258
urbanisation 4, 12, 49, 143
- in developing countries 50
- in industrialised countries 50
urbanisation, degree of 4, 50
urgency of CO_2 problem 152
US Energy Programme 80
useful energy, definition 340
upwelling water 90, 97, 182

variety of energy supply options 166, 209ff
vegetation
- period 169
- statistics 84
Vikings 184
volcanic eruptions 3, 34, 38, 136
- climatic effects 135ff

warm periods 32, 151, 157, 314
- unipolar 190
warming
- CO_2-induced 118
- detection 117ff
- troposphere 133
warming effect 150
warning of climate change 38

warning signals, for CO_2 effect 120
waste heat 28
- amount 254
- climatic effects 12, 144ff
- definition 340
- reduction 147
- use 24, 147, 254ff
water
- availability 177
- consumption 177
- cycle 15, 26, 139, 176
- deficit 15, 180
- demand and climate 176
- engineering projects 144
- power 165, 224, 330
- resources 177, 240
- resources and climate 176ff
- rights 180
- supplies and climate 15, 30, 175ff
- vapour 3, 126
water hyacinths, CO_2 storage 18, 207
waterways, lack of water 16, 180
wave energy 165, 224
weather
- changes 43
- definition 33, 340
- effects on energy supply 329ff
- extremes 198
- fluctuations 38, 43
- prediction 43
Weddell Sea 90
well-being, general 247
West Antarctic, ice sheet and ice shelf 185
wheat yields and climate 15, 171ff
"willingness-to-pay" method 158
wind energy 165, 224, 330
wine yields and climate 175
WMO, World Meteorological Organisation 28, 44, 243, 340
wood
- carbon source 68
- energy producer 68
World Climate Conference 28, 44
- Declaration of 28
- Programme 44, 240ff
world coal
- production 5, 54
- resources 54
- trade 5, 54
world economic depression 79
world energy
- conference 56
- consumption 145
- demand 6, 58ff
- models 59ff
- scenarios 6, 59ff, 226ff
world population growth 4, 46ff
World Weather Watch (WWW) 76
Worldwatch Institute, Washington D.C., U.S.A. 46

yield reduction, through climatic effects 169ff
yellow fever and climate 195

zero growth rate, fossil fuel consumption 100, 262
zero growth scenario 21, 228, 229
zonality of circulation 41

Prof. Dr. Wilfrid Bach

Wilfrid Bach received his Ph.D. in atmospheric sciences at the University of Sheffield, England in 1965. He has served on the faculties at universities in Canada, the U.S. and Switzerland. Since 1975 he has been Director of the Center for Applied Climatology and Environmental Studies and the Department of Geography at the University of Münster in Germany. In 1983 he was appointed Dean of the Faculty of Mathematics and Natural Sciences. He is the author of more than 100 scientific papers and the author/editor of 'Atmospheric Pollution', 'Man's Impact on Climate', 'Interactions of Energy and Climate', 'Food-Climate Interactions', 'Renewable Energy Prospects', 'Least-Cost Energy: Solving the CO_2 Problem', 'Carbon Dioxide. Current Views and Developments in Energy/Climate Research'. His expertise as a research scholar has been sought by German, U.S. and Swiss Government Bodies as well as International Agencies.

GPSR Compliance

The European Union's (EU) General Product Safety Regulation (GPSR) is a set of rules that requires consumer products to be safe and our obligations to ensure this.

If you have any concerns about our products, you can contact us on ProductSafety@springernature.com

In case Publisher is established outside the EU, the EU authorized representative is:

Springer Nature Customer Service Center GmbH
Europaplatz 3
69115 Heidelberg, Germany

Batch number: 10761599

Printed by Printforce, the Netherlands